Blocton

WELCOME TO
WEST BLOCTON

BLOCTON

The History of an Alabama Coal Mining Town

Charles Edward Adams

The University of Alabama Press
Tuscaloosa

Manufactured in the United States of America

Typeface: Garamond

∞
The paper on which this book is printed meets the minimum requirements of American National Standard for Information Sciences—Permanence of Paper for Printed Library Materials, ANSI Z39.48-1984.

Library of Congress Cataloging-in-Publication Data

Adams, Charles Edward.
Blocton : the history of an Alabama coal mining town / Charles Edward Adams.
p. cm.
Originally published by the Cahaba Trace Commission in 2001 with a new epilogue from the author.
Includes bibliographical references and index.
ISBN 978-0-8173-1777-5 (trade cloth : alk. paper)
1. West Blocton (Ala.)—History. 2. Company towns—Alabama—West Blocton—History. 3. Coal mines and mining—Alabama—West Blocton. I. Title.
F334.W47A33 2012CIP
976.1′82—dc23
2011052588

Publisher's Note: *Blocton: The History of an Alabama Coal Mining Town* was originally published by the Cahaba Trace Commission in 2001. This new edition from The University of Alabama Press corrects some errors in the first edition and contains a new Epilogue from the author.

Frontispiece: Welcome to West Blocton, Harvey Pope, intersection of Main Street and Old Highway 5 (Truman Aldrich Parkway). (Courtesy of Wanda Pope Miller)

In Memorium

In honor of West Blocton veterans who defended our country, protecting our freedom, and our underground miners who served their country at home working in Blocton area coal mines. In memory and recognition of West Blocton town officials, past and present, for serving the community unselfishly, promoting its welfare and preservation.

Mayor Jesse "Jabo" Reese & Council Members
Greg Blake, Myrtle Jones, Frank Norris, Paul Province, Jr. & Sam Snelson

The faults of our brothers we write upon the sand, their virtues upon the tablets of love and memory. Living or dead, an Elk is never forgotten, never forsaken. Morning and noon may pass him by, the light of day sink heedlessly in the West, but ere the shadows of midnight shall fall, the chimes of memory will be pealing forth the friendly message, "To our absent brothers."

Blocton Lodge No. 710
Benevolent & Protective Order of Elks

In memory of Jack Bowling and Geraldine Clark, my dearest and closest friends and classmates at West Blocton High School from the seventh through the twelfth grades and for the remainder of their lives.

Roy Mason

In memory of Harry James Adams who served his country during World War II; was a coal miner and UMWA member for thirty–five years; was a loving, caring, generous husband, father, grandfather, and great–grandfather; was proud to be a citizen of West Blocton, and would be so proud of his brother, the author of this book.

The Adams Family

In grateful appreciation to the men and women of Bibb County who served in the armed forces of our country.

Bibb County Heritage Association

Mary Davie was a gifted pianist who spent much of her life teaching piano and voice to countless young people in West Blocton. She served her church, First Baptist, as pianist, and played for Sunday services at the Bibb County Nursing Home where she was a resident until her eyesight and health prevented it. She was a caring, loving mother, and I am proud to use this space for a memorial to her.

Robert N. Davie, M.D.

Libby Jeanne Israel Caplan *and* **Elmo Israel Ellis** *revere in memory our beloved parents, Samuel B., and Bertha Seletz Israel; our devoted brother, Frank Seletz Israel; and all former members of the West Blocton Israel family.*

To honor and remember with love my parents, Tip and Margrette McCool, lifelong residents of Blocton and West Blocton, who taught our family that "home" is a very special place.

Debbie McCool Cobern

In memory of my brother, Richard Howard Moon, Private, United States Marine Corps, World War II, killed in action on the Japanese Island of Iwo Jima, February 26, 1945.

James Daniel "Dan" Moon

In honor of our West Blocton High School classes, 1947, 1949, and 1951.

Sue, Ann, and Patsy Pickett

In loving remembrance of Macon Pope, Sr. and Harvey Pope, devoted brothers, fathers, and Christian men who preached the gospel in and around West Blocton for a combined total of more than ninety-five years.

Wanda Pope Miller, Macon Pope, Jr., Faye Pope Hickman, & Mary Ruth Pope Dennard

In memory of our parents: Theo Blake Roycroft Acker and James M. Roycroft; and W.C. and Ruby Johnson Doss.

Kenneth B. and Helen Doss Roycroft

In honor of my parents, Edna Gilbert Thrasher (1916-) and George Martin Thrasher (1913-1992) for their love of (West) Blocton.

Helen Raye Thrasher, M.D.

In memory of W. R. "Will" and Lennie Lawrence Young, our grandparents, who were among the first settlers in Blocton. Will arrived with his parents at age 5 and Lennie a few years later. Their love for each other was obvious to all who knew them.

George Howard Young, Jean Young Ivey, & James W. "Jimmy" Young

THIS PUBLICATION WAS MADE POSSIBLE THROUGH SIGNIFICANT FINANCIAL DONATIONS FROM THOSE WHO PROVIDED THE PRECEDING MEMORIALS.

TABLE OF CONTENTS

	MEMORIALS	v
	PREFACE	ix
	FOREWORD	1
CHAPTER I	Nineteenth Century Beginnings	5
CHAPTER II	Early Churches and Synagogue	51
CHAPTER III	Tennessee Coal and Iron Company Era	64
CHAPTER IV	The Bloctons: 1901 Until World War I	82
CHAPTER V	Over There and Over Here	130
CHAPTER VI	The 1927 Town Fire	170
CHAPTER VII	The Great Depression and Rebound	191
CHAPTER VIII	Fighting Hitler and Tojo	217
CHAPTER IX	Coal Mining's End in West Blocton	238
CHAPTER X	Town Centennial and Beyond	303
EPILOGUE		313
APPENDIXES	I. Mayors of West Blocton	319
	II. Young and Company's 1910 Business and Professional Directory of Blocton and West Blocton	320
	BIBLIOGRAPHY	323
	INDEX	326
	LIST OF CONTRIBUTORS	341

PREFACE

In 1983, Dr. Rhoda Coleman Ellison asked me to read galley proofs of her book's chapter titled "The Coal-Mining Towns," in *Bibb County, Alabama: The First Hundred Years, 1818–1918.* It was the most fascinating and interesting information I had seen on my home town, West Blocton. I started to believe her findings should be expanded to produce a more focused town history. To that end, I gathered material on the subject, greatly aided by a belated mastery of computer word processing.

Marie Bankhead Owen's *Story of Alabama,* compiled in 1949, lists West Blocton in its index. One would expect at least a short paragraph on the community, however, the referenced page had a mere one sentence entry: "West Blocton: Interior town and post office in northern part of Bibb County northeast of Centreville." Well, sirrah, with this tome that is now disassembled. The fleshed out, revisionist version follows.

I have been advised that using "first" as an illustrative point is risky in recording history. Despite the warning, this work claims a number of "firsts" for Blocton and West Blocton, and I believe all are correctly affirmed.

Blocton and West Blocton, like Birmingham and other Alabama industrial centers such as Anniston and Gadsden, first appeared on the map after the Civil War. This chronicle starts in 1883 when the Cahaba Coal Company was formed by two Yankee entrepreneurs. Some say the founders of this town were carpetbagger coal mine developers, but that does a disservice to the men who started Blocton. On the contrary, these innovative industrialists came South, gained reputations as good citizens and stayed through their productive years; one until his death.

I spent my first eighteen years in West Blocton. It is part of my roots. My childhood and teen years spanned the Great Depression and World War II, and I remember the time vividly. I listened to family and neighbors' stories. Some of my recollections are recorded in this book.

Many Bloctonites had British Isles ancestors. My paternal grandparents were immigrants who arrived in Birmingham, Alabama in 1876 from Bristol, Somersetshire, England. My maternal side reaches directly back to a Scottish forefather who arrived on this country's shores in his early youth. He fought in the American Revolution and the War of 1812. Another maternal ancestor came to Alabama from the Carolinas shortly after Indians were forced to give up claim to the Alabama Territory. A maternal great grandfather was an infantryman for the Confederacy. So, on my father's side, I am a second generation American and maternally, my family reaches back to the American Revolution.

The stories of the trials and tribulations of ancestors, of patriotism and responsible citizenship were a part of my upbringing. Coal mining was the basis of family livelihood and scourge of its existence. In 1926, my father was maimed in a mine explosion that put the family at risk, financially and psychologically. My connection with the life of a coal mining community, through its brutality and tragedy, is shaped by personal experience.

I have given Rhoda Ellison recognition for my motivation to write this history. My parents were also major inspirational factors. They were Blocton and West Blocton residents from their arrival there in 1913 until the ends of their lives. My father, as well as being a miner, was a self-trained musician. He was involved locally in the organized labor movement and was active in politics. He played his brass instruments for patriotic, civic, and religious occasions, and he belonged to local fraternal lodges. As president of the union local, he was victimized and fired during the 1920–21 national strike.

My mother was a homemaker, a church-goer, a member of the Order of the Eastern Star, and a patriot. She fired my father's hand gun into the air at 11:00 a.m., November 11, 1918, and then joined a joyous war's-end celebration on Main Street. She flew the nation's flag from our front porch on national holidays when no one else in the neighborhood did. For as long as they lived my older sister, Dorothy Amelia Adams Caffee, and brother, Harry James Adams, Sr., related their recollections of growing up in Blocton and West Blocton. They and another sister, Fannie Mae Adams Eady, contributed financially to my education and otherwise constantly encouraged me to "be somebody." Being "somebody" is elusive and relative. Perhaps this history of a small Alabama town makes a modest contribution to that status.

I depended on the several local weekly newspapers (1891–1931) for much of the material used. From the early 1930s through the 1980s, I relied heavily on the *Centreville Press*. I am particularly indebted to the University of Alabama's Hoole Collection staff, specifically, to Joyce Lamont, the director of that unit until her retirement. The managers of the Gorgas Library's interlibrary loan office were patiently helpful in retrieving materials from the Alabama Department of Archives and History, and the student aides in the microfilm reading area were always ready to provide help. I am also indebted to the Samford University Library's staff where I searched for photographs and Tennessee Coal and Iron Company records.

Assistance with other photograph gathering came from friends Martin and Sandra Everse, both of whom also encouraged this project. They were among the first readers of the unedited manuscript. Their suggestions were incorporated and made the finished product better. They were fully engaged in the book's completion. It is not an exaggeration to say that the book may never have made it off my computer's hard drive without their assistance.

Specific sources are identified in the bibliography. Special thanks go to those who have responded by letters and those supplying information by phone or in personal conversations.

Perhaps the most valuable contribution of all was by Connie Zeanah Atkinson, a cousin in the history department at the University of New Orleans who applied her magical and scholarly gift of editing to my amateur writing efforts. Connie has West Blocton connections, she spent her early childhood there. Without her blue pencil this volume's readability would be diminished. When unexpected teaching and administrative duties restrained Connie, the rector and chaplain of Canterbury Chapel at the University of Alabama, Reverend Kenneth L. Fields, once a college English teacher, came to my aid by editing the last three chapters. I am deeply grateful for his erudite and generous help.

And now, a final statement of gratitude. This publication was made possible by a grant from the Cahaba Trace Commission and the many contributors that matched those funds. These generous people were mainly former West Blocton residents from Pennsylvania, South Carolina, Georgia, Florida, Washington, Colorado, Connecticut, Texas, and California as well as Alabama. There were also many donations from friends and family.

To those who deserve a place in this history but for some reason were not included, I offer the following quotation from The *Book of Ecclesiasticus* in the *Apocrypha*:

> There are some who have left a name, so that men may declare their praise. And there are some who have no memorial, who have perished as though they had not lived; they have become as though they had not been born, and so have their children after them. But those were men of mercy, whose righteous deeds have not been forgotten; their posterity will continue forever, and their glory will not be blotted out. Their bodies were buried in peace, and their name lives to all generations.

The whole of the verse quoted begins, "Let us now praise famous men . . ." When I read, or hear read, these lines close to All Saints' Day, I think of my maternal grandparents who are buried in unmarked graves in Mt. Carmel Cemetery in West Blocton. They are remembered now by a memorial marker placed there in 1995 by their grandchildren. They died in hard times when no funds were available for tombstones, the struggle to provide basic needs for the living came first. Many old coal mining community families can easily relate to this reality. "Their bodies were buried in peace, and their name lives to all generations."

Charles E. Adams
Tuscaloosa, Alabama, 2001

Familiarity with local traditions and local history is essential to a pride in one's home whether it be the nation, state, county or community. A love of the section in which one resides grows and develops with increasing knowledge of its history.

Rucker Agee
Alabama Review, July 1952

FOREWORD

Hometown Revisited

Town of my childhood days
Where those, now dead,
Are bound to me by gesture,
Word and phrase,
Bound to me by memory
That had been forgotten until now...
And yet so much,
In degree entire,
A part of me,
The town and I make whole
Its history.

Hannah Kahn

Geological formations millions of years old contain a series of coal seams that underlay the hills and valleys of the Cahaba Coal Field. The Blocton Basin, within this field, is named for the company town that once was the most prominent coal mining center in the southwestern reaches of the field. Dr. Phillip Lamoreaux, former Alabama state geologist, believes that Sir Charles Lyell, preeminent English geologist of the nineteenth century, likely explored part of the Cahaba Field on his second American visit in the late 1840s. Lamoreaux believes there is evidence in Lyell's writings to suggest he may have visited the north Bibb County area that was to become known as the Blocton District. Other prominent early visitors were Professor Michael Toumey, Joseph Squire, and Dr. Eugene Allen Smith, all distinguished geology scholars and explorers. Reports by these scientists led Truman H. Aldrich and Colonel Cornelius Cadle to the field and to founding their famous enterprise, the Cahaba Coal Mining Company.

Blocton was first called Gresham. The *Bibb Blade* of the town of Six Mile used this name in a February 1884 issue. But by March 1884, the name "Blockton" was regularly identified in the paper's pages. When the *Blade* editor announced a new post office for the town on March 26, he could not remember the name given by Colonel Cadle for the post office site, but Blockton appeared in subsequent issues. According to Ellison's *Place Names of Bibb County*, "Gresham" may have been named for the family of Mrs. E. H. Gressham buried in nearby Caffee Cemetery. Although that is an honest deduction, it does not seem probable. Incidentally, there are five places named Gresham in the United States but none have a known direct connection with this place name.

My research has revealed a far more likely theory about the early place name, "Gresham." Colonel Cadle was the town's first postmaster so it was his responsibility to give the post office a name. Perhaps he assigned a name in haste to meet a deadline. The following is the likely reason the former Union soldier chose "Gresham."

General Walter Quinton Gresham, 1832–1895, statesman and jurist from Indiana, served under Union

General W. T. Sherman during the Civil War, as had Lt. Col. Cornelius Cadle. As a matter of record, these two served in closely allied units in September 1863 in an expedition from Natchez, Mississippi, to Harrisbourgh, Louisiana, as recorded in *War of the Rebellion: Official Records of the Union and Confederate Armies*. Incidentally, the stately mansion "Rosalie" in Natchez, a Mississippi State Society of the Daughters of the American Revolution National Historical Landmark and visited annually by thousands of tourists, was the Union Army headquarters during campaigns in the area. A photograph and biographical sketch of General Gresham is prominently displayed in "Rosalie's" main hallway.

General Gresham organized a company of volunteers in 1861 and rose to command a division. Cadle and Gresham doubtlessly had known each other as comrades-in-arms. Prior to the war, Gresham had been an advocate of the Abolition Movement, also an interest of Cadle's. In 1884, Gresham was considered a possible favorite son candidate for President of the United States on the Republican ticket. Perhaps Cadle wanted to honor General Gresham by naming the new town in Bibb County for him.

Another possible motivation for naming the town for the general is that Gresham was serving as U. S. Postmaster General in 1883, the year the Cahaba Coal Company was organized and its first mine slopes opened. However, "Gresham" was discontinued as the town's name after that one known appearance in the Six Mile paper. It could have been an overruling by Aldrich who chose to use a less controversial, unique name. Possibly, he recognized it was enough to have a couple of Yankees as top bosses without having the new deep South town named for a Federal general. Aldrich also knew that during the last days of the war, Federal troops destroyed Tannehill and Brierfield furnaces, burned the textile mill at Scottsville, occupied the village of Centreville and burned the Cahaba River bridge there, and they also burned the University of Alabama some thirty miles to the northwest. Feelings in the area might have run high against commemorating a Federal war hero.

So it is conceivable that Colonel Cadle, who had postmaster duties in the early months of the company's founding, briefly used the name of a wartime compatriot for the post office name and his decision was revoked by Aldrich. Incidentally, Gresham served a month as Secretary of the Treasury under President Chester A. Arthur and as Secretary of State under President Grover Cleveland, serving from March 1893 until his death in May 1895.

Blockton, the name eventually given the town in late February 1884, is attributed to the Cahaba Coal Mining Company's president, Truman H. Aldrich. He did so, as the story goes, after a block of coal weighing approximately one ton was taken from the company's Number 2 mine. Large chunks of coal were not uncommon when the new seams were opened. Wayne Cline's *Alabama Railroads* says, "These mines [of the Cahaba Coal Mining Company] were so productive that single blocks weighing over a ton each were removed from them. Aldrich dubbed these enormous lumps of coal 'bloc[k]tons,' and the settlement surrounding the mines thus came to be known as the town of Bloc[k]ton." In December 1884, the company employed a flat car to ship a five-ton specimen mined in "Blockton" to the New Orleans Exposition.

The Six Mile paper in September 1884 reported that "Blockton contains 100 houses constructed and built by the Cahaba Coal Mining Company. One hundred more are to be built. The population is about 1,000. It has a union church, schoolhouse [one building] and Odd Fellows Hall. The town is well supplied with water from both wells and springs." Almost overnight this new town was the largest community in Bibb County.

This central Alabama county had been created in 1818, a year before the state was admitted to the Union. Its economic base had been largely agricultural, with short spans of industrial growth. A significant textile enterprise was located in Scottsville dating from 1836. Primitive iron forges had been built in the county's northeastern sections, some as early as the 1830s. Notable blast furnace installations were constructed at Brierfield and Tannehill, the latter property on the Bibb County line in its northernmost section. In late March 1865, these furnaces were, like Scottsville's textile mill, victims of Wilson's Raiders, a large Federal cavalry force. A postwar development was the Edwards furnace at Woodstock, which operated from 1880 until 1890.

Some early coal mining had occurred east of Blocton in the Piper area. Employing only very primitive mining methods, this enterprise used slave labor from Mississippi during the Civil War. The coal from the Piper area mines was used in the iron making process at Brierfield and other nearby early forges. Joseph Squire's *Report on the Cahaba Coal Field, 1890*, a Geological Survey of Alabama publication, indicates that the first underground mining in Alabama was done in this area around 1856. Confederate government demand for coal gave new impetus to mining in the Cahaba Field, mainly at mines in Shelby and Bibb Counties. Squire's handwritten autobiography reflects that in the late 1870s he surveyed the southwestern portion of the Cahaba Field to the vicinity of Scottsville in Bibb County.

An attempt to revive the iron making industry in post Civil War Bibb at Brierfield was undertaken by General Josiah Gorgas, who had served as Chief of Ordnance for the Confederacy. After four years of financial struggle, Gorgas gave up the effort in 1869 to become associated with Sewanee University in Tennessee. The Brierfield installation continued under various hands until 1894, when it was "blown out" for the last time.

Coal mining in Bibb County, in its primitive state, had diminished considerably at war's end. The severely depressed economic times of the post Civil War South made it difficult for industry of any sort to prosper in this central Alabama region. Ellison's fascinating county history recounts the post war struggle. During the period the lumber industry made a significant industrial contribution in communities like Randolph, Bibbville, and Greenpond. Ellison related that "none of these industries of the last decades of the nineteenth century made as much impact as the coal mines in the northern section of the county. For half a century after 1880, they were to bring new life and hope."

Birmingham's coal and iron development in the mid-1870s jolted the state into a new era of industrial development. Blocton, a decade later, added its high grade, vast coal tonnage to the industrial scene. It and other nearby mining towns created a thriving and successful industrial community in the formerly agrarian Bibb County. The coal mining era was to last almost seven decades.

West Blocton celebrated its centennial in 1984. Though its mining economic base is gone, the town has survived in spite of devastating strikes during the rise of the organized labor movement, the end of coal mining, a ruinous fire and the Great Depression. One former resident living in Virginia suggested the history of West Blocton should be subtitled, "The Town That Refuses to Die."

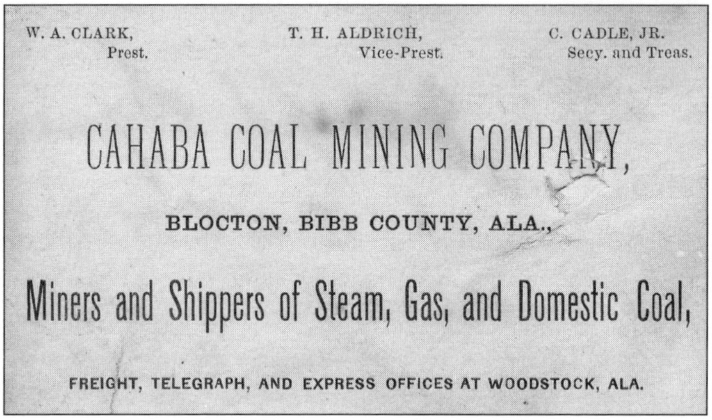

W. A. CLARK, Prest. | T. H. ALDRICH, Vice-Prest. | C. CADLE, JR. Secy. and Treas.

CAHABA COAL MINING COMPANY,

BLOCTON, BIBB COUNTY, ALA.,

Miners and Shippers of Steam, Gas, and Domestic Coal,

FREIGHT, TELEGRAPH, AND EXPRESS OFFICES AT WOODSTOCK, ALA.

Cahaba Coal Mining Company Business Card. (Courtesy of Ken Penhale, Helena, Alabama.)

Uriah Smith and Family. An early land developer, Uriah Smith is credited with naming the town, "West Blocton." (Courtesy of Cahaba Lily Activity Center, West Blocton, Alabama.)

CHAPTER 1

Nineteenth Century Beginnings

Blocton, the Cahaba Coal Company town established in 1883 but not named until early 1884, no longer exists. The town once boasted over 400 company-built dwellings and other structures. In 1889, company records listed the population as 3,600. There were churches, a water system, schools, company store, and other enterprises as well as a two-man, company-sponsored, police department. Blocton was politically important enough to send John C. Jones as a delegate to the Alabama Constitutional Convention in 1901. In 1892, the sizable holdings of the company founded by Truman H. Aldrich and associates were sold to the Tennessee Coal, Iron, and Railroad Company (TCI). In 1907, that company became a part of United States Steel, one of the nation's industrial giants.

Blocton is one of the many extinct Bibb County towns that grew and prospered in the late nineteenth century and through the second decade of the twentieth century, fading between the end of World War I and the 1960s. Blocton ceased as a company entity in July 1928, though one could still receive and post mail at the Blocton Post Office until 1935. The TCI company store, business office, white and black schools, medical dispensary, along with the one remaining large mine, Blocton Number 10, all closed in a major company retrenchment a year before the infamous stock market crash that ushered in the Great Depression. Waterworks on Caffee Creek, the county's first filter plant dating from the mid-1880s and installed by the Cahaba Coal Company, was destroyed by dynamite charges. If the installation was offered to West Blocton for a price, it is not on the records.

Blocton is one of the many extinct Bibb County towns that grew and prospered in the late nineteenth century and through the second decade of the twentieth century, fading between the end of World War I and the 1960s.

West Blocton, an incorporated municipality since 1901, was once called Suttletown after an early enterprising merchant, Julien Suttle. It was briefly called Smith Town after Uriah Smith who, in the mid-1880s, drew up a modest street layout and sold commercial and residential lots. Uriah called his configuration, approximately eight blocks from north to south and some ten blocks from east to west, a "Map of West Blocton." Smith's "blocks" are not standard measures; they are some yards less than the one-tenth mile regulation city block. Uriah Smith is credited, by so labeling his map, for naming the town.

West Blocton along with its sister town, Blocton, grew rapidly in the 1880s and 1890s into a bustling commercial, service, and residential center. For several years, advertisements in an early local weekly newspaper, the *Blocton Courier*, carried the business locations as "Suttletown," "Smith Town," and "West Blocton," and some added "West End" to their ads. Proprietors wanted to be absolutely sure prospective customers knew exactly where businesses were located.

No demarcation line existed between Blocton and West Blocton except on maps showing West Blocton's town limits. Often, the two towns are linked in the minds of those living in the area. The community of West Blocton is frequently, though incorrectly, shortened to the old company town's name, Blocton. Shortly after the company town's creation, the Post Office Department changed the town's spelling from Blockton to Blocton. The reason given was the frequent confusion with a similarly named south Alabama town, New Brockton. Ellison reports that William Luther Pratt succeeded Cadle as Blocton's second postmaster. Pratt, from River Bend, had attended Alabama Polytechnic Institute (Auburn University), but left before graduation. He eventually read law (studied while observing and assisting with cases) in attorney J. M. McMaster's office in West Blocton. Pratt also was an early teacher in the company school and after passing the bar exam, practiced law. He became Bibb County's probate judge in the late 1890s, an office he held for more than forty years.

West Blocton owes its beginning to a group of northern industrialists who had made a detailed, engineer's-eye study of the state's geological reports and maps prepared by Professor Michael Toumey, Alabama's state geologist from 1848 until his death in 1857. They also studied the later reports of his successor, Eugene Allen Smith, who refined and expanded the State Geological Survey from the mid-1870s well into the 1900s. Geologist Smith's survey crew, operating from a horse-drawn, converted ambulance, made detailed reports on the mineral resources of all Alabama counties. A volunteer assistant, Truman H. Aldrich, and his associate Cornelius Cadle, accompanied Smith on many of his expeditions into the mineral districts. One of the areas surveyed was the hills and valleys of northern Bibb County known by geologists and cartographers as the Cahaba Ridges. The agrarian populace of Bibb County dubbed this rugged landscape, the southernmost foothills of the Appalachian mountain chain that reaches from Maine to Alabama, "The Uglies." The *Bibb Blade* reported in September 1883: "Everything is on a boom in the Ugleys [*sic*]."

There is ample evidence that Native Americans once wandered over this area in significant numbers. Ellison reports that Creek tribesmen had made the banks of the Cahaba River their home for around three centuries. Her *Bibb County, Alabama: The First Hundred Years* recounts excellent, detailed research on indigenous peoples. It is likely there was a small Creek town on the Cahaba near the Bibb-Shelby County line. Indian hunting parties ranged near the Blocton area as evidenced by the large number of stone artifacts that have surfaced in plowed fields. One of the most productive sites was searched carefully on Sunday afternoon excursions in the 1940s by William "Billy" Smith, his wife, Nina, and children, Kennedy and Cherry. The senior Smiths, now residents of Florida, have an extensive collection of stone implements collected from near Ashworth's Crossing about three miles north of the old Blocton depot site. Another collector in the area was Robert Majors of the Hebron community. His artifact sites were near the many springs that dot the landscape of the "Uglies." These collectors and others have found a variety of weapons and tools including arrowheads, spear points, knives, hatchets, nutters, and pottery shards. Folk tales claim the Indians used coal

State Geologist Eugene Smith and Survey Crew. One of Dr. Eugene Smith's prized possessions was this South Bend, Indiana Studebaker wagon in which he criss-crossed the state surveying the mineral wealth of Alabama. (Courtesy of Alexander Sartwell, Geological Survey of Alabama.)

in their campfires from outcroppings on the Cahaba River and north Bibb creeks. "Cahaba" is a corruption of the Choctaw name for the river, *Oka Aba*, meaning "the river above." The enterprising mining corporation organizers of the mid-1880s named their Bibb County company after the river.

Truman Heminway Aldrich and Colonel Cornelius Cadle were among those who organized the Cahaba Coal and Mining Company. Ellison's history records the company's founding:

> Four men, some of whom might be labeled builders of carpetbagger coal towns, formed the Cahaba Coal and Mining Company. Signing a petition of incorporation at Woodstock on June 27, 1883, they were Truman and W. F. Aldrich of New York State, and Cornelius Cadle, Jr. and Charles Turner, both residents of Selma. These four were later joined by W. S. Gurnee, a banker and financier, of New York State. They declared a capitol stock of one million dollars, composed of 10,000 shares of $100.00 each.

As company president, Aldrich acquired the land from Henry Fairchild DeBardeleben in a $400,000 deal to begin building his new coal mining world in the unpopulated forest of northern Bibb County. Research by Robert E. Praytor (1998) verifies that from April through July 1883, Bibb County deed books show Aldrich recorded purchases of tracts from George W. Tillery; Elias Callahan; J. C. Reach; Uriah Smith; W. G. Morrison; R. D. Smith; S. G. and M. J. Daniel; and also property from Swann and Phillips, no given names recorded.

Aldrich assumed the post of president and treasurer of the new company, and Cadle was vice president and manager. The other founders were named directors. Later added as directors were A. L. Tyler and Samuel Noble of Anniston who became associated with the company to assure the availability of Blocton coal and coke for their Anniston furnaces. John Grote was also locally associated with the company management. These individuals made up a formidable list of post Civil War industrial developers in the state. Ethel Armes' *Story of Coal and Iron in Alabama*, a 1910 publication sponsored by the Birmingham Chamber of Commerce, contains substantive and interesting biographical information on these men. It also provides interesting details on Blocton and the other Bibb County mining towns. Armes, a talented writer and researcher, came to Birmingham in 1902 having previously held a position on the *Washington Post*. Incidentally, Armes' book made no mention of the strikes and the evolution of the organized labor movement. Chambers of Commerce rarely have recognized organized labor, and it was the policy in 1910 to ignore the presence of labor unions.

Truman H. Aldrich. (Courtesy of Alexander Sartwell, Geological Survey of Alabama.)

Truman Aldrich, a native of Palmyra, New York, was an 1869 civil engineering graduate of Rensselaer Polytechnic Institute of Troy, New York, and was one of the first mining engineers and technically trained men who ventured into the post Civil War South. Armes credits him as heading the list of the first big coal operators of Alabama. She stated that no scientific or genuinely practical methods were applied to coal mining in Alabama until the early 1870s when Aldrich came into the field. He arrived in Alabama in 1871 to enter the banking business in Selma in partnership with Cornelius Cadle.

Armes provided a short biographical sketch of Cadle. His home state was Iowa where he grew up in Muscatine though he was born in New York City, May 22, 1836. He was in business partnership with Aldrich in Selma and in the development of mines at Montevallo in the 1870s. Cadle was twenty-five when the Civil War began. He served in that conflict with the Eleventh Iowa

Volunteer Infantry. He participated in all the battles and campaigns of the Army of the Tennessee, commanded by Gen. Ulysses S. Grant, from the battle of Shiloh, where Cadle served under General McClernand and later under Colonel (later Brig.-Gen.) Marcellus Crocker. During one of Cadle's war exploits, he and two other officers in his company placed their unit's flag atop the capitol building in Jackson when that city surrendered. Cadle was wounded at the siege of Vicksburg. His family history relates he was "struck in the right cheek by a minie ball that passed through his right ear, and glanced off at the back of his neck." Considering battlefield surgical techniques of the era, he must have carried a disfiguring scar for the rest of his life.

Immediately after the war, Cadle was connected with the Freedman's Bureau, stationed in Montgomery, and in 1867, he was appointed receiver of a bank at Selma. He spent some years associated with Aldrich at Montevallo and later in other mining enterprises in Jefferson County. He left Alabama in 1881, spending a year in Colorado involved in mining interests and returned to Alabama in 1882 to join Aldrich in planning for the organization of the Cahaba Company. After the company's consolidation with the Tennessee Company in 1892, Cadle remained as a manager until January 1893. Other details about Cadle are found in Owen's *Dictionary of Alabama Biography*. In 1895 he was serving as chairman of the Shiloh National Military Commission, a position he held for fifteen years. He was injured in a storm, called a "cyclone," that damaged the park in October 1909. He resigned because of disability, according to family history. The records of the Shiloh National Park do not reflect this injury and indicate he may not have been in the vicinity at the time of the storm. In 1998, the park historian took issue with Owen's material and the family genealogical report on Cadle that had been published in 1915. Cadle's health failed progressively, and he died after an extended illness in January 1913 at Cincinnati, Ohio. Cadle married Lucy Ann Barrus in Providence, Rhode Island in September 1867. There were no children. No mention is found of his spouse during the years Cadle was in Blocton.

Cornelius Cadle. (Courtesy of Shiloh National Park Archives and Musser Library, Muscatine, Iowa.)

One can only speculate how some Blocton citizens and native born county leaders felt about this former Yankee colonel. Many local workers, businessmen and professionals were Confederate veterans. Bibb County had suffered from raids of Federal troops during the last few months of the war. When Blocton appeared on Alabama's maps, it had been barely two decades since the war's end and less than a dozen years since the official end of Federal occupation.

However, Cadle won the hearts of many locals by sponsoring an amateur baseball team and financing a finely uniformed, nineteen piece, company musical ensemble which was known as Cadle's Silver Cornet Band. Cadle had an interesting war-time experience with band organization while under General Sherman in the campaign known as "the March to the Sea." The unit in which he was serving as adjutant captured a full set of brass musical instruments while foraging in a Georgia town. His recollection of recruiting musicians to play for a parade review was reported in a 1903 publication by Cadle, *Sketches of War History*. It seems the hastily assembled band's performance of one of the only two tunes the band had mastered before a military gathering in 1864 was not worthy of being called on for future corps parades.

It is assumed that Cadle founded the company sponsored Cadle School for African Americans which was abandoned in the early 1950s. No concrete evidence has been found to confirm the source of this school's name, however, no other prominent Cadle family resided in the area when the institution came into being. Because of Cadle's compassion for former slaves and their families, his interest in the Abolition Movement, and his service as an official in the

Freedman's Bureau, it is only reasonable to conclude that the school was named for the Cahaba Company mine official.

Truman H. Aldrich had applied his civil engineering training to railroad building in New York and New Jersey shortly after finishing his degree at Rensselaer Polytechnic Institute. This prepared him for his involvement in mining development in Alabama. Ironically, he is reported to have said about his acquisition of the Montevallo mines in 1873, "I knew just about as much about practical coal mining then, as a horse about holy water." In the state's Warrior Coal Field in 1878, along with Henry F. Debardeleben and James W. Sloss, other notable Alabama mining industrialists, Aldrich organized the famous Pratt Coal and Coke Company and served as its superintendent. Later he developed mining operations at Henry Ellen and Dudley, now extinct mining towns. After a brief foray into politics, he operated the Hargrove mines from the mid-1890s until 1903.

By the time Aldrich and Cadle opened mining operations in Bibb County, they had acquired formidable knowledge about modern mining machinery and techniques. The Blocton mines were opened applying the latest developments in 1880s mining engineering technology.

When Aldrich undertook in 1883 to connect his mining operation in northern Bibb County with the Alabama Great Southern Railroad at Woodstock by constructing an eight-mile railroad, his practical railroad building experience in the North and later at Montevallo and other Alabama locations proved invaluable. The railroad line's easiest, relatively level route was along Caffee Creek's valley. Still, several earth cuts requiring the movement of millions of square yards of soil caused engineering difficulties. The major cost, however, was incurred because the meandering creek had to be bridged ten times, causing the project's cost to exceed $160,000 (approximately $20,000 a mile), a huge sum for the 1880s. The line was completed in record time. The projected completion date was January 1884, however, the first coal shipment over the rail line was in early February.

The *Bibb Blade* reported in September 1883 that 400 hands were working on the railroad and 50 at the mines. The road builders were organized into squads, each under the management of an experienced foreman. The chief engineer was a Major Barker.

The steam engine owned by the company was built at the Baldwin Locomotive Works in Pennsylvania and was of the Mogul pattern. The Baldwin Company was founded by Matthias William Baldwin, a philanthropist with an interest in supporting the welfare of African-Americans. His abolitionist sympathies led to the boycotting of Baldwin locomotives in the South in the days before the Civil War. William Baldwin died in 1866, and the boycott had faded by the 1880s. By the time of the founding of the Cahaba Coal Company, Baldwin locomotives had been used in Alabama but, according to railroad historian Wayne Cline, were not very popular. We can assume that Cadle's association with causes supporting the welfare of African-Americans may have led to the choice of the Baldwin locomotive. Two other locomotives of the same make and model were acquired later as the company grew. This is an interesting speculation about motivations behind management purchase decisions in Blocton's early days.

While the railway was under construction, rapid development occurred at the site of the mine slopes. Wagon roads were scant in this heavily forested area of Bibb County. A road of sorts had been in use south from Woodstock to the site of J. W. Thrasher's grist mill on Caffee Creek. An improved roadway, perhaps traversing the route to the mill, was extended south to the mine openings. It later became known as the Blocton-Woodstock Road. The improved route was cut east of the creek bed and roughly parallel to the railroad. This road extended north of Woodstock toward Jefferson County and was known in that stretch as the Blocton Pike. Passage of mule and oxen-drawn wagons over this hastily formed route south from Woodstock was possible in the months immediately before the rail line was completed. It is identified today by many in the area as the Old Woodstock Road. On contemporary maps it is labeled the Griffintown Road. Over this new wagon road came basic mining equipment necessary to start the several slopes.

With the railroad's completion, heavy mining machinery arrived, including steam engines for hoisting mining cars from the mine depths and huge ventilation devices, major necessities for modern, late nineteenth century mining. Included in the machinery inventory were two 100 horsepower, steam-driven hoisting engines to raise coal cars from the mines manufactured by C. P. Williamson Company of Birmingham. Installation was accomplished rapidly with

Blocton Mine Number 2. (Courtesy of Will Reese, Montgomery, Alabama.)

mines one and two (later called Blocton Number 1 and Blocton Number 2) opening in quick order. Coal was shipped in February of 1884, just eight months after the incorporating papers had been signed. The *Bibb Blade,* reporting on the community's development while referring to Birmingham's newly acquired sobriquet, proclaimed, "This [place] bids fair to become another Magic City."

The railroad connected the company's mines with the Alabama Great Southern line at Woodstock and thereby with the Blue Creek extension of the Birmingham Mineral Railroad at Blocton Junction depot near Woodstock. Within a few years, as recorded in Joseph Squire's *Report on the Cahaba Coal Field,* another railroad was completed that entered the Blocton Basin from the east side, coming from Montevallo to Gurnee, and from Gurnee to Blocton. It was constructed by the Brierfield, Blocton and Birmingham Railroad Company. This gave the Blocton Basin connection with the Alabama Great Southern Railroad, the Birmingham Mineral System, and the Louisville and Nashville Company's main line and the East Tennessee, Virginia and Georgia main line by means of the Selma, Rome and Dalton Division, which were three of the most important railroads in the state. This information comes from the writings of Joseph Squire, an English immigrant who arrived in Alabama just before the Civil War. His cited report provides information about the area's early railroads. Credited with being an effective explorer of Alabama's coal fields, his works, according to prominent Alabama historians, made fortunes for the capitalists who listened to him.

The Ken Penhale document collection in Shelby County contains correspondence between Squire and Truman H. Aldrich spanning several years concerning land and mineral investments (coal seams in particular), preferred railroad routes, and property rentals. Amounts considered in the correspondence range from $20 house rentals to expenditures of $50,000.

This new Bibb County town, Blocton, less than a decade old and carved out of a deep virgin forest, as documented by Squire, was connected to the outside world by modern rail lines. Massive railroad bridges were necessary to cross Caffee Creek and small streams to the southeast of Blocton. The creek's deep gorge as it nears the Cahaba River demanded major trestle installations. Among the bridges were seventy foot high Bluff Trestle, a massive steel, sandstone and concrete structure; Longhorn Trestle, a long curved wooden-pile and concrete pier construction spanning a small stream in a deep ravine; and an impressive span across the Cahaba River. The cost of the lines to the eastern rail connections greatly exceeded the eight-mile line constructed by Aldrich's company from Woodstock south to the mines. The last major hurdle in completing the rail link to the east was the Cahaba River trestle, completed in mid-summer 1890. The Brierfield, Blocton and Birmingham Railroad, later known as the "Old BB&B," was ready for operation.

In 1898, the Georgia, Mobile and Ohio Railroad completed its line connecting Tuscaloosa with points east. The railroad thus came to Centreville and created a boomtown atmosphere in that community. Later that year, a line connecting with the GM&O was extended north from Eoline to Blocton, giving the north Bibb mining towns a rail link with the county seat. It also opened the way for development of other mines at such places as Moffat, Red Eagle, and Lucile. This rail connection was in use until its abandonment in 1945.

Joseph Squire. Many notable industrialists consulted with Joseph Squire to tap his unparalleled knowledge of the Alabama coal fields. Truman H. Aldrich, Henry DeBardeleben, and J. W. Sloss were frequent visitors to Squire's Helena home. (Courtesy of Ken Penhale.)

The Great 1896 Train Wreck

Connecting Blocton with Gurnee and Montevallo, the 110 foot high and 1,300 foot long trestle spanning the Cahaba River was the scene of a dreadful train wreck on December 27, 1896. Twenty-eight passengers plunged to their deaths when local train Number 40,

Cahaba River Trestle. Two days after Christmas 1896, twenty-eight people were killed when the train they were riding plunged through this bridge and crashed into the rocks and river below. (Courtesy of Will Reese.)

consisting of the engine, baggage, and two passenger cars heated by coal stoves, crashed through the trestle. The train, after falling into the river gorge, quickly caught fire trapping the injured. The number of dead varies depending on differing newspaper reports. The first accounts listed twenty-two fatalities. A tenth anniversary article declared twenty-eight deaths. Reports said only four on the train survived. When word of the accident reached Blocton, a train was dispatched immediately carrying every physician from the town to the scene. Listed in the *Birmingham News* were Drs. Robinson, Ray, (T)rigg, Jones, and Meadow. Accompanying the doctors was a flatbed car crowded with miners who volunteered to help. The paper said, "The doctors went to work manfully for the injured and had them carried to a warm place and where they could be given some little attention. The men who came over from Blocton went to the work in the debris made by the fire. They worked hard to rescue all that could be rescued."

Twelve-year old Harry Adams and others from Belle Ellen heard the wreck. He and several friends ran the mile and a half distance through the woods to the accident site. For the rest of his life, Adams remembered the screams of the injured and the terrible condition of the burned bodies at the scene. Belle Ellen mine superintendent Llewellyn Johns accompanied by several others, offered aid from the Belle Ellen community. Until a major train tragedy near Mobile in the early 1990s resulted in the loss of forty-seven lives, the Cahaba trestle wreck stood as Alabama's worst train disaster.

Railroad officials, at the time of the wreck, claimed that the horrible tragedy was the work of train wreckers. Those who arrived at the scene believed there was sufficient evidence to show that one rail had been moved entirely. The Louisville & Nashville Railroad offered an unprecedented $10,000 reward that was never collected. A story observing the tenth anniversary of the wreck stated the cause was never determined. As a result of compromise payments and in the case of the families of Tom White of Birmingham, the dead engineer, and L. W. Martin of Blocton, who sought court action, the L&N Railroad paid out an amount between $75,000 and $100,000.

Blocton was shocked. Stories circulated of those who would have perished had they not missed the train and those who were spared because they had changed plans at the last minute. The most disturbing news for Blocton was that there were eleven residents killed and three injured. The newspaper report said, "Many are the homes made sad in Blocton, Birmingham and Helena, funeral procession after funeral procession have wended their way through their streets to the cemetery where the last rites were paid the remains of the loved ones; some so charred and blackened that identification was almost impossible." A town accustomed to accidental deaths and injuries in the mines found this tragedy one of the worst experiences in its young history. Its occurring so close to Christmas added to the sorrow.

The incomplete list of those killed from Blocton included: Bruce Phillips, L. W. Martin, Miss Ada Powers, D. J. Powers, Mrs. Emma Powers and two children, Mrs. R. Little, mother and sister of James Gardner, and Andrew Bryson. Three of the injured survivors were from Blocton.

While out of historical sequence, this story is one of the events in the early days of Blocton that left its mark for years to come. Locals reported seeing the wreck's remains in the water for decades following. As late as the days of World War II, scrap metal from the train wreckage was removed from the river's bottom to be used in the country's war effort.

Blocton Growth

The Cahaba Coal Mining Company included prominent Alabamians on its payroll. In 1886, for example, James W. McQueen was joint telegraph agent for the Alabama Great Southern and the Cahaba Coal Mining Company stationed at Blocton and Woodstock. By 1910, McQueen had become vice president of the Sloss-Sheffield Steel and Iron Company in Birmingham. In Blocton, he helped manage complex communication and transportation needs of the new enterprise. It was a busy place. The *Blocton Courier* carried schedules of more than twenty trains daily into Blocton in 1892, a significant indication of the boomtown atmosphere that must have prevailed in the early history of the area.

One of the contemporary extant reports on Blocton's early development appeared in the *Shelby Sentinel* of July 21, 1886:

James W. McQueen. The early chronicler of the Alabama coal and iron industries, Ethel Armes, called McQueen "one of the big guns of the Birmingham District." He began his career in Blocton. While there he met and married Lydia Edwards, the daughter of the famous ironmaster Giles Edwards. (Courtesy of Virginia Green, Homewood, Alabama.)

Alabama Coal Mines-Blocton

(From the Chattanooga Tradesman)

by Henry E. Colton

One of the new enterprises of Alabama is the mines of the Cahaba Coal Company which are located nine miles south of Woodstock station on the Alabama Great Southern Railroad. The capital invested here is almost entirely from the Northwest, the principal owners being Mr. Underwood of Kansas City, and Mr. Clark, of Muscatine, Iowa and the general manager is Col. Cadle, also formerly of Iowa. The capitol invested is about $750,000 and the company owns 12,000 acres of coal land, eleven miles of railroad, standard gage laid with steel rail, 150 houses for hands, two slope openings into the coal seams, with complete stores, offices, etc. The number of men employed averages 300 and $12,000 per month is paid out as wages. The coal output at present amounts to 450 tons per day, but the mine is being so developed that it will reach 600 tons per day by next winter. Such is about the present status of an enterprise that was only commenced late in 1883. The Cahaba Coal Company opened on two seams of coal by slopes; later one of these slopes was abandoned. The location here is just at the apex of a slight roll in the coalfield, and there are seams in a few hundred yards of each other dipping both to the southeast and northwest. The ore at present worked dips to the southeast, at the rate of 13-1/2 deg. at the outcrop, decreasing to 9 deg. eight hundred feet down the slope. This slope is in the whole 875 feet long. It is one of the most complete in any section, is 18 feet wide and 6 feet high clear of the timbers, has a double trackway for hauling out coal and a manway. The mine is ventilated by a way or slope along side this manway, at the outlet of which is a fan, 10 feet diameter 3 feet face, and with paddles 3 feet in width, in summer [sic]. The water collects in a sump and is drawn out by a steam pump with 16 x 10 cylinder through a four-inch pipe.

The coal is now from four side entries on each side of the main slope, a distance of 200 feet being between them. The pillars left between the rooms are 10 feet thick, the entry pillars 15 feet. All coal in rooms is shot down, and the miners are paid 55 cents per ton for lump coal; good men mine 8 to 10 tons per day. In January one man mined 176 tons, and two mined 163 tons in 15 days.

The coal seam is as follows: From the bottom up, coal five feet, rash 6 inches, coal 10 inches, slate 18 inches, coal 18 inches, slate 3 feet, coal 10 inches. The 5 feet and the 10 inches of coal are mined, the next slate is left for a roof, and is not a good one. The coal is run over a screen after it comes from the mine and there all the rash which accidentally gets in is sifted out. As sent to market it is a very hard clean coal, in irregular shaped lumps. It is highly esteemed as a steam and grate coal.

Two miles northeast of this mine the company [is] opening another slope on the same seam [this likely refers to No. 3 mine]. At this point the seam is: coal 5 1/2 feet, rash 6 inches, coal 1 foot, then slate with no coal above. Hence this is likely to be a much better mine to work.

Domestic coal is sold on the cars at Woodstock at $2 and $2.25 per ton, steam coal at $1.25 and $1.35. The Alabama Great Southern Railroad is a great consumer, also much is sent to the New Orleans

& Northeastern, also to Morgan's [presumably J. P. Morgan] Texas railroads.

The entire business is managed by Col. Cadle, who commanded a regiment of the Union army during the late war, and he is evidently a man of good executive talent. In Mr. P. A. Smith, the engineer in charge of all the outside surveys and inside plannings of the mines, Col. Cadle has an efficient assistant. He is an Alabamian by birth and a young man who by modesty and careful study in the field acquired knowledge which will ever make him valuable to mine operators. The entire arrangement of the mine is excellent and the slope one of the best I have seen.

Coke Ovens

A major ancillary enterprise of the Cahaba Coal Company began with the mine development. Still visible are the remains of 467 beehive coke ovens constructed under Aldrich's supervision. Four rows each approximately a quarter-mile long occupy several acres just to the east of West Blocton. The structures represent one of the few remaining vestiges of old Blocton. The ovens originally supplied coke for steel mills in Anniston owned by Cahaba Coal Company directors, Samuel Noble and A. L. Tyler. An 1899 photo shows the ovens in operation. Some thirty men can be seen performing various duties in the installation's operation. In May 1998, a historical marker was dedicated reflecting a brief history of Blocton and the coke ovens, a project sponsored by the Alabama Historic Ironworks Commission. Significant park development at the site was begun in 1997 and shows potential for making the remains of the coke ovens a prominent addition to the Cahaba Trace, a scenic and historical route through the Cahaba River Valley established by the Alabama Legislature

Details of the coke ovens of Blocton can be found in records of the Tennessee Coal, Iron, and Railroad Company transcribed by W. B. Allen in August 1932, in the Birmingham Public Library's Department of Archives and History. Plans were in place from the Cahaba Coal Company's beginning for the erection of coke ovens. The first phase of the ovens was completed in March 1888, and the ovens were expected to be operational by August of that year. However,

Blocton Coke Ovens. A view of the Blocton ovens from the north end. The men in the foreground are standing next to a stack of large refactory bricks continually needed to repair the ovens. (Courtesy of Tennessee Coal, Iron and Railroad Company Annual Report, 1900.)

Jeremy Prather. (Courtesy of Howard Prather, Birmingham.)

a local strike intervened in the plans and the first 140 ovens were not placed in operation until around March 1889. Soon after, 150 more were ready to produce coke. The quantity shipped during the year was 17,802 tons, 100 tons per day to the Eureka Company at Oxmoor and the same quantity to the Birmingham Furnace and Manufacturing Company at Trussville. The Pioneer Mining and Manufacturing Company's successful use of coke made from the company's coal, and a favorable report by John Fulton, a recognized authority of the time on the qualities of coke for blast furnace purposes, led the company to make preparations for the erection of 185 additional coke ovens. However, slightly fewer were completed than had been planned. These last ovens were ready for firing in June 1890. The total built was 467 with an output capability of 600 tons per day. Records indicate the facility operated until *circa* 1904.

For the construction of the ovens, huge boulders for use as buttresses were cut from solid sandstone outcroppings in the area as was the stone used earlier in pier construction along the company's railroad crossings of Caffee Creek. The hard sandstone is from the Pottsville formation, the same geologic strata containing the area coal seams. The exact location of the site from which the stone was quarried is unknown. More than a million fire bricks specially cast for the oven construction were shipped by rail from the Birmingham area.

The first superintendent of the coke oven operation was Lewis Minor of Connellsville, Pennsylvania. Jeremy Prather, a North Carolina native and Confederate Army veteran, is one of the known stonemasons who worked on the ovens' construction. His grandson Howard Prather of Homewood reported that this pioneer stonemason is buried in an unmarked grave at Mount Carmel Cemetery in West Blocton. Jeremy Prather was the father of James and Zeb, both of whom worked in the local mines and, in 1898, they served in the army during the Spanish-American War. They and Albert Gilbert are the only known veterans of that war to have resided in the Blocton area. Neither Prather saw action, however, spending their service months in Montgomery. Zeb and James are also buried in Mt. Carmel Cemetery. The Prather family placed memorial stones for Jeremy and his wife in the family plot at Mt. Carmel in October 1997, and a government supplied tombstone was added in 1998. At the placement of a Mt. Carmel Cemetery historical marker in November 1997, special tribute was given to Jeremy Prather's Civil War record and his role as a stone mason who helped build the Blocton coke ovens. Three bricks from the coke oven ruins were cemented in the ground near the Prather foot marker during the November 1997 dedication ceremony. The name of another brick mason who worked on the coke oven construction appeared in the local paper during the town's Bicentennial Celebration. This was Reed Martin, Sr.'s grandfather, Lucian "Dump" Martin.

Local residents have used materials from the coke ovens site for years and in the 1980s some of the sandstone endwall material was used to reconstruct the double furnace at Tannehill Ironworks Historical State Park. Many underhouse enclosures and yard retaining walls in West Blocton are constructed from brick from the ovens. The side and back walls of a building on Main Street occupied by a pharmacy in 2000 (the former Gillespie/Collins Funeral Home) are made of brick from the site. In 1997, Tulon E. Desmond of Birmingham said that after the company left Blocton, his father contracted the sale of brick from the ovens site as agent for TCI, so the removal of materials was an orderly and official business enterprise, not a scavenging operation as some have reported. An advertisement appeared in the local paper in 1933 that offered a million bricks for sale. Tulon Desmond and W. A. Jones were listed as agents.

The coke ovens, later operated by TCI, supplied coke to the company's Oxmoor and Bessemer iron furnaces. Coal from TCI's Blue Creek mine as well as from the Blocton mines was used in the coke making process. Information on the ovens is found in Squire's *Cahaba Coal Field Report*, TCI Company's *1900 Annual Report*, and in the Allen transcribed material on the Cahaba Coal Mining Company cited above.

An interesting recollection of the ovens appeared in a 1946 edition of the *Centreville Press* written by S. L. Johnson of Oxford, Alabama, who spent his boyhood in sight of the ovens:

> When I first saw the coke ovens as a boy they were a great show to me. From our home [on Smith Hill] they could be seen at night - long rows of flickering light, over which hung a black pall of smoke. In still, cloudy weather the smoke covered all the territory around Blocton. I soon succeeded in getting acquainted with the crew that ran the small engine out on the track over the ovens, and they allowed me to ride with them and in that way I could see the whole layout. The row was called a battery, and was built of stone and brick with the interior divided into cells or hives. There were huge doors on each side near the bottom but these were sealed up after starting the fire.
>
> When the coal had been made into coke, the fire was allowed to go out and the bottom doors of the hives were opened and men reached long steel hooks into the ovens and pulled the coke out onto the earthen ramps and water was thrown on it to cool it, and then using great coke forks they threw it into the coke rack cars and it was carried to the furnaces.
>
> I have seen whole trains of coke going out of Blocton and I suppose some cars moved out each day. These men knew their work, and it was interesting to watch them opening an oven and pulling down the coke. They often worked long hours and their families sometimes helped them and brought their meals to them where they worked.

This article was read by Martin Everse, Tannehill Park Director, during ceremonies dedicating the West Blocton Coke Ovens Park and the West Blocton Improvement Committee's historical marker in June 1996.

Official records of coke production in Alabama indicate the last year of significant coke production in Blocton was 1904. Technology in coke oven design made beehive ovens obsolete. One design used in the Birmingham District was the Semet-Solvey ovens that provided extraction of byproducts such as amonia, tar, and surplus gas.

Johnson's writings in the *Centreville Press* provide more insight into the early days in Blocton. In the June 23, 1949, issue he recalled the Number 1 Mine:

> Blocton grew up around No. 1 mine. The railroad came down the creek valley and the depot was built only a few hundred feet from the mine entrance. The railroad yards, the depot and the mine tipple occupied a level area between the creeks and at the base of a hill. As the mining operations expanded the depot building was enlarged and when I first saw it the office and freight rooms, all included, were over one hundred feet long and a great volume of freight passed through it daily.
>
> The mine entrance faced southward and the tipple extended across the valley at the rear and to the west of the depot and the empty railroad coal cars were pushed up a spur which extended up the creek. [This extension crossed the creek, and the railroad bed is still visible upon close examination of the undergrowth.] These [cars] were let down by hand under the tipple and as a boy I watched many times these cars being loaded from the strings of little mine cars which were drawn up from the mine by the hoisting engine. Old No. 1 was a great mine about 1895 when I first saw it running.
>
> The company offices were built on the hillside just north of the depot and the footpath to No. 1 Town and Suttletown went up the hill past the offices. No. 1 Town was the first town and the first houses for the miners were in a row far over the hill and near Suttletown. The mine Superintendent's home stood on the highest point, [this is one of the earliest found references to this structure] and on the right of the road to Suttletown was a row of nice homes occupied by the switching engineer and the mine inspector and others.

> The Baptist, Methodist and Presbyterian Churches and the school were all in No. 1 Town and also the homes of the doctors and the preachers. Dr. Twigg [sic] [should read, Trigg] was a company doctor, and Dr. Peacock arrived in the Nineties. Our Methodist preacher was named Len F. Whidden.
>
> No. 1 mine was said to have a lot of water and a great and constant volume of cold, clear water poured from a six or seven inch pipe from the mine into a ditch which crossed the road west of the mine and into the creek. [It was a] great spot for barefoot boys to wade and play on a hot summer day. [A stream still flows from this abandoned mine entrance that has been back-filled to comply with government regulations.]
>
> A few hundred yards northwest of the depot and mine was the great mining company mule lot and barns. To a boy who had been reared on the farm this lot was a great attraction. I had spent lots of time feeding farm animals. The wagon road to Suttletown passed to the south of the tipple and to the right of the mule lot and up a draw and through No. 1 Town. A high fence enclosed about an acre of land and inside were long sheds with feeding troughs. From fifty to a hundred mules were kept here, the greater part of which were used in the mines for pulling the coal cars to the base of the chain slope. Others were used with delivery wagons and heavy hauling.
>
> The lot keeper was Lige Kemp, who was assisted by his son in feeding the mules. I would watch them pour whole sacks of corn and oats down the long troughs, and break for the mules many bales of hay. The mules were well fed, but it was terrible to see the cuts and gashes on them as they came from the mines. Some were injured in accidents, some by the harness, and some were beat up by the drivers. Drivers were often discharged on account of extreme cruelty to the mules.
>
> Near the barns was the harness shop. The harness maker was Mart Riley who had an endless job of replacing, repairing and making new harnesses. Even so, he had lots of time to talk and gas with Lige Kemp and the wagon drivers.

The S. L. Johnson *Centreville Press* series contains interesting further description of the Number 1 mine as he remembered it from his youth. He described a very early miner's lamp (it was shaped like a small tea pot, a wick protruded from the snout) that was worn on their cap visors. It used oil that gave poor light and smoked very much. The oil, said the writer, was hard to put out but when it did go out for one reason or another, the miner carried a supply of matches. Next to breathable air, light was an essential in the depth's blackness. One company marketed this fuel as "Miners' Sunshine." Rather than having a consistency of oil it was more like lard and had to be packed into the lamps by hand.

Mining fundamentals and basic operations were well described by Johnson:

> All mining was by hand. With a long metal coal auger the miner bored a hole in the face of the coal seam at his working place in such a direction and slant that when he had placed the charge of blasting powder in the hole and fired it, the blast sent down a pile of coal just where he could most handily shovel it into the little coal car which the mule driver brought up from the main entry way. These little four wheeled cars carried about one to two tons, and when the miner filled the car he hung his brass tag with his number on it attached to a leather strap on a nail on the inside of the car at the front end. Then the mule driver came with his mule and carried the car back to the main entry and there it was herded along, with other cars from other rooms, to the foot of the line where the trip was made up for hoisting.
>
> Up on top, and under the tipple house, was a large room in which was located the hoisting engine, for a mine like old No. One, was a beautiful and efficient piece of machinery, and the engineer who operated it was a very skillful man. No one was allowed in this room while the mine was running, which I liked to look in at the window and watch him. He sat on a high seat where he could look down at his engine, and the big drum on which the large wire rope a mile long was reeled and unreeled, and he

> listened for the signals from down in the mine telling him when to start up his engine pulling the trip of five to ten cars from the bottom to the top of the tipple.
>
> When the trip ran over the hump at the top of the tipple he stopped his engine and the tipple workers halted the trip of cars with sprags [sic] made of hickory which they thrust into the car wheels [spokes]. Then they pushed the cars one by one over the weighing scales where the weight of the car and the number on the miner's tag was entered into the scale book. After that the car was pushed on to the tipple and its contents poured into the railroad coal car on the track below.
>
> When old Number One was running it was a busy scene around the old tipple. For more than twenty years coal poured out of this wonderful mine and up over that old tipple and hauled away for the world to use.

Johnson's memoirs of his boyhood days in Blocton and the way things were in the town's early years are vividly detailed. It is important to quote his works at length to give readers of a later generation a better understanding of those times. Among his last columns was one on July 5, 1951, which had more to say about the Number 1 Mine and other matters of community life. The subject was the company shop and miners' recreation:

> This was the place [the company shop] where the supplies for the mines were fashioned. In the blacksmith shop all the iron parts for the little coal cars were made, wagons repaired, mules shod, and miners' tools sharpened. Two or three smiths and their helpers were always busy there.
>
> Then there was the part where the mine slope rollers were made. Down the long slope from tipple shed to bottom of the slope in the mines, a long wire rope ran up and down in the center of the track, and unless supported, in some way, would have worn out soon by dragging on the earth and ties. And so about every fifteen or twenty feet a round wood roller, made of gum wood, about two feet long,

Hauling Mining Props. Underground, a miner's life depended on the strength of massive timber roof supports or props cut from trees. Horace Roland and Moodie Martin were hauling props to an area mine, *circa* 1915. (Courtesy of Billy Mack Roland, West Blocton, Alabama.)

was mounted on iron spindles at each end, the spindles being housed in bearings on a lug fastened to the ties.

The wire rope [cable] soon cut a groove in the rollers, and some time in a few days cut all the way through, then the repair man came with a new roller to replace the old. This went on every workday. B. J. Edmonds was the roller maker for a number of years.

There was a man for every job around these old mines. There was a timber called a mine prop. It was cut in the woods the right length for the height of the coal seam in the particular mine it was cut for. It was used to brace up the overhead rock after the coal was removed. To make up for any difference, if the prop was too short, the miner had a "cap board," a short pine board about five inches wide and six inches long. They would be used to be wedged in at the top of the prop that was found too short.

The miner worked long hours, and sometimes very regular, but when he did have a holiday, he got lots of fun out of it. All of them did not drink liquor and beer, but some of them had beer parties, and I can recall seeing the little, dumpy, five gallon kegs of beer being unloaded from the train from Birmingham, and later this same keg might be seen on a wagon loaded with men going to some quiet place in the woods.

The main diversion was baseball. Every body could enjoy and have part in this. Old Number One Mine almost mustered a ball team in the Nineties. It was called the Blocton team and its fame reached far and wide. Some times an excursion train would carry the team to Bessemer, Montevallo, Aldrich, Randolph and other places.

The players were: Joe Smith, first base; H. H. Farrah, second base; Fred Lawley, third base; Bob Burton, right field; Louie Farrah, left field; Albert Bracknell, center field, Charley Fuller, short stop; Pat Finnen, pitcher, and Joe Sansing, catcher and DeWitt Edmonds, utility man. All these worked in No. One mine except Finnen who worked in No. Two and Joe Smith, who was a trainman.

Ball games were played in a rugged way in those days. These men had rugged occupations, and when they went on the playing field they played a "rugged" game. It is not meant that they were not clean sportsmen, but the other team would know after a game, win or lose, that it had not been coddled on a feather bed.

There was a Fourth of July game about 1896, and a team came down from Bessemer. There was speaking at the speaking stand under trees on the high ground east of the creek at the edge of No. Three Town, and perhaps dinner on the ground. The ball game went on so long, a double-header, that the train had to be held for more than an hour so that the crowd and team could go home.

Some times there would be a hot fist fight around a game that might furnish as much excitement for the fans as the game. These fights were more often engaged in by the fans than the players.

There was no paid admittance to games in those days, but on big occasions the fans would provide big tubs of ice-cold lemonade and soda pop for the players. The soda pop bottle had made only a recent appearance. The bottle was shaped about as it is now, but it had no metal cap. It had a hard rubber stopper held in place by a ring shaped wire. When the bottle was filled the stopper was brought up by force and closed the opening and the gas inside made it tighter. When you wanted to open the bottle you would strike the wire ring a hard blow and force the stopper down. You had to be ready to drink, for the red pop would soon be gone in foam if you waited.

The W. B. Allen file in the Birmingham Public Library on the early days of the Cahaba Coal Company's history and how it related to the Tennessee Coal and Iron Company reveals details about the late 1880s developments in the company town.

Blocton Mine Number 7. (Courtesy of Will Reese.)

During the fiscal year ending in March 1888, four new mines were opened. They were Number 3 and Number 4, both slopes, and Number 5, a shaft mine, and Number 6 another slope. According to a brochure prepared for the May 1888 visit of the American Institute of Mining Engineers, six mines had been completed. By the summer of 1889 two more mines, Number 7 and Number 8 were in operation. Railroad tracks were completed to these openings and to the coke ovens adding six miles of lines to the company's rail installations. During 1889, 20,000 tons of coal were shipped to the Export Coal Company at Pensacola and from there to Havana and other Cuban ports. The coal industry had introduced international trade to Bibb County, a "first" for the county, and in coal production, the first in the state. According to Armes, this development was a milestone in the history of Alabama's coal and iron industry. Her chronological table entry says: "1889–1890. Export trade of coal with Gulf ports, West Indies, and South America inaugurated by T. H. Aldrich."

Dr. Stewart Lloyd of the University of Alabama observed in 1957 that, "Aldrich personally taught householders, churches, stores, etc. how to use Cahaba coal instead of wood for heating, and more important, pushed its use on the railroads running west from New Orleans. Through his Export Company, later merged with the Cahaba Company, he supplied coal to Central and South America, and bunker coal to ships at Mobile and New Orleans."

The *Centreville Press* editor called the series of memoir-like articles by S. L. Johnson, "A Brief History of Bibb County." The entry for June 10, 1949, told about Number 3 Town and other locations:

> In the old Blocton, the road ran east from the L. & N. Depot and passed under the No. 2 railroad track, and past the large meat market; across the road on a bridge, then over the tracks of the E. T. V. & G. [East Tennessee, Virginia, and Georgia] Railroad and you would be in No. 3 Town. This was a

large community, laid out in streets running east and west. The road ran east to about the middle of the town and then turned at a left angle to go up on Smith Hill.

The town was built on an abandoned farm field and sloped from the side of Smith Hill to the south edge of the town near the No. 3 track. The houses were nearly all of frame construction, four rooms and one story. There were no shade trees, very little shrubbery or flowers, but most of them had a garden and a cow.

The town was built to house the miners who worked in the No. 3 mine, the entrance of which was a half mile east of the town, and up a long draw between two steep hills. The houses were occupied by native farmers, turned to coal miners, and a few Bulgarians. Water for the houses was drawn from a hydrant in the street. Water pipes brought the supply from water tanks on the hill above the coke ovens [just to the northeast of the 1925 Baptist Church building]. No. 3 was a large mine and all the workers could not be housed in No. Three Town so many of them lived in Little Italy, Stacks Hill, Smith Hill and Thomas Town. Little Italy was the exclusive Italian community built in a deep draw between Smith Hill and Stacks Hill. Some of the houses were built on a hillside so steep and high up that sometimes another house would be built under it. Cows and goats were kept under the hillside houses. When they [the Italian residents] held a wake it lasted sometimes two days and nights, and they could be heard up on Smith Hill where we lived. Then when the funeral moved out to their church [St. Francis of Assisi Catholic Church] on Gunlock Hill near the company store, the men walked in a company behind the hearse. Their customs were so very different from our own that I can recall walking up and down on the railroad track in front of their town to see what could be observed going on. They were good citizens for I never heard of them giving any trouble.

Houses on Stacks Hill to the left of the track, and near the mine entrance, were built of logs and chinked with clay. The miners who lived in these houses had a short walk of only a few hundred yards to the mine entrance. One family kept pea fowls and their cries were like a siren. They could be heard day and night on Smith Hill.

Thomas Town was about a half-mile north of the mine and on the road to McCulley Hill and Belle Ellen. Fifteen or twenty families made up Thomas Town and these worked in various places. Bill Seales and his family lived there, and he and his sons followed the business of supplying timbers for No. 3 mine. These timbers were called "props" and were cut from pine trees at the proper length for propping up the overburden as the coal was removed.

Many miners lived on Smith Hill and as a boy I could see them going by our house before daylight with lighted lamps on their caps. Our neighbors were miners: Jim Greene, Jim and Leonard Stone and a Mr. Lawley. Stone saved his money and opened a store near our house. Other neighbors were John Gholson, Mrs. Holifield and Mims Keith. About a block north there was a Henderson family who farmed the lands of Rufus Smith. They were Mormons and entertained traveling representatives of that faith who traveled by twos and lodged with whomever gave them shelter

The switch engine would have to get a flying start back on the yards to push a string of empty coal cars up the hill to No. 3 mine, and in the winter we could witness this run from our house on the hill, and it never ceased to fascinate me. The mine tipple was up at the head of a long hollow between the hills, and to get the train of cars up there the engineer would get a long running startup north of the depot, and with the fireman shoveling coal and the whistle blowing they would go up around the hill at the foot of No. Three Town and up through Dago Town, and if they had a good run, would push the cars past the tipple and spot them, and then bring down the train of loaded cars. If they had too heavy a train, or if the track was slippery, they would have to go back to town and run again, and the second run was the most fun to me because then they made most fuss.

> One of the engineers was named Dill and one of the switchmen was named Smith another was Will Stamps and one was named O'Rear, the father of Mrs. T. C. Tuggle. One fireman was Tod Broadhead and another fireman was John George. Smith lived near the coke ovens and I plowed his garden for him.
>
> Near the edge of No 3 Town stood the Odd Fellows Hall, a two story building and when I passed there at night there was always a light up there and I wondered what they did to the people who joined. When later in life I joined my first lodge, and was being shown the initiatory [sic] mysteries, I thought of the lighted windows in that hall.

In a subsequent column Johnson identified other residents of Smith Hill: John Woods, Love Riley, T. T. Harris, Gaz Caldwell, Monroe George, and W. R. Young. Also, there was Jim Greene, Frank Lawley, Cornelius Stone, Bail Holland, the Widow McElroy and her son, Thomas. He recalled that T. T. Harris ran a dry goods store and that H. C. and W. B. Reynolds opened a general store on Smith Hill. Workers there were Forest Colier and Watson Hollifield. Other residents were Bud Perry and Clay Fulton. Another store was that of J. R. Thomas after whom Thomas Town was named since he had built several dwellings there on the northern edge of Smith Hill.

The S. L. Johnson series in the *Centreville Press* also gave details about mine development during Johnson's youth. In August of 1949 he wrote:

> Blocton had a Number 5 mine, and it was a vertical shaft, located up and on the east side of the creek and north of the point of junction of the L. & N. and the Southern railroad tracks. The coke ovens were a short distance southwest and the roundhouse was a few hundred yards northwest. Number 5 mine was operated for some time and much coal was taken from it, but I came along too late to see it in operation. About 1896 it was being kept in a standby condition. Above the ground there was the cage enclosure, the boiler, and engine rooms and the tipple shed, all in good condition when I knew them.
>
> A man was there continually in charge to fire the boiler to make steam for keeping the pumps going, the air fans, and the lifting cage when needed. There was a beautiful engine which the man kept spotless, and as a boy I liked to be around when a party came and wanted to be sent down in the cage. The men going down would put lamps on their caps and when they stepped into the cage the man would send them down; one cage coming up as one went down. When I asked him how deep they would go down, he would reply, "Plenty deep." Later, with the other mines, Number 5 was abandoned.
>
> Number 4 mine was up a hollow about a quarter mile northeast of Number 5, and was abandoned when I first saw it. All the above ground structures had been removed, and at that time there was only a black hole in the hillside. There were a number of small houses near the mouth of the mine and Negroes [sic] lived in them. My information was that No. 4 mine had been worked out with a large output of coal in the late Eighties.

In Johnson's June 9, 1949, column he described the company store (this was after the days of the Cahaba Company).

> For more than twenty-five years the company [TCI] ran a great store in Blocton, at a point almost in the center of the cluster of towns which made up the whole of Blocton. This was a large building, perhaps sixty feet wide and a hundred feet long, and two stories high, and generally crammed with goods from bottom to top. Everything the miner needed could be purchased in this store, either for cash or with the company's "checks." These checks were issued to the man for any amount of time he had in it. A small discount was charged to the man for issuing, but many of the miners drew their time

up to the last minute and spent it in "the big store," A few thrifty workers allowed their pay to accumulate until payday, and then drew the cash and proceeded to shop around on the "Hill" or in Suttletown for the best bargains.

The big store was operated by a manager and usually a half-dozen clerks, and two or three delivery men. The deliveries were made by mule teams and on large, heavy wagons. My father, after moving from the farm to Blocton about 1895 drove one of those delivery wagons about twenty years. In that time he came to know all of the people in all the houses. The store manager for most of that time was Samuel Morgan.

The store stood about three hundred yards southeast of the L&N Depot on a hillside and facing north, and along side the railroad track to No. 2 mine. Just west of the store was Walker's bake shop. The post office stood slightly to the northwest of the bake shop, and in between [was] the road to Buck Town, the exclusive Negro town. On the bluff of the branch which runs between the depot and the store site, and facing south, stood the doctor's office. Between this building and the post office a road ran back northwest to Frank Dexter's livery stable.

When all the mines were in full operation all this area was a busy scene.

Cline's *Alabama Railroads* (University of Alabama Press, 1996) contains information on the markets for Blocton coal, "Aldrich's company shipped millions of tons of Cahaba coal to a multitude of railroad and industrial companies throughout the country, in addition to a growing number of Birmingham District furnaces. The Cahaba Company supplied steam coal to all but one of the railroads operating out of New Orleans, with the Southern Pacific consuming five hundred tons a day in its New Orleans-San Antonio run."

By 1890, the company town of Blocton had increased in population to 3,600 people, nine-tenths of whom lived

Company Store, Blocton. Coal mining companies operated their own general mercantile stores, and the prices found there were often a source of discontent. Miners often referred to them as "pluck me" stores. (Courtesy of Will Reese.)

Lawrence Family. The Lawrence family was one of the first to move into the new company housing at Blocton. (Courtesy of Jack Lawrence, Atlanta, Georgia.)

in the company's houses. The Cahaba Coal Company report indicates there were 404 company owned buildings including dwellings and structures for support enterprises. Photographs show there were varying sizes of company houses for miners. Some had as many as five rooms while others were smaller. External walls were mostly board and batten while interiors were plastered. Plastered interior walls were not common in rural Alabama homes so this may have been an influence of the two northern town founders who recognized the importance of insulation during cold months. There were no basements and each house was served by an outdoor privy, bathrooms being unknown. A tin bathtub near the kitchen stove or in front of a fireplace served during cold months. Some of the management-level homes were of clapboard construction as the superintendent's home, one of the last examples still standing.

The *Alabama Sentinel* of April 2, 1892, had a comment on the Cahaba Company's housing. The living quarters are described as model in contrast to other "stark" company housing. "Each family had its own house enclosed by a picket or other fence, with lots of room for the cultivation of garden truck. The houses . . . are lathed and plastered inside, rents reasonable and fair, coal fuel $1.50 a ton delivered, and water for drinking and domestic use supplied by the Cahaba Company in abundance. This is as it ought to be elsewhere in the coal belt." Unfortunately, the *Sentinel* continued, it was seldom as it ought to be elsewhere in the coal belt.

From this report it may be observed that credit is due Aldrich and Cadle for the comfortable company housing arranged for their employees. Also reported by the *Sentinel* was a company plan for its workers to take part in the "Savings Department." Workers could deposit amounts from $25 up to be managed by Birmingham banks and the company to produce interest rates from 4 to 6 percent depending on length of deposit. The paper observed:

> Under the control and management of a responsible firm like the Cahaba Coal Company, such a scheme is a positive benefaction to the working man, encouraging, as it undoubtedly does, the saving habit in men who, in many sad instances, are liable to squander their surplus earnings, if the same is permitted to remain in their custody. They might take surplus cash to Birmingham and Bessemer, 'get

half seas over' and turn their cash over to a saloon keeper for safe-keeping finding nothing left when asked for its return.

S. L. Johnson's remarkable memory produced many essays in the *Centreville Press* about his youth in Blocton. The following appeared on May 26, 1949. It provides information about the early days found nowhere else:

> I recall having seen Blocton for the first time about 1888. In going from our home near Rocky Church, we passed through River Bend, and then on through the pine forest for about six miles and came to Number Two Town. This town was located near the opening of the mine. There were about forty houses and these were created by the company for the occupancy of the bosses, foremen and others connected with the mine. It was intended to be a model town, for those days, as a company owned village for workers.
>
> The town was laid out in two streets, or four rows of houses, and these lay from the southwest to the northwest, with the main road running through one of the streets. The houses were well constructed of good material and some of them were two stories high. The people who lived in them were, in the main, high class workers. I recall that my father carried produce to sell to the people. As I understand it, many of them were from Wales, where they had followed mining before coming to this country. They were interesting people and had customs very much different from our own. There were no stores, all were served by the big company store just to the north. There were no churches nor a school. [Johnson's memory failed him on this matter. By 1886, the Methodists had built their first church in Number 2 Town.]
>
> The town was near the edge of a large tract of abandoned farm land that reached to the company store, and just a short distance to the south was the mine entrance. All day long and almost every week day in the year, the people heard the throb of the exhaust of the hoisting engine bring the little cars of coal up from the mine to the tipple, and also hear the rumbling of the coal being dumped into the train cars. The switch engine would go over every day, and sometimes twice, and bring a long train of loaded cars of coal to the yards, and it was wonderful coal, large lumps often weighing a quarter ton.
>
> No. Seven mine was located east of No. Two, about a mile and a half distant [this distance recollection by Johnson is exaggerated]. It was called a drift mine, so level that the coal was brought to the entrance with mules, where the little cars were carried by a small locomotive to Number Two tipple and loaded on the train cars.
>
> The railroad track to No. Two passed over the hill at the rear of the company store and crossed a hollow over a long and high trestle, which was later filled in to the top with waste from the mines. When the M&O Railroad was completed in 1898, the track to No. Two was continued on to Eoline, and much coal passed over this line. Lucile and other mines were opened along that line.

Blocton was not incorporated. In one of its areas a water supply could not be obtained so a small waterworks system was provided. Two policemen were employed by the company and good order was maintained. The Cahaba Coal Company's 1888 report also said the railroad between Blocton and Woodstock carried 31,436 passengers and 427,000 tons of freight including coal and coke during the year.

These developments led to the formation of a new voting precinct in May 1888. The *Bibb Blade* reported:

> The Blade is informed that the commissioners court at its last session, which was held on the 2nd Monday in April, granted a new petition to form a new beat at Blocton. This place has, heretofore, been divided between beats No. 1 and 9 but the rapid increase of population has made it necessary to form a new one. We are not informed where the lines will be located, but are informed by our present tax-collector that about 400 votes can be polled within the beat. [Women, most Blacks and foreign-born citizens were not eligible voters in this era.]

Eventually two beats, ten and eleven, were formed following appropriate court action.

The prominence of the Blocton mining operations did not go unnoticed at the University of Alabama campus, and its geology students were frequent visitors in these early years. The area was also well known by the state geologist, Dr. Eugene Allen Smith, stationed in Tuscaloosa who had a professional relationship with Truman Aldrich. The following is an entry under "Locals" in a student publication, *Alabama University Monthly*, dated June 1887. The editor was Cadet O. D. Street who later served as a probate judge in Marshall County and still later practiced law in Birmingham. In the 1920s he married a prominent West Bloctonian, Mary Curd Allen.

> In the early part of last month the Senior Scientific Class made its first Annual Geological Excursion for the purpose of obtaining some practical geological knowledge. Hereafter each year this class will make a similar trip to various parts of the state. This year the scene of its labors was the Cahaba Coal Fields in the vicinity of Blocton, Bibb County and the great anticlinal valley separating the Warrior and Cahaba fields, which valley can be easily traced as far north as Pennsylvania. The class succeeded in getting complete sections of this valley both in the vicinity of Woodstock and at Birmingham. The boys enjoyed themselves finely, and at the same time derived an immense practical benefit from the trip. Dr. Eugene A. Smith, Professor of Geology and State Geologist, assisted by Prof. W. A. Langdon, was in charge of the party. We owe many thanks to the people of Blocton for the many kindnesses and courtesies extended to us by them. Especially do we wish to thank Mr. T. H. Aldrich, President of the Cahaba Coal Mining Company, for his kindness in granting to the party a free pass over his road from Woodstock to Blocton; to the gentlemanly and affable Mr. Latardi, who shall ever retain a warm place in our remembrance for his many courtesies and the kindly interest he took in us, and for carrying whom so deep down into the mine, from which he issued again puffing and blowing in good style, we have not ceased to upbraid. We must also tender our thanks to young folks of the town for their earnest endeavors to make our stay a pleasant one, and for the nice dance they also gave us. Our thanks would not be complete without a mention of our most genial and obliging host and hostess, Mr. and Mrs. W. H. Douthit. During our short stay with them we came to feel ourselves perfectly at home, and we shall ever cherish a pleasant remembrance of them and of the few days we spent under their roof. Taken all in all, we do not recall any place more pleasant than this little mining town. We must not forget Mr. J. U. Ray, Jr., who attended the University two years ago, but is now conductor upon Mr. Aldrich's Mineral Road. We, alone of our party, were not previously acquainted with him, but it did not take long to get acquainted, and to be sure a jollier, better-hearted fellow than Ury is seldom again soon found. He also shall not be soon forgotten. All of our party unites in the wish to be able to visit Blocton again.
>
> One of the geological party happened to a very sad accident during his stay at Blocton. The young people of the town gave to the party a dance. Mr. R., who contemplated cutting a big swell carried his pumps along. When the dance began he slipped them on and placed his shoes under the corner of the house. With merry foot he tripped the "light fantastic" for some hours, but when the time to go home was announced, the shoes oh! Where were they? Great was the young man's grief when he learned they had fallen a prey to the rapacity of some miner.

Further evidence of academic and professional interest in the new mining development in Bibb was the visit by the American Institute of Mining Engineers during their May 1888 meeting. A brochure was issued by the Cahaba Coal Mining Company describing the seams, mines, analyses of the coal and coke, location of the operations and other data. The seams were identified as the "Underwood" and "Woodstock." The publication said the company had built 325 houses for its employees and had employed 750 men.

Robert E. Praytor, historian and pastor of the First Baptist Church of West Blocton, has located maps in court-house records showing Blocton street lay-outs north of the railroad and along Caffee Creek. The "model community" shows north-south, east-west streets: Tenth, Ninth, Eighth, and Cadle streets on the hill, and Front, Aldrich, Anna, Spring, and Madison Streets were between the creek and the Old Woodstock Road. Church, school, community garden, and tennis court sites were included in the plans. All evidence of this configuration has disappeared, obliterated by erosion, timber harvesting, and structural abandonment.

A conflicting report in a special edition of the *Blocton Enterprise* in August 1910 reports that surveyor William A. Harvey, a native of Auburn, was employed to lay off the streets of Blocton (perhaps this should have read West Blocton). He later entered the mercantile business and was given credit for erecting the first brick building in West Blocton, the drug store built in 1900 bearing his name. He also was owner of the first electric plant in town called the Hydraulic Light and Power Company that supplied electric power to several downtown businesses. Harvey was active in civic affairs serving as a town councilman and a member of the school board and the dispensary commission. He was married to Minnie Pratt, daughter of Leonard C. Pratt of River Bend.

The Organized Labor Movement in Blocton

During the decade of the 1880s, organized labor in its infancy came to the Alabama coal fields. Blocton was one of the more active areas in the movement. It is reported that the Cahaba Coal Company in its early days had built a structure to be used jointly as a "union church, and school house." The upper floor was for the use of the Knights of Labor. This national group's passion for anonymity has left the barest of records of their activities, likely because of the need to protect leaders and membership from the retribution of mine owners.

According to Ward and Rogers' *Labor Revolt in Alabama* (1965), many of Alabama's strikes in the 1880s probably were inspired by the Knights of Labor. This book tells of the first strike action in Blocton. The basis for discontent was the practice of the owners lowering wages when profit margins failed to meet their expectations. In some areas of Alabama when miners struck for retention of wage levels or better wages, the mine companies would employ convict labor. However, there is no evidence that convicts were used in Blocton although they were employed at a later date in the nearby mining towns of Belle Ellen and Lucile. A report on the Cahaba Coal Company appearing in the December 17, 1884, *Bibb Blade* emphasized that "none but free labor [was] employed." The complaints, low wages and convict replacements during strikes, were major reasons for unionization. In the mid-1880s when the Knights organization faded, other groups formed. The National Federation of Miners and Mine Laborers, organized in 1886, took up where the Knights had not performed effectively. State miners voted overwhelmingly to affiliate with this new union.

Blocton's First Strike

Blocton miners recorded their first strike under the leadership of the Miners Federation. In October 1888, miners at Blocton struck against a reduction in wages. It was this strike that interrupted the building of the company's coke ovens. There were other strikes at the time at the Pratt Mines and in Walker County. All three strikes ended in failure. The *Birmingham News* printed the following in their October 22, 1888, edition:

> STRIKE AT BLOCTON
> TWO THOUSAND MINERS QUIT WORK TODAY
> AND THE MINES SHUT DOWN
>
> The entire force of miners, numbering nearly 2,000 men [this number is an exaggeration], went out on a strike at Blocton Mines this morning. An official announcement to this effect has been made by the authorities of the Cahaba Coal Mining Company. The mines are now shut down and will remain so until the matter is settled. The trouble grew out of the announcement that wages would be reduced 10 per cent. The miners refused to accept this reduction, and consequently quit.
>
> All the miners quit and left very quietly, there being no trouble of any kind, nor is there likely to be any. A large number of them came into the city [Birmingham] this morning. It is very probable that a conference between the miners and officials of the company will be held at an early [date].

Company bosses selling the most desirable work areas was a major factor in worker dissatisfaction, and the practice was attacked by Blocton miners in 1899. Researchers Brown and Davis recorded the following in *It Is Union and Liberty* (1999):

> Corrupt practices in the distribution of cars and rooms were scaled back (as a result of 1888 strikes). In 1899 the men at Blocton No. 3 mine forced the resignation of a mine boss who sold rooms to the highest bidder. One "Enos" recounted the episode in verse:
>
> There was a boss at No. 3
> Who with his men couldn't agree,
> And so the Blocton miners struck
> And said this boss must light a shuck!
> This smart boss was on the move
> To try his innocence to prove,
> But all his efforts came to naught
> When men turned up whose room [they] had bought;
> They to the truth did firmly stick,
> And this poor boss turned quite sick,
> Then to the office he straight went
> And in his resignation sent.
> It was accepted and there will be
> No more rooms sold at No. 3
> A better chance we now will stand,
> For our new boss is a very fine man.

On October 30, the *News* reported that the strike was still on with no prospects of a settlement in the near future. Miners were reported to be in Birmingham looking for work with the railroad.

Alabama: the History of a Deep South State by Rogers, Ward, Atkins, and Flynt (1994) states the reason for labor unrest: "Short of slave labor, it was difficult to imagine a more exploitative system than the coal industry. Miners' wages, always pitifully low, were now reduced even further in the vise of economic depression. The miners were cheated on the coal they mined, cheated on the rents they were charged, cheated at the company store, and summarily fired at the hint of complaint or union activity." This observation was in connection with the major strike of 1894, which will be treated in pages to follow. The Aldrich operation in Blocton has not been characterized as this extreme, nevertheless, the statement is generally applicable to the state's labor history.

Jesse Cole Persinger, the Methodist preacher serving Blocton at the time, writing in his eightieth year, told of the 1888 strike in his autobiography. He remembered the troublesome times: "The church was growing . . . and everything seemed to be very hopeful until the early spring (the septuagenarian's memory failed on details, the strike was in the fall according to Rogers, *et al*) when there was some agitation among the miners. Everything was shut down in the way of work in the mines. On and on it lasted; good men and their families were tied there with no way to make a living. The labor troubles were caused by hot heads and foreigners." Reverend Persinger did not remember, it seems, the wage reduction or other reasons for the miners' work stoppage.

In his life story, Persinger gives some insight into the problems a single man encountered in the coal mine boom town and to the crowded conditions in boarding houses. He wrote of his difficulties in finding suitable lodgings:

> There was no such place [in Blocton] as the good one I had at Woodstock. Nearly every family had a spare room to let out. There were two or three big houses the company had for boarding houses, but they were not desirable places for a preacher to board. I must have moved four times. Once I had a room with a teacher, Charley Cross. We were great friends. He left to go west and eventually

> became a preacher. I lived for a time in a room adjoining the doctor's office but later moved in with Brother Self's family where I had been taking meals. He was a bridge gang foreman and boarded some of the men who worked for him. It was a crowded house, one room housed five men. Mrs. Self, an elegant lady, said the only space she had was a bed I'd have to share with her brother-in-law, the only one in the house with a bed to himself. I was desperate enough to agree to that arrangement. While the men were away at work, I did the studying my profession required. I swore to find a wife so I would not be alone.

Those who know mining towns appreciate that another reason for discontent and concern was the lack of mine safety. Horror stories abound on the subject, some more poignant and sad than others. Two items in the April 15, 1893, *Blocton Courier* are representative: "Geuseppi Graziono, an Italian, working in Number 5 mine was caught under falling rock last Tuesday morning at 10:30, and killed. He leaves a wife and two children in the old country." And in the same column, "Louis P. George was instantly killed by a falling rock in Number 1 mine. It took eleven men to pry up one end of the rock, in order to get Mr. George's body out. The deceased was a young man and well liked by all who knew him. To his heartbroken young wife and other relatives we offer our deepest sympathy." In the December 23 *Courier*: "On last Thursday two white men, Joe Schober and Willie Patterson, and a Negro whose name we did not learn, were very badly burned by an explosion in Number 2 mine. The fire was caused by ignited coal dust. Face and hands of the men were burned but they will recover." In February 1894, a fire in Number 2 mine took the life of J. T. West, the mine pumper who was overcome by gas and smoke while trying to escape. West left a large family.

The January 10, 1895, *Blocton Journal* reported a gruesome account of an accident in the Number 1 mine:

> Last Saturday morning, a little after seven o'clock, a Negro by the name of Frank Sheppard, who was employed at No. 1 mine to keep up the rollers on the slope, was accidentally killed by a trip of loaded cars. The particulars, as far as we could learn, are as follows: The Negro was at his work on the slope, and got out of the way to let the trip pass. After the cars passed, which were being hoisted to the top, the Negro resumed his work, thinking everything was all right. He had occasion to cross the track, and in order to do so, had to crawl through a small opening. Just as he had gotten his head through, four of the cars, which came uncoupled before reaching the top and ran back into the mine, caught him, and he was instantly killed. A hole the size of a man's fist was knocked in the side of his face and head, scattering his brains in every direction.

The *Montgomery Advertiser* of April 24, 1897, carried a mine disaster story from Blocton:

> A disastrous explosion occurred in No. Two mine at this place this morning about 3:30 o'clock. Four Italian miners were burned to death. Their clothing was burned off them and the skin was peeled from their faces, fingers, arms bodies and legs. They presented a most ghastly appearance. The dead men worked on the night shift and also in adjoining rooms. All were young men except Charlie Talliali, who was about 50, and all were unmarried. It seems that no one else was in the mines in that section at the time of the explosion and it is not known just how it happened. The force of the explosion was terrible. Props were blown down in the rooms and the pillars of the entry were knocked down for quite a distance. No. 2 has always contained gas and great caution is exercised in mining there. Had the accident occurred in the day time the dead would have numbered into the hundreds for there is a large force that works on the day shift.

Ellison, in an article in the fall 1986 *Alabama Heritage* magazine, recounted the plight of the miner: "For the men who worked the mines life was hard and dangerous. During the early years, especially when working conditions were more primitive, the Bibb County mines claimed many lives." The *Blocton Messenger* of July 3, 1903, described the miners' condition. "The coal miner's wages shouldn't be measured by other classes of labor. The miner risks his life as no other man does. In the morning when he enters the dark hole that carries him away from the light of the world, he knows he is risking his life."

Accidents were recorded almost every week in the local paper. The State Inspector of Mines Biennial Reports began in 1892. These reports are also an official source of recorded death and injury in the states' mines. Ellison summed up the circumstance succinctly: "Accidents in the mines resulted in demands for more and better safety measures, and in the consideration of the danger, higher wages."

Ward and Roger's book reports the organization of the United Mine Workers of America (UMWA). The first convention in Columbus, Ohio, in January 1890, placed Alabama in its District 20. By the end of June, local unions had been organized in Bibb County. One of the first strikes by Blocton miners after affiliating with the UMWA started in December 1890 and lasted until middle January 1891. The strike was reported in the records of the Cahaba Coal Mining Company compiled in 1932 by W. B. Allen: "The strike was caused by an effort of the United Mine Workers of America to gain control of the mine labor and to dictate wages and rules to the mine operators. The strike was rendered unsuccessful in its object by the introduction of Negro labor which indicated to the striking miners that they would be permanently unemployed unless their demands were withdrawn." There is no record of Negro laborers being sent to Blocton, however, there was an incident involving Negroes connected with a mail train recorded in other pages. (See Chapter III).

The *Bibb Blade* gave a terse report of the 1890 strike: "Blockton is on a big strike. They were getting 45 & struck for fifty cents two weeks ago and from what I hear will stay struck as the Co. refuses to raise or compromise. Strikes has [*sic*] proved to be a curse in southern states."

To show that nothing in human relations is constant, Birmingham's *Labor Advocate* reported an interesting Blocton incident in August 1901. The paper's article is paraphrased by Daniel Letwin in a University of North Carolina Press publication, *The Challenge of Interracial Unionism*, as follows:

> Both the extent and limits of inter-racialism came to light at Blocton, where the meeting hall of the white local burned down in 1901. To assess the situation, the black local and the white local held a joint meeting in the latter's hall. There, one participant reported, the black miners offered "enthusiastic" support for the funding of a new hall for the white local: "Brother Starr, a Hercules in size and oratorical power, said that the colored miners would be as proud of an appropriate hall for the white miners as they would for themselves." A magnanimous gesture of solidarity across the color line in support of a racially defined local—here was an approach that defies sharp dichotomies of race and class consciousness.

Blocton miners' dedication to the labor movement and support of miners nationwide is evidenced in a *Blocton Courier* item of August 1892. The famous Homestead strike in Pennsylvania had dominated the headlines for most of July. The story follows:

> There was a mass meeting here on last Friday night, for the purpose of taking steps toward relief of the Homestead strikers. There were a number of prominent gentlemen from Birmingham, who entertained the assembly with speeches. The meeting was largely attended, and a committee appointed to solicit subscriptions for the benefit of the Carnegie locked-out men. Blocton is always to the front when it comes to sending assistance to those who need it.

Cahaba Coal Company Sold to TCI

The end of the Cahaba Coal Mining Company in Blocton was announced in the summer of 1892, with transfer papers signed in December. By then the company had extended its holdings to a new mining operation, the Excelsior Coal Company at Gurnee east of the Cahaba River. There was a sizable community at Gurnee. Some totals give the number of structures, miners' houses, and support buildings for both Blocton and Gurnee. Robert Praytor's research shows the number to be 575 structures. Approximately 404 of these were in Blocton. In a complex stock deal, the company's holdings in Blocton and Gurnee were acquired by TCI. According to Armes, "with this move, the Blocton mines became a part of the largest coal and iron company in Alabama and the third largest in the nation."

An announcement appeared in the August 15, 1892, *Blocton Courier* calling a September 16 stockholders meeting of the Cahaba Coal Mining Company. The fourth item of business to be transacted said, "To consider and vote upon any proposition which may be submitted at said meeting, to sell or otherwise dispose of the property and business of said Company." Approval was endorsed for sale of the company and consummated shortly after the meeting. Final transactions in the transfer of the company to TCI occurred on December 19, 1892 (Robert Praytor research).

Truman Aldrich and Cornelius Cadle, Cahaba Coal Company organizers and managers, remained in Blocton for a short transition period, Aldrich as superintendent and Cadle as manager. The January 13, 1893, issue of the *Blocton Courier* reported a special event honoring these men on their departure. The gathering was held at Blocton's Odd Fellows Hall. A large assemblage was present, said the paper's editor, with the hall beautifully decorated. The Cadle Silver Cornet Band opened the program, followed by remarks by a Mr. Wilkins praising both gentlemen. A banner over the stage read, "Long Life to Mr. Aldrich and Col. Cadle." Each was presented engraved silver services. Wilkins told the honorees and the assembly that best wishes were extended from the whole community and that included the poorest laborer to the highest skilled worker. Cadle and Aldrich made generous responses. Cadle remarked, "I am proud of Blocton, of the work that has been done in the last ten years, and all of you who have been associated with me in that work. We have had loyal help from all. I know that all the people of Blocton are my friends. I thank you sincerely for this beautiful testimony of your regard." Aldrich made a more lengthy response including the following:

> In leaving Blocton we regret to part from the many friends we have made here, and the associations that have come to us by the starting of this place. In my new position, I expect to watch over Blocton and to try and have the world do justice to it. I brought the present managers here the other day and showed them this property, and we heard nothing but kind words and praise, and that there was no place in Alabama like Blocton, and the people of Blocton have made it what it is today.

The editor ended his report by saying there was much enthusiasm and good feeling displayed.

Little is known about Aldrich's personal life in Blocton. His relatives now living in Birmingham and North Carolina have not been able to provide more than very basic information. Aldrich was married twice. His first wife was Anna Morrison of New Jersey, and he probably met and married her while he was involved in railroad building in New Jersey and New York. He fathered three children in that marriage Marie, Georgia, and Herbert. It is believed the couple and their three children lived in Blocton. The *Blocton Enterprise* in a 1910 special edition refers to Mrs. Aldrich as active in fundraising for the Episcopal Church when she was a resident. Special suppers and other programs were held as early as 1887 for this project. The church was not built until 1900 (it was consecrated in 1906), eight years after the Aldrichs' departure. According to the *Blocton Enterprise*, August 1910, Mrs. Aldrich was a resident of Gurnee when the church was built. The family made a generous donation to the building fund. The first Mrs. Aldrich died in 1913, and Aldrich remarried in 1914. His second wife, Helen (Leonard) Harris, a widow, had four daughters so he became stepfather to her children. After a short time in Birmingham as TCI Company head, he had a brief involvement in politics serving a partial term as Congressman representing a district that included Jefferson and Shelby Counties.

He returned to Bibb County soon after the turn of the century and managed the Hargrove mines located a short distance southeast of Blocton. Aldrich was appointed postmaster at Birmingham by President William Howard Taft in December 1911, and he served in that position until December 1915. One of Aldrich's daughters was married to a brother of Mrs. Taft. During World War I, he was one of the first "Dollar-a-Year" men, serving the government for that "pay" in whatever professional capacity he was needed.

Aldrich was a collector of fossilized shells and wrote many scientific papers on the subject. His collection of more than twenty thousand shells and shell fossil specimens was given to the State Museum of Natural History at the

University of Alabama in the early 1930s. (His interest in this study began in his teen years.) He was associated with that museum and its director, Dr. Eugene Allen Smith, and later with Dr. Walter B. Jones. Because of his prominence in the field, the University conferred on him the honorary doctor of science degree in the early 1930s. Dr. Smith said he was one of the four most eminent tertiary paleontologists of all time, along with T. A. Conrad, Isaac Lea, and English scientist Sir Charles Lyell (Lyell was one of the early geologists to examine the geological formations in the Cahaba Coal Field during his 1840s visit to North America).

Aldrich died April 28, 1932, at the age of 83. He is buried in Birmingham's Elmwood Cemetery. West Blocton honored Aldrich in 1995, renaming what had been known as old Highway 5 the Truman Aldrich Parkway.

John H. Adams, born in Birmingham, England, came to Birmingham, Alabama in 1880. He was subsequently associated with its industrial development, holding a number of responsible management positions in the district. His biographical sketch appears in Armes' book. Adams counted Aldrich as a friend and colleague. A poem from the Walter B. Jones collection, composed by Adams, gives an insight into how highly esteemed Aldrich was held by his contemporaries:

Truman H. Aldrich
An Acrostic*

Through quiet vales o'er hills where mighty forces clashed,
Revealing earth's great secrets by erosion gashed:
Unfolding for him to read full page of hidden stores,
Man's gifts from God—Coal, Iron, Gold, and other ores,
Amidst the "Silent Places" oft alone he wrought;
Nay, not alone: His tireless brain forever caught
Higher and wider visions, that cheered him on and on,
And nerved his heart when Hope was almost gone.
Life held so many things for him, no task too small;
Deep in some darksome mine, a tiny "Shell" would call
Rich veins of thought he loved to pass to youth,
In turn to have them prove Creation's wondrous truth.
Can Alabama, as it calls the "Roll of Really Great,"
His name forget? He gave the best he had to serve his state.

(*"Acrostic" A poem or series of lines in which certain letters, usually the first in each line form a name, motto or message read in sequence. The first letters in each of this poem's lines spell Truman H. Aldrich.)

It is speculated that Aldrich built the residence now known as the Superintendent's House that has for the past several decades been the home of the Pickett family. The house has been altered significantly inside and out. Once a two-story home, a former resident reported to the Picketts that a Victorian styled observation tower once adorned the front of the structure. Evidence of a stairway is visible today in the main hallway, and the older fireplace mantles in the house are vintage 1880s design.

Though the exact date of its construction has not been precisely determined, the house is historically significant since some of Alabama's most prominent industrialists have lived there while serving as company superintendents. It has been confirmed that Peter B. Thomas, who succeeded Aldrich as the top local company official for TCI in early 1893, lived in this house. Thomas had served as the mining superintendent for the Cahaba Coal Company since 1885. His daughter, Edith (Thomas) McGowen of Birmingham, reported to owner Suzanne Pickett in 1976 that she and her three brothers were born in the house. On a document acquired in August 2000, the birth of her older brother, Darius Allen Thomas, was shown as August 1886. This is the earliest date on record for the house's existence. If it could be proved that Truman H. Aldrich built and lived in the house, it would be even more significant. Researcher Robert

Praytor has concluded that Aldrich constructed this house in 1883. His evidence includes the fact that Aldrich's brother William is reported to have built a similar home in Aldrich, near Montevallo, that also had a Victorian observation tower. The house was mentioned by S. L. Johnson. In his March 29, 1951, column about the rarity of houses being painted in the 1880s and 1890s he remembered: "At Blocton, the company superintendent's house was painted. It stood on an elevation back from the road, and I supposed it was the finest house in the land as I passed it when I was a boy."

Truman H. Aldrich. An early historian wrote that "the name of Truman H. Aldrich heads the list of the first big coal operators of Alabama." In scientific circles, his life-long interest in fossils also gave him a reputation as an outstanding paleonlogist. Here he is pictured in south Alabama collecting fossils along a stream bank. (Courtesy of Frances Haynsworth, Birmingham, Alabama.)

The change of the coal company ownership in 1892 was the end of an era in the young community's history. The local weekly, the *Blocton Courier*, boasted of a population close to 5,000. That may have been an inflated count, but it was not far off. There were 404 dwellings and other company structures in the company town. At the same time, a number of homes were being built in West Blocton along with some sixteen businesses on Main Street. If the population count reached as far as Lowerytown and Primitive Ridge, included Smith Hill, Stack's Hill, and Little Italy, it is possible that the population was that large. The census of 1890 was lost in a Washington, D. C. fire so we do not have its numbers to verify editor McCumber's count.

This new town needed identity and community spirit, an element recognized by Aldrich and Cadle. Efforts in this undertaking were manifested by the financing and promotion of a company musical group. Blocton's first band, Cadle's Silver Cornet Band, first mentioned in a *Blocton Courier* story in early December 1891, was one of Colonel Cadle's contributions to the community. A New Orleans trip for local miners and about fifty ladies was sponsored by the Cahaba Coal Company and their musical group. Five coaches, one sleeper, and a baggage car carried 300 excursionists over the Louisville and Nashville Railroad via Mobile to the Crescent City. The trip was reported in the New Orleans *Times-Democrat*. The band's musicians on the trip were: J. W. Wallwork, bandmaster; John Auton, E. Lewis, J. Robertson, M. D. Freeman, T. B. Ellison, James Waugh, J. Duncan, Dave Lloyd, Sam Waugh, M. E. Lennigan, R. Bryin, J. Barclay, D. Thomas, L. L. Carmichael, W. A. Chase, A. Bryles, James Richardson, and W. A. Thomas. When they struck up their music, these nineteen bandsmen in their colorful uniforms must have been a sight and sound that gave much pride to the Bloctonites on the excursion. The party left Blocton at 4 p.m. on Thursday and, after stops in Montgomery and Mobile, reached New Orleans on Friday morning. They left on Sunday morning and had dinner in Mobile. Since it was Sunday, noted the Blocton paper, there was no beer.

A larger train was required for another trip in June 1892. This was a shorter haul but perhaps a more entertaining and emotional one since local pride was a factor. Blocton's company sponsored baseball team, under its manager Colonel Cadle and Superintendent Peter B. Thomas, had experienced a good spring season. Cadle and Thomas thought the team unbeatable. The *Blocton Courier* reported that company officials bet $50 on the Blocton Nine. Through Aldrich's connections with the University of Alabama, a game was scheduled to be played on campus. The company owners thought their team could easily best the University Cadets since it was their first year to field a

Superintendent's House. (Courtesy of Cahaba Trace Commission, Brierfield, Alabama)

baseball team. There was considerable promotion in the local paper. The announcement on June 10 said, "Excursion to Tuscaloosa. The Blocton Club will cross bats with the University team. Special Train, leave 7:00 [a.m.] return 7:00 [p.m.]. See Tuscaloosa sights." Cost for the round trip to Tuscaloosa was one dollar. Five hundred fans boarded the passenger cars at the Blocton Depot on that Saturday morning. To encourage the players and to entertain the fans, the Cadle Silver Cornet band in their spiffy uniforms made the trip to Tuscaloosa.

The write-up of the game and trip appeared in the following week's *Blocton Courier*, June 17, 1892. The news was not good, a different "Mudville" had struck out. One fan identified only by initials W. D. T. penned a poem, which the editor printed. It was along the theme of a popular period piece, "Casey at the Bat." It carried a revealing title:

Blocton Not "In It."

BY W. D. T.

The Blocton base ball nine
thought they were very fine
But they found the other day
They were not "in it."

They went to Tuscaloosa
To play the University
And there is where they found
They were not "in it."

Blocton took the bat
And gave the college boys the field
But they very soon found
They were not "in it."

Cadle's Silver Cornet Band. This photograph of the band was discovered in the attic of the Duncan family home in Oneonta in the mid-1990s. (Courtesy of Harmon W. Frederick, West Blocton, Alabama.)

They never made a round
'Til the U. had them drowned,
And so they found that
Blocton was not "in it."

Then there was Billy Sheean
He was always known to win
And even Superintendent Pete
Was sure the boys would beat.

So he bet fifty dollars on the game,
And of course he lost his money;
And us boys all felt funny
To see our noble nine lose their fame

So the boys returned to town
And gave the saloons a round,
And they drank the Globe bar perfectly dry.

And at 7:00 got aboard the train,
While their hearts raged with pain
To think they had muffled every fly.

Poor boys, they did try,
But they muffled every fly,
And it looked as though
They never would be "in it."

They were Acorns, Reddy and Young,
Of victory they had always sung;
But that was a time when
They did not begin it.

For they played the fatal game,
And had lost their base ball fame,
And decided they never had been "in it."

The local favorites lost 20–0. The editor, in a sour grapes mood, wrote that the Tuscaloosa transportation system's mule drawn rail cars were not fit for man nor beast and criticized the town's restaurants and hotels for running out of food by lunch time and being unable to serve supper. He suggested that should there be a repeat contest, the Tuscaloosa food establishments should be given advance notice that they were not dealing with a one-horse town and so should be prepared. He did admit that the campus was a friendly and attractive place. He also reported proudly that Cadle's Silver Cornet Band had serenaded an appreciative downtown audience from a hotel balcony.

There is no extant Tuscaloosa paper from the date of the game, but the *West Alabama Breeze* reported that the Blocton fans attending a game with the University Cadets, had spilled over into Northport. It said further that "the saloons in Tuscaloosa were so crowded that some parties had to come to Northport to get a drink. Blocton must be a dry town because the visitors tried to fill up for the summer." Some of the team may have been among the Northport visitors and that might explain their ineptness at bat that day. The baseball series between the University and Blocton continued over several years with no better results for the local team. The 1896 *Corolla*, the University's yearbook, said in its listing of the year's events: "May 30: Alabama 9, Blocton 1; nice 'scrap' afterward."

Company Baseball Team. Pictured were many early residents of the Blocton area. (Left to Right) Front row: Charles Young, Len Auton, Pat Finnen. Middle row: Charles Lint, Charles Cummins, Edward Davis, Mick Nash, Joe Smith, Forney Abbott. Top row: George Bainbridge, Tom Gordon, Tom Finnen, J. G. Moore, Charles Fuller, Bill Thomas. The boy behind the group was Marlon McCool. Two years after this photograph was taken, Pat Finnen was shot through the heart in the Bell Top Saloon in Blocton. (Courtesy of Darrell Finnen, Nashville, Tennessee.)

Number 2 Blocton Company Band. This photograph was given to the University of Alabama by a member of the Duncan family living in Pensacola, Florida and a relative of a band member. Several bandsmen appear to be former members of Cadle's Silver Cornet Band. (Courtesy of William Stanley Hoole Special Collection Library, University of Alabama, Tuscaloosa.)

Cadle's Silver Cornet Band announced later in the *Blocton Courier* that it "was open for engagements during the presidential campaign. They will serve at political meetings when requested. For terms, apply to the secretary, James Robertson, Lock Box 372, Blocton, Alabama."

The only contemporary local rival of Cadle's band was the Suttletown String Band. The *Courier* of December 2, 1892, identified the Suttletown members as T. J. Vickery, Fayette Findley, Charles Ellison, and S. E. Vaughn.

Following the Cadle ensemble in the community was the Blocton Number 2 TCI Band. This group was not mentioned in the weekly newspapers. However, it may have been "the band" mentioned during the years before the Blocton Home Band was organized in 1913.

In January 1893, the *Blocton Courier* reported the organization of a Philharmonic Society composed of Blocton's best talent. On March 1, plans were announced for its entry in a Birmingham contest. The group was still active in April 1894 when the *Courier* reported that Odd Fellows Hall was filled to capacity for their concert. The Philharmonic Society, the paper noted, was composed of "our Welsh brethren." The proceeds from this night's performance were to benefit the Welsh Congregational Church. The proceedings were pronounced "excellent."

The town's sporting scene was further enhanced in April 1893 by the formation of a bicycle club. The *Blocton Courier* reported that Dr. A. E. Meadow had acquired one of the town's first two-wheeled vehicles and several townsmen, seeing his success in riding over the hills without breaking any bones, decided to join him in purchasing bicycles. The gentlemen named as members, with the doctor, of the new sports group, were Messrs. Cunningham, Clark, Norris, Spraggins, Forney, and Melvin. Meadow reported to the paper that he planned to ride his wheel to the next session of court in Centreville, an eighteen-mile journey over a very rough road. The editor praised the innovative cyclist declaring "it takes worse than rough roads and distances to back the Doctor out from an undertaking when he makes up his mind."

Bicycles were first reported in Blocton in the February 13, 1890, *Centreville Enterprise*. Its reporter from Blocton wrote, "Bicicles [*sic*] are all the go with our boys, but some of them have trouble in making them go. The bycicle [*sic*] is an untamed animal to the uninitiated, at least that is the avowed opinion of the unfortunate would-be bycicilists." It happens that the two-wheeled conveyance was so new to Bibb County that not just riding it but also spelling it was not easy.

Bicycles continued to be both a pleasure and a hazard for a number of years. Ellison reported that the vehicle's popularity in West Blocton created a need for the town council to enact an ordinance on March 3, 1903: "Be it ordained by the mayor and councilmen of West Blocton that any person who rides a bicycle on the streets at a greater speed than three miles an hour, shall on conviction for each offense be fined not less than one-cent nor more than ten dollars or may be imprisoned in the town jail or sentenced to hard labor for the town."

Dodging speeding bicycles was not the only thrill available to residents of 1890s Blocton. S. L. Johnson, in a *Centreville Press* column from December 2, 1948, recalled what excitement a circus brought to town in 1896:

> My father saw John Robinson's show, which we now call a circus, and which traveled in wagons over the roads through the South in the Eighties. No circus came to Bibb County over the railroad until Blocton grew up and had a large population, and I never saw a circus until I was a large lad, and that was in Blocton. We were living in Blocton in 1896 and a big circus came and showed on a field just east of No. Three Town and alongside the railroad track which ran up to No. Three Mine. [This is approximately where the old Number Nine Blocton Mine was located. Near this site the ground has recently been graded for youth baseball fields].
>
> When it was billed for the date to show, I began to wonder how I might see it. Never having seen a circus, and for it to set up just under the hill from our home, was just too much for a boy of my age to think of missing. I took a stand at the rear of the big tent after it went up where I might have glimpses of the animals. This was in the morning before the afternoon performances. There was no water in this area of the show grounds, so water for the animals had to be carried in large pails from a hydrant at the end of the street where people on that side of town got water. Two heavy set Irishmen with large pails, which held about four gallons each began to carry water and they spied me and a pal, who had joined me, and they at once conscripted us and gave us another pail. When full, it was too heavy for one to carry, so we put a pole under the handle and it seemed we carried water for almost a half day. The elephants would suck up our pailfull and then look wistfully at us for more, and back we would go for more. The Irishmen would not let us lag and it was hard work, but we thought it was great fun to get inside that animal tent. There were some monkeys in a cage and they were comical and when we tarried to watch them the Irishmen would knock us along with their pails back towards the hydrant for more water. They were very rough towards us we thought but now I know they just pretended. When we asked what we might expect as pay they said we were seeing enough already for pay, but at last when the task was ended they gave us a red card each and told us where to be when the show opened. We had no dinner, but were at the right spot when the show opened and showed our cards. We were admitted and told to show the cards again inside and when we did they gave us seats inside the railing around the ring. We sat on some boxes and thought we were the privileged of the earth. We could see other boys we knew back in the tiers and we could imagine how they looked with envy to where we sat almost in the path of the performers. We saw everything and were dazzled by the many acts and the funny stunts of the clowns. Toward the end of the show our two Irishmen appeared at the edge of the ring and started building a bonfire with bursted *[sic]* crates and barrels. They called for us to join them in bringing stuff in from the rear. We did and again knew the boys burned up at seeing us out there in the middle of things. While we were bent over the fire putting on fuel a clown cried out "FIRE" and in came running more clowns with our old water pails full of water and dashed it up and over us and the fire. We were drenched from head to toes and it was a very cool day. Then we knew the other boys had the laugh on us as we made our way out. When we appeared at our house wet to the skin, it was hard to explain where we had spent the day, for it was not usual for one to come home from a circus as wet as we were.

When children play on the baseball fields planned for the area around the old Number 9 mine, they should be aware that over a hundred years ago two young Blocton boys had an unforgettable day in a circus tent right on that spot attending Bibb County's first real railroad-traveling circus.

The author remembers the last circus to play in West Blocton in the early 1930s. How exciting it was to see two elephants make their way up the hill on Florida Street toward the water tanks pushing circus wagons en route to the TCI Park where the one-ring circus tent had been set up. A bridge on the regular route could not bear the weight of the elephants and their loaded wagons, thus the unusual route by way of another road.

Fortunately, Blocton was served by local weekly newspapers from 1890 until 1931. Ellison reported the first was the *Labor Advocate* established in March 1890, but no issues have been found. There is a reference to it in the March 28, 1890, *Bibb Blade*: "The *Labor Advocate*, Blocton's paper, was circulated on the streets yesterday. This is the first issue, and it is printed at present in Birmingham." In Centreville's *Blade-Enterprise* of April 10, 1890, J. R. Tankersley was identified with this publication. He was reported to be in town to attend county court. The paper was proclaimed to be Blocton's "latest newspaper effort." This statement implies there may have been other weekly newspapers that failed to prosper. There was a paper called the *Labor Advocate* published in Birmingham during these years, but no connection between the papers has been found.

The *Blocton Courier* opened in West Blocton in November 1891. Its editor was Thomas McCumber. Its pages and those of other papers to follow have provided historians with information available from no other sources. The *Blocton Courier* was followed by the *Blocton Intelligencer*, the *Blocton Times*, the *Blocton Messenger*, and starting in 1906, the last weekly in the town, the *Blocton Enterprise*. The last paper ceased local publication in 1931. It was continued by editor Jim Oakley of the *Centreville Press* for several decades, printed at the Centreville plant.

A remarkable event of interest to pugilistic fans was recorded in the *Courier*'s issue of September 2, 1892. It was titled:

> SULLIVAN & CORBETT
>
> By special arrangement with Associated Press the people of Blocton will get the detailed telegraphic report of the Sullivan and Corbett fight by rounds. These reports will be read in Odd Fellows Hall on the night of the fight, and all present will know the result at once. Tickets will be sold at the low price of fifty cents. This is the first instance of this kind Blocton has experienced and the people will owe Mr. Frank Hally for the arrangement. This is a spirit of entertainment that should be encouraged. Remember the date September 7, admission—fifty cents.

This historical sporting event went twenty-one rounds and lasted an hour and twenty-three minutes with Corbett the victor. Those fifty-cent tickets were a bargain. In a later paper editor McCumber reported there was a pretty good crowd and in this item spelled the promoter's name, "Holley."

Early 1890s Businesses and Institutions

The newspaper serves as a catalogue of the community as it was at the time. For example, in December 1891, a *Blocton Courier* column titled "OUR BUSINESS HOUSES AND WHAT THEY HANDLE; CHURCHES AND SCHOOLS" presented a business and institutional picture of the Blocton, West Blocton, and Smith Hill communities. Names of merchants, clerks, and other community leaders provide an important record. It is interesting to find that there were so many businesses in the company town, and this article also indicates there were two railroad depots in operation. The list follows:

> CAHABA COAL MINING CO.'S STORE, general merchandise. Here you will find everything that is usually kept in a general store. This store does an immense business, in fact the largest of any in the place. The Manager, Mr. W. F. Tyler, is a polite, affable gentleman, and is the right man in the right place. The clerks are all gentlemen and they treat all customers with respect and courtesy. WILLIAM

Blocton Company Store. The company store was a popular gathering place. A young couple like Claude Dinkins and Etta Reach Pritchett could meet there without breaking the era's strict rules of courtship. (Courtesy of James Ammons, McCalla, Alabama.)

WALKER BAKERY, Mr. Walker knows how to be entertaining - if you want something, he will wait on you; and if you want nothing, he will talk with you. He has a fine soda fount, and will dispense hot drinks in the winter. H. E. REYNOLDS, fruits of all kinds - next door to the post office—There you will always find Mr. S. W. Reeves, with his stencil outfit, ready to make you a key or stencil plate. H. G. McBRIDE WATCHMAKER, near the Company's Store. He gives special attention to repairing fine work. N. R. STACK, BOOKS AND STATIONERY, near E. T. V. and G. Depot. Here you find the very latest periodicals and most popular books [one source says foreign newspapers were sold by this firm]. JOHN GILBERT POOL ROOM, near L. & N. Depot. Keeps the latest newspapers and magazines; also cigars and cider. D. A. ABSTON is next to the pool room and is always busy making and mending shoes. JOHN QUINN in the same building takes your scalp for 25 cents and shaves you for a dime. Mr. Lamb now has charge of the BLOCTON HOTEL, and we predict for him success. The hotel is situated near the company's store and the L. & N. Depot. J. H. STRICKLAND keeps everything in the way of fish, oysters and meats. He spares no pains to please his trade. A. D. BELCHER, in No. 2 town, furnishes all kinds of fresh meats. He has a good trade and knows how to stand in with his customers. W. D. COX, PHOTOGRAPHER - tent near E. T. V. and G. Depot. He will snatch your beauty if you will call on him. SAM MORRISON, at No. 2 Mine, is a first class blacksmith, and does a great deal of work. [This ended the Blocton commercial listing. Thirteen businesses are identified.]

SMITH'S HILL is on top of a hill overlooking the company's property, and furnishes a picturesque view of almost the entire town. There is an immense amount of business done here. The business houses are as follows: W. L. CLARKE, DRY GOODS AND NOTIONS. This store is managed by Mr. T. T. Harris, and is kept in a manner which does much credit to the manager. D. J. FAGGARD, the fat man - dry goods and groceries. Mr. Faggard is only a short time with us, but he has so far proven to be an honest, straightforward gentleman. MRS. R. E. WALTERS, MILLINERY AND LADIES' NOTIONS. Mrs. W. Keeps a first-class store, and spares no pains to please customers. C. C. COLLIER, DRY GOODS AND GROCERIES. This is a well selected stock, and Mr. Collier does a handsome business. JOHN R. THOMAS, GENERAL MERCHANDISE. Mr. Thomas has a large building, and it is well filled. He is a pleasant gentleman, and takes pleasure in showing customers around. MARK DENNIS, GROCERIES, this is a nicely arranged little store, where one can find anything in the family grocery line. LOWREY AND THOMAS, BAKERS. These are young men just starting out, but we predict they will succeed. Mr. L. is a good baker, and they will endeavor to please their customers. ROBERT HOLLAND, supplies the hill and a large portion of the town with good beef. [These eight businesses ended the listing for Smith's Hill.]

SMITH TOWN - WEST BLOCTON This is a beautiful portion of Blocton. The land is somewhat level and the town is situated about one-half mile west of the post office - hence the name West Blocton. The business houses are: J. E. J. WALTERS, FURNITURE AND UNDERTAKING. Mr. Walters keeps a full line of furniture, and everything pertaining to the furniture trade. This is the only installment house in town. E. N. COTTINGHAM & CO., two stores, one - general merchandise, and one clothing, hats, and shoes. The general store is presided over by Mr. J. L. Suttle, who is adept at pleasing his trade. Mr. Cottingham has charge of the other store and both do an immense business. W. L. CLARK, DRY GOODS. For the manner in which this store is kept, much credit is due Mr. W. W. Perry, head clerk. This store is under the supervision of Mr. T. T. Harris, and he knows how to display

Blocton's Busy Corner. (Courtesy of Will Reese.)

> goods. F. P. SCHLOSSER, WATCHMAKER. Keeps a nice line of watches, clocks and jewelry on hand, and makes a specialty of fine repairing. W. H. DOUTHIT. This is a new store and in the most handsome building in the place. Mr. Douthit will keep a full line of dry goods and groceries. J. B. DAVIE, DRY GOODS AND GROCERIES. This is a nice store and well assorted stock of goods, with polite and energetic clerks to wait upon customers. DRUG STORE. This store is in the hands of Dr. W. F. Black as pharmacist and Bush Black as head clerk. The store is owned by Mr. J. H. Flinn and Dr. W. F. Black. They keep a fine line of drugs and patent medicines. COMERFORD & CASEY. Staple and fancy groceries. They keep a nice line and are nice young men. K. CLEMENTS. FRUITS AND PRODUCE. He is around every day with his delivery wagon. W.B. BRACKNELL. FRUITS AND PRODUCE. He always has country produce on hand. W. D. THOMAS. THE BUTCHER - supplies the town with meat in nice style. ELLISON SNIDER, does the blacksmithing and woodwork. J. E. GLOVER, Makes the shoes, and he is a good one. TILE BOWLING ALLEY, well we have one and it is in operation. And last but not least is the COURIER. Office over the drug store. Here you can get all kinds of printing except book work - and one dollar pays for the Courier for a year.

The *Courier* editor thus ends the list of Main Street businesses in Smith Town - West Blocton. At least one overlooked was the DOUTHIT HOUSE HOTEL, known to have operated since the mid-1880s. Also in business during the era was the MANRING HOTEL. There may have been others. It was the age of horse drawn conveyances, wagons and buggies. There must have been a livery stable or two to serve the wagon and buggy trade. The editor continues his roundup column with the following information on sawmills, churches, schools, fraternal "secret orders," railroads and a comment about the town's population:

> W. B. MARTIN, SAW MILL furnishes the lumber for the town. He also deals in general merchandise. DEAN & KING, about two and one-half miles west of Blocton, operate a large saw mill, and ship a great deal of lumber.
>
> CHURCHES: Besides the business houses we have eight white and two colored churches, in which are represented Baptists, Primitive Baptists, Methodists Cumberland Presbyterians, Christians, Episcopalians and Catholics.
>
> SCHOOLS: The Blocton High School, Prof. W. Holdrige, principal, is in a flourishing condition, and the roll is continually increasing. Mr. R. W. Jones has a good school at the Odd Fellows' Hall, and his list also continually increases. Miss Dovie Hayes is building up a magnificent school at Mt. Carmel church.
>
> SECRET ORDERS: We have four secret orders all in a flourishing condition. They are: Odd Fellows, Knights of Pythias, Masonic and Red Men.
>
> RAILROADS: Blocton is connected with the outside world by three railroads - E.T.V.&G., A.G.S. and L. & N. - Birmingham Mineral. Over these roads we have a great many trains daily. All trains out of Blocton make close connection with other roads for all points north and south and at Birmingham for all points of the compass. The E.T.V.& G. connects at Birmingham Junction for all points east and west. In railroads, we have three of the best in the world.

The population of Blocton is composed of an honest, industrious people. Here almost all nations work side by side in perfect harmony, and the best of feeling always prevalent. The foreigners are especially generous and kind, and few of them would be willing to give up American freedom for their native homes.

Editor McCumber's list of businesses, a total of thirty-six, churches, railroads, and his comment on the community's diverse population provides us with a time capsule documentation of the area as it existed in December 1891. The new town in Bibb was in its seventh year.

Schools

The earliest reference to Blocton's school was the *Bibb Blade's* reference to the "union church and school house" built by the founding company. The building in Number 2 Town Blocton later was identified as "the old hall' and as the 'Tabernacle." The *Birmingham Age-Herald* of June 4, 1918, carried a lengthy story about Blocton and West Blocton schools. Its opening paragraph gave a terse history of the earliest school in the company town. "The first school at Blocton was taught in an old fraternal lodge. Later a small schoolhouse was built and when it burned down the present building was erected."

Documentation is available from the pages of the January 22, 1892, *Blocton Courier* in an item titled "Blocton School Notes." The school in the Old Hall had been outfitted with patent desks. The principal was W. Holdridge, Ph.B., and the assistant teacher was Miss S. V. Wallace, A.B. The school was reported to be no place for pupils who did not like to study. Latin classes were doing good work, and map drawing was creating much interest in the geography classes. In spite of inclement weather, a good attendance was indicated. Next to the item, an ad for the school appeared: "BLOCTON HIGH SCHOOL, Male and Female." The term had opened in December 1891. It offered a "thorough course in English, Latin, Mathematics and Sciences." While it is not clear, this short-lived institution appears to have been a private school. No reference is found following its 1892 spring term announcement.

The *Blocton Courier* carried a brief item in April 1893 stating that the white public school had been removed from the Baptist Church in Number 1 Town to the Old Hall in Number 2 Town. This is evidence that the public school was still in a developing stage. On September 22, the paper announced that the Blocton High School would open on the first Monday in October at the Old Hall. Dr. J. M. Meadow was principal. In December 1894, the paper announced a concert to be given by the Number 3 (Blocton) school. A permanent place for educating the local children had not been settled.

The educational institution to be known as the TCI School was in place by September 1894. The *Blocton Journal* reported that "a meeting of patrons and district school trustees was held in the Number 1 School House for the purpose of arranging about starting the school and selecting teachers." As evidence of local interest in the school, the building was too small to hold the assembly and the meeting was moved outdoors. The decision was made to have the trustees select the teachers. J. G. Moore, the company superintendent at Blocton, was chairman. Others having leadership roles in the meeting were William Dickey and Sam Hartley.

An inventory of the school scene in Blocton appeared in Centreville's *County Press*, December 9, 1897. The reporter was Trion Harris, who had joined with William Dickey (both were trustees in the Blocton School District) and Superintendent Capt. R. H. Pratt, to make an official inspection of the schools. He reported four white schools and two colored. The visit started at Number 2 School in Number 2 Town where Miss C. B. Schooler was in charge, assisted by Annie Moore Whitten. "These accomplished young ladies were surrounded by a crowd of intelligent boys and girls." When the gentlemen arrived at the Number 1 school the children were at recess. They observed the pupils' re-entry at the tap of the bell. The children formed a line in two's and marched to piano music to their respective rooms. The scene reminded Harris of his soldier boy days at the "City of Oaks" (a reference to the University of Alabama when it was still a military school).

The school had 140 pupils enrolled. S. L. Callen was the principal, assisted by Tula Harkey in the primary department. The music department was presided over by Miss Issa Phillips, an accomplished musician. Next, the visit extended to Campbell Town where the school was presided over by Capt. Pratt's niece, Lillie Pratt. Then the delegation visited the Smith Hill School. The trustee reported, "This school is located on one of the highest points in Blocton, overlooking 'beautiful Italy' in the valley some hundred feet below. This school's principal was S. F. Verner, nephew of Prof. W. H. Verner of Tuscaloosa. His assistant was Jennie Harris, "one of Bibb's brightest and sweetest daughters, [and] we found their department overflowing with bright happy faces." The report ended with Harris saying that he and Mr. Dickey visited the colored schools and found them in good condition and all doing good work. Typically for the times, no mention was made of the teachers' names in the black schools. The schools were reported to have six-month terms. Public money was supplemented by patrons' fees to make up the teachers' salaries, about fifty cents per pupil monthly, a wage considered by

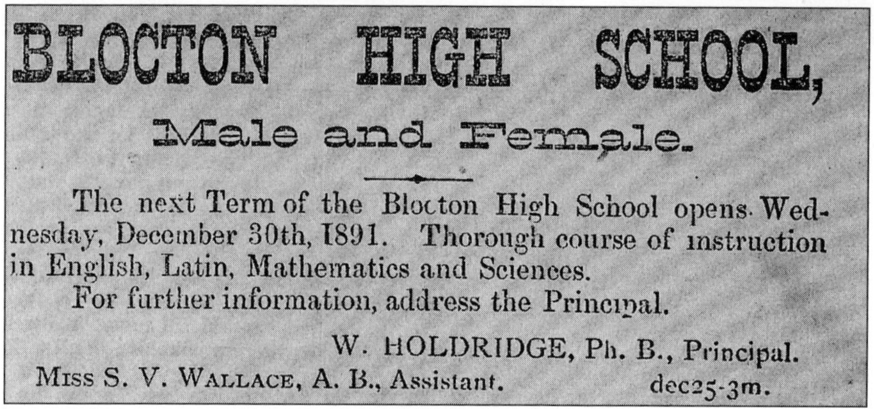

Blocton High School Advertisement, *Blocton Courier*, 1892.

TCI Blocton School, Number 1. (Courtesy of Alabama Historic Ironworks Commission, McCalla, Alabama.)

the trustee to be "up to date on this line." These early schools were to serve the community until the Catholic parish opened a school in 1901 and the West Blocton School opened in 1903. Evidence suggests the company schools served the earnest scholar well. The following are examples of success stories and school evaluations.

Ruben H. Wright, son of a local attorney, attended local schools, first at the company school and later in West Blocton. In June 1911 the local paper announced that his friends should rejoice that Ruben had graduated at the University of Alabama and won the law prize, a twenty-dollar gold piece. Wright practiced law in Tuscaloosa and subsequently served as circuit judge. He was prominent in state legal education programs and well-respected for his professionalism in the state's legal community. Wright's achievement is a remarkable success story of a locally educated scholar.

Polglaze Grave. In the Mt. Carmel Cemetery, the worn and weathered monument of the senior Polglaze reads: W. M. Polglaze, Born in Penzance, England June 1, 1847, Died Mar. 26, 1902. At Rest. (Courtesy of Cahaba Trace Commission.)

Richard Polglaze was born in Blocton in 1885 to an immigrant father, William, who was born in Penzance, Cornwall, England. His mother, Leah Shamblin Polglaze, was a native of Tuscaloosa County. William had been a tin mine worker in Cornwall so the change to coal mining was an easy adjustment. The Polglazes were one of Blocton's pioneer families. Richard attended the schools provided by the company, entering primary grades about 1891. He was a born engineer, reported nephew Claude Stabler of Greenville, Alabama, and with the fundamentals provided by Blocton schools and self-study, entered Auburn University (then named Alabama Polytechnic Institute) around age twenty. He worked at the water plant full time in the City of Auburn and attended college when he could. His father was killed in a runaway coal car accident in a Blocton mine in 1902 and is buried in Mt. Carmel Cemetery. Richard then had to support his mother and two sisters. He graduated in 1913 in civil engineering. His first job was superintendent of Gadsden's waterworks. In 1917 he married Lillian Dodge, whose parents, incidentally, were from Devonshire, England, adjacent to his father's Cornwall birthplace. In 1923 Polglaze and Frank C. Basenberg formed an engineering firm in Birmingham. They were highly respected for water plant construction and operation. Many of the 1920s era waterworks serving Birmingham were products of their design. In 1939, as World War II approached, they were contracted to plan the expansion of the water system for Fort McClellan. Perhaps childhood visits to the company waterworks on Caffee Creek had been an inspiration for Polglaze's interest in his profession. The Blocton coal company school gave a respectable academic start to his remarkable story of success. Richard, maybe relating to his father's birth place, Penzance in Cornwall, England was an avid fan of Gilbert and Sullivan operettas and was especially well-versed in singing tunes from *The Pirates of Penzance*.

Another 1913 Auburn graduate, also in engineering, was Aden Castagnoli, a resident of the Italian community and graduate of the company school in Number 1 Blocton. Some of his elementary school years may have been spent at St. Francis Catholic School. The editor of the *Blocton Enterprise* said the town should take civic pride in the success of this

Cadle School. (Courtesy of Alabama Historic Ironworks Commission.)

scholar who had attained high honors in mining engineering, noting his immediate appointment to an instructorship at Auburn. While it may be apocryphal, former Italian community residents have said that Castagnoli was one of a team of engineers commissioned to reconstruct the foundation of the Leaning Tower of Pisa. By 1913, the Italian populace of Blocton had begun to have better relations with the mainstream social structure in the area. The recognition of this outstanding scholar in the local paper is evidence of that changing attitude.

Samuel B. Israel, born in Poland in 1884 when that country was a part of Czarist Russia, arrived in West Blocton shortly after his father William "Wolf" Israel established a business in 1892 on Main Street. Young Sam's early education was obtained in the company sponsored schools. His siblings, attesting to his intellect, told that he worked their math homework in his head on the walk home from school each day. Israel left school at an early age to help his father in the family store. He later obtained training at Massey Business College in Birmingham. During his entire adult life, Sam Israel was one of the town's most energetic and devoted civic leaders. Former store employees and fraternal lodge associates were not hesitant in identifying him as a genius. He was recognized statewide for his Masonic ritual knowledge that he doubtlessly considered a mental exercise when reciting page after page of the written and unwritten material. Israel's son, Elmo I. Ellis of Atlanta, reported that his father gave the local school credit for giving him a good, fundamental education start. (Elmo changed his surname to Ellis to satisfy the radio broadcast profession's use of euphonic names for announcers).

An oral history project conducted by Willie Jo Parker in 1983 included several interviews with African-Americans that made references to the company school's contribution to local education. The institution was called Cadle School, named, as recorded earlier, for one of the Cahaba Coal Company founders. Among those reporting to Willie Jo Parker were Ruth Johnson Cutts, Rebecca Payne Parker, and Frank Jackson. The latter won the TCI scholarship in 1926. Jackson finished high school at Tuskegee Institute, earning a B.A. degree from Alabama A & M. He returned to teach at Cadle School and served as principal at West Blocton Heights School (elementary and junior high grades). He subsequently worked for the U. S. Postal Service. In 1983, Jackson was a resident of Birmingham. Effie Heard reported that her oldest children attended the TCI School. In 1987, she was one of the oldest residents of West Blocton having been born in 1897. Mem Wells, a seventy-six-year-old West Blocton resident in 1983, said of the Cadle School, "The learning was second to none. You could make your way through the world with it." Children walked from as far away as Frog Level to attend the school's nine grades. Parker reported that local residents called the school's location on the Old Woodstock Road "Number 3 Quarters."

One of Cadle School's most well-known and successful scholars was Louis Charles Goodwin. He was one of those determined youngsters who lived in the Frog Level vicinity, several miles from the school location, but found a way to make the daily three mile trek to the Blocton company school. Goodwin finished high school in Birmingham at Daniel Payne. He earned a B.S. Degree from Alabama A&M and served in the Mediterranean area during World War II. He later earned an M.A. and Ph.D. from Columbia University in New York. He held a number of important assignments at institutions of higher learning including registrar at Alabama A&M, a professorship at Grambling University in Louisiana, and eventually was made dean of Grambling's graduate school. Goodwin was a frequent visitor to his family home near West Blocton and was admired by his friends and relatives. His untimely death at age 60 in 1980 was a shock. He is buried in the Caffee Cemetery just north of West Blocton.

Another person of achievement, identified by Parker as having been born *circa* 1918 in the area, was Cora Brown who has the distinction of being the first African-American woman elected to the Michigan Legislature. Later, Brown was appointed to an upper management position in the U. S. Post Office by President Eisenhower.

The Community's Cosmopolitan Composition

Editor McCumber's mention of foreigners in his roundup story was a reference to the cosmopolitan makeup of the town's population. The industrial leaders in Jefferson County had for years advertised for workers in European newspapers. When these foreign nationals came as individuals to Alabama, they were easily assimilated but when they were brought in by mining companies through labor brokers, they were a challenge to local, native-born miners who vowed that "they are not going to take our places in the mines. We came here first and propose to stay." The volume by Rogers, Ward, Atkins and Flynt, *Alabama the History of a Deep South State* relates this problem in the chapter, "The Rise of Industry."

A detailed description of the Blocton foreign population is told by Dr. Rhoda Ellison in an article titled "Little Italy in Rural Alabama" that appeared in the fall 1986 edition of the *Alabama Heritage* magazine.

As related by Ellison, a culture clash had begun in the mid-1880s when the first immigrants arrived. Some were Slavs, others were Poles, but the majority were Italians. Many had come from villages in the hills above Bologna, south of Florence, and from the Italian colony on the island of Corsica. In their native land they had been shepherds, farmers or gardeners. They had been lured to Alabama by coal mining company recruiters looking for cheap labor, and later, by letters from relatives and friends who preceded them to the Bibb County hills. While the larger cities of the country all established their "Little Italy" communities or ghettos, Blocton's Italians, after suffering the horrors of an Atlantic crossing in the stench and heat of steerage accommodations on major liners, and experiencing the indignities of the immigration process on Ellis Island and other ports of entry, found themselves placed in a stockade at their job site to work off their cost of passage. In Blocton this enclosure, called the "Dog Pens" by locals, was near the Number 3 mine where most of these immigrants worked in the early days. In a hollow between Blocton and Smith Hill, with steep hillsides, the Italian immigrants settled into their small town version of "Little Italy." They built their homes there, opened stores, a restaurant, and boarding houses. It did not take long for the natives to label it "Dago Town" or "Dago Hollow." Eventually, the Italians themselves referred to the place as "The Hollow." Since they found themselves ostracized from the community's mainstream fraternal lodges and other social activities, they soon developed their own clubs and community activities.

In the underbrush covering the forlorn area today, the community outdoor oven still remains where the Italian women baked bread. Since they were probably not welcome in the Main Street bowling alley, some of the residents constructed a boccie court where the national game of Italy, a variety of lawn bowling, was played. Later, some of the more ambitious Italian citizens established a social and literary club called the Prosperity Society of Blocton. The women formed a social club, which they called "Yolanda," and the men formed another, "Il Figli d'Italia" (Sons of Italy). These groups are said to have had weekly meetings.

Little Italy, Blocton. (Courtesy of Will Reese.)

The Fraternal Lodges

The fraternal groups that apparently did not welcome the Italians as well as other minority groups were a significant part of the community's social life. These were the secret orders identified by editor McCumber in his listing of community assets. The Order of Odd Fellows functioned in Blocton from the earliest days. Its hall is mentioned in the *Bibb Blade* in one of its first reports on the new town. The Blocton group was known as the Ridgley Lodge. Later a West Blocton chapter was organized and was named Mitchell Lodge after John Mitchell, a renowned labor leader. The Free and Accepted Masonic Lodge Number 474 was founded locally in 1889. The *Centreville Enterprise* of January 2, 1890 reported:

> The Blocton Masonic Lodge have [sic] secured their charter from the Grand Lodge of the State of Alabama. [Dated December 15, 1889] They held their first meeting under the charter on Monday evening, the 15th inst., when the following officers were elected: Worshipful Master, Samuel Hartley; Senior Warden, Morgan Jenkins; Junior Warden J. L. Davenport; Secretary, Dr. J. M. Meadow; Treasurer, T. Banks, Tyler [a Masonic office]; Milton Rally [probably Riley]; Senior Deacon, William Walker; Junior Deacon, Mit Cundiff; Stewards, Dr. B. Jones and Seth Thompson. This lodge is in excellent condition and promises to be among one of the most substantial of the order in the State. It is composed of some of Blocton's best men, men of character and worth whose influence is felt among their fellow men.

Another fraternal group was the Order of Red Men. Cahaba Tribe Number 27, organized in 1892. Ironically, the bylaws permitted only white men as members so no Native American would have been welcome in the tribe's wigwam. The fourth of the orders was the Knights of Pythias. The Benevolent and Protective Order of Elks, Lodge 710, was chartered in 1901. This lodge, still active, will mark its one hundredth anniversary at the start of the twenty-first century.

The members of these orders took their lodge rituals seriously. In some cases the rites extended beyond the meeting

Little Italy, Blocton. (Courtesy of Will Reese.)

halls. A Knights of Phythias wedding held in March 1894 and recorded in the pages of the *Blocton Courier* is evidence of the practice of using the orders' solemn rituals.

Perry-Wood

Married at the First Baptist Church, on Sunday evening, 18th inst., at three o'clock, by Rev. Huff of Cottondale, Mr. Walter W. Perry and Miss Alice, daughter of Rev. F. M. Wood, all of Blocton. Attendants - Mr. W. O. Broadhead and Miss Jerusia McMath. The church had been beautifully decorated by friends of the contracting parties; and long before the appointed hour, the house was filled to its utmost capacity, standing room being at a premium. The Uniform Rank K. of P., of which Mr. Perry is a member, took a very interesting part in the ceremony. The division marched into the church in two platoons, formed on each side of the church, facing inward. The bride entered on the arm of Miss Jerusia McMath, maid of honor, while the groom came in with his best man, Sir Knight Broadhead. The bride was beautifully and tastefully dressed, and the groom was attired in full dress uniform of a Sir Knight. Miss Carrie Thompson, of Nashville, Tennessee, presided at the organ, and as the beautiful Mendelssohn Wedding March pealed forth its sweet strains, it was truly a lovely sight.

The bridal party marched to the center of the platoons and formed the base of the double triangle, while the Rev. Huff stepped forward, about three feet, facing the party. The division, under command of Capt. Hill, executed to perfection the double triangle - he and Col. Wadsworth forming the apex. After the ceremony, the bridal party marched out under crossed swords followed by the Uniform Rank. Thus ended one of the prettiest weddings Blocton has ever seen.

Incidentally, W. W. Perry anticipated the need for a domicile for him and his bride by building in 1892 one of the first homes "on the hill" overlooking the town's developing Main Street some two blocks south. The house now is occupied by the Louie Sewell family.

A major public recreational facility, the Park of the Four Orders, under construction in Blocton in 1892, was a joint project planned and financed by these fraternal groups. The park was the site of many patriotic celebrations, ball games, political rallies, band concerts, circuses, carnivals, and other events. A circus advertisement in an issue of the *Blocton Enterprise* indicated that Annie Oakley, the famous woman sharpshooter from Buffalo Bill's Wild West Show, performed there in 1911. The park was later improved and maintained by the Tennessee Company and became known as the TCI Park. A pavilion once stood at its north end. This covered, wooden structure that served as a bandstand and dance floor, was in use during World War I and is believed to have been removed during the early 1920s. Later, it was the practice field for the local high school's first football teams. Will H. Carroll who spent his boyhood in Number 1 Blocton company housing said he used to go to the practice field every day. He reported it was especially convenient for practicing players from Piper and Coleanor who rode the train. When they heard the engine whistle blow for Ashworth's Crossing it was time to make a mad dash to the depot to catch the 4:39. Today the park acreage is a residential area.

Early lodges in the African-American community were identified by Willie Jo Parker's oral history survey of older citizens in 1983. Masonic Philanthropic Lodge No. 41 was chartered on June 24, 1890. Worshipful Master of that early fraternity was F. E. Thomas; Senior Warden was A. Boyd; and Junior Warden, A. Davidson. A metal-sided building built by Matthew Cleveland has stood since *circa* 1900 serving as the lodge's meeting place. It was remodeled under the lodge's Worshipful Master Felix Parker in 1977. Other prominent lodge leaders were I. S. Dooley, Will Hill, Edward Parker, and Percy Black. Today the lodge still holds regular meetings.

Other lodges in the African-American community included Odd Fellows, Blue Ridge Lodge No. 3333 chartered in 1890. Leaders were F. E. Wells, George Bumley, and George Wildo. It is uncertain when this chapter became inactive. The community also had a Knights of Pythias lodge, the only leader remembered is Alex Mixon. When it dissolved is unknown. The Order of the Eastern Star, Orpha Chapter, was organized in 1905 and chartered on August 24, 1907. Mrs. Kellog was the first worthy matron. Other leaders have been J. P. O'Riley, Janie M. Balasco, Anna Coleman, Emma Dooley, Anna Blake Duncan, Mamie B. Lacey, L. W. Goodwin, and Willie Jo Parker.

CHAPTER 2

The Early Churches and Synagogue

Mt. Carmel Church Singing School, May 1915. (Front row, left to right) unidentified, Annie (Creel) Burns, Theo Lewis, Sarah Forman, Kathleen Yeager Pounds, Estelle Harper, Allie (Harper) Smith, Jarome Frost, Martin Cook, Hector Foman, Freeman Henderson. (Second row) Savanna Ashworth, Anna (Barns) Cook, Lela Mae (Mills) Creel, Ina (Grady) Owens, unidentified, Willard (Owens) Deerman, Edna (Hartley) Rakestraw, Mattie (Price) Williams, Bama (Ashworth) Dailey, Martha Martin, Myrtle Dyer, Vernon Dyer, Beatrice (Dyer) Thrasher, Mary Ellen Frost. (Third row) Reanie Henderson, A. J. Grammer, Ada Reach Thrasher, Mattie Reach, Dealia Reach McSpadden, Nora Sullivan, Floria Lawrence Hall, Ora Lee Yeager Gleaton, Verna Harper Bice, Shirley Cook, Amos Martin, Curtis Martin, Carl Dyer. (Fourth row) Rev. Richard Henderson, Rev. Sam Henderson, Nealey Cook, Clinton Gleaton, Luther Reach. (Courtesy of Kathleen Pounds, Woodstock, Alabama.)

Mt. Carmel Baptist Church

To the northwest of the new town of Blocton, serving a small farming community, was a Missionary Baptist Church constituted in 1869. First named Burmah or Burmee, for a nearby community and later named Mt. Carmel, it has the distinction of being the oldest church in West Blocton. The following appeared in the *Centreville Press* July 24, 1984, which printed several church histories in response to the West Blocton Centennial Church History Sunday.

> Be it remembered that on Friday before the 5th Lord's day in August 1869, we few scattered members of the Baptist Church of Christ in consistency to a church of the same faith and order thought it advisable to become a constituted church in order to try to keep house for the Lord.

Mt. Carmel Church. The second Mt. Carmel Church was constructed around 1900. (Courtesy of Glenda Simpson, West Blocton, Alabama.)

> A Presbytery was chosen and we met at the time specified by Elders Joab Langston, W. H. Williams and Henry Griffin and after being examined by them, respecting our faith and ability to keep house for the Lord was constituted a church of the Baptist order.
>
> Mt. Carmel's first building was that of log construction located on the western edge of the town of West Blocton on old Highway 5. She was admitted to membership in the Tuscaloosa Baptist Association in 1870–1872. In 1872–1877 she became a constituent member of the Mud Creek Baptist Association. In 1877 Mt. Carmel became a constituent member of Harmony Baptist Association and in 1902 she became and still is a constituent member of the Bibb Baptist Association.
>
> Around the turn of the century a new building was erected and dedicated to God. It was added to from time to time to accommodate a growing membership. The present church building erected in 1970 is the third structure to carry its name. The first pastor was Reverend. Joab Langston. Other early preachers were: W. G. Hubbard, G. W. Freeman, W. W. Carlton, S. A. Henderson and A. L. Nichols.

Mt. Nebo Primitive Baptist, another one of the early churches near the new towns of Blocton and West Blocton, is reported to date from 1875 at Primitive Ridge, a sawmill community at that time, about a mile and a half to the southwest. No details on this church have been found.

The first residents of the new coal town may have attended worship services in these two earliest churches, Mt. Nebo and Mt. Carmel.

The Methodists

The first informal congregation in Blocton was a Methodist Mission sponsored by the First Methodist Church of Tuscaloosa. The small mission was established in December 1883, and its minister was Reverend Rudolphus Strider Hullett. The oldest church records identify Hullett and six members received by letter. They were Edward Donaldson, William M. Owen, E. Paralee Strickland, Isabella and Mary J. Donaldson, and Lula A. Turner. These early members were probably associated with the railroad and mine construction. The oldest church register says on its cover in bold print,

"Church organized July 6, 1884." Although records have not been found to provide details about the early days of the congregation, it is known that the R. D. Carver was the Methodists' first full year appointee.

Reverend J. W. Spencer was assigned to the Methodist mission in December 1884. The first entry mentioning the Blocton Mission was in the 1885 volume of the *Journal of the North Alabama Conference*. It reported forty-one members, a church building valued at $200, and that the preacher had been paid $49.80. An appropriation of $200 to support the mission had been granted by the Tuscaloosa District Domestic Missions Fund. One year later, the 1886 *Journal* reported 218 members, a church building valued at $600, and a Sunday School with 100 scholars. The spectacular increase in the church's membership in one year is evidence of the rapid growth of early Blocton.

Among the members listed in the 1885 Church Register were Matilda Pritchett and her son, Pitman. Older church members in 1987 reported that Pitman Pritchett and four other men were the leaders in the church's 1894 formal organization.

The autobiography of Dr. Jesse Cole Persinger, the fourth preacher to serve the Methodist congregation, gave information about the early years of the Blocton Methodist Church. It also provided insight into the early days in Blocton. Excerpts from this personal memoir appeared in the March 2, 1988, *Centreville Press*. At first, Persinger had his primary assignment in the Woodstock community with only part time duties at Blocton, but the Blocton church grew so rapidly that he was moved there full time. About his experiences he wrote:

> We held our services in The Hall, a building owned by the mining company and located in Number Two Town; the lower floor was used for school, church and social events. [This is the building the *Bibb Blade* editor identified as the union church and schoolhouse.] The upper floor was for the

Methodist Church, Number 2, Blocton. (Courtesy of Robert Majors, West Blocton [Hebron], Alabama.)

miners' organization, the Knights of Labor. On Saturday nights the hall was used as a rendezvous for gamblers and drinkers. I have gone there for services on Sunday morning and thrown out the cards and whiskey bottles before the congregation could come.

The first church building in the community was erected by the Methodists during Persinger's tenure in Blocton. The building, in place in 1886, was located in Number 2 Town Blocton to the south of the company store and post office. Church records from 1904 show the exact coordinates of a lot seventy-five feet by one hundred feet conveyed to church trustees by Truman H. Aldrich of the Cahaba Coal Mining Company. Trustees listed in the transaction were Dr. A. E. Meadow, W. B. Martin, H. H. Self, Leonard Auten, John Mertz, M. Y. Hayes, and B. F. Jones. It is likely this group included church organizers that joined Pitman Pritchett in the 1884 church founding. The Cahaba Coal Company was generous in providing land for churches. It is reported that land as well as modest church structures were provided for African-American congregations of Methodists and Baptists during the town's first decade.

By 1897, the Methodist congregation had begun to make plans to move their church from Number 2 Town since most of the members now lived in the Number 1 vicinity. On New Year's Day 1900, a "Church Moving and Building Committee" was formed. The members were Dr. A. E. Meadow, T. T. Harris, J. J. Howard, and J. V. Porter. The preacher was W. H. Wells. By July in the same year, the new sanctuary in Number 1 Town was in use.

Methodist Church, Number 1, Blocton. (Courtesy of Patricia Clinner Deerman, West Blocton, Alabama.)

Blocton Methodist Intermediate Girls' Sunday School Class, 1927. (Author's collection.)

A 1910 photograph of the church taken for a special edition of the *Blocton Enterprise* shows the simple beauty of this Victorian, Carpenter Gothic structure. It had a majestic steeple at the left front and interesting fish scale shingle decoration on the front gable. This edifice was to serve a large congregation for thirty-five years. For a number of years around the turn of the century, this church's membership numbered second only to the First Methodist of Tuscaloosa in the district. It was to hold that distinctive position again around World War I.

The Roman Catholics

The Tuscaloosa Catholic Parish, St. John's, supported the organization of a mission church for Blocton in 1886. Reverend John Cassidy traveled to Blocton every second and fourth Sunday to offer Mass in a private home; identified by some sources as the residence of the John Finnen family. While it is uncertain exactly when the church was built, it was in use before the end of the 19th century. The church, named in honor of St. Francis of Assisi, was constructed under the direction of Father McCafferty and is reported by church historian Rose Gibbons to have been a crude 20' x 30' structure located on Gunlock Hill just to the north of the company store. A more substantial facility was soon in place as the Catholic population grew with the town. The mining management had joined Birmingham industrial developers in placing advertisements in European newspapers for workers to immigrate to jobs in the district. As a result, a number of Italian workers found their way to the Blocton mine operations. It was this sizable group that formed the membership base of a mission of the Roman Catholic Church.

In 1901, Reverend James S. Downs became the first resident priest and through his efforts a room was added to the church for use as a rectory (living quarters). In 1903 he built a two-story, six-room parish house and also that year a three-room school was added to the church. Benedictine Sisters from Sacred Heart Motherhouse, Cullman, staffed the school that enrolled 125 students.

The church on Gunlock Hill remained the place of worship for the Catholic Parish for more than sixty years. In 1950, a new church was built in West Blocton and consecrated as St. Thomas Aquinas by Bishop Toolin of Mobile. The parish was declared extinct in 1973 and the building sold to the congregation of the Faith Baptist Church. The altar and religious figures were distributed locally and to other Catholic churches in the diocese. In the 1990s, the church bell that called Blocton and West Blocton parishioners to mass for more than eighty years was given to the Montevallo Catholic congregation for use in a prayer garden. It carries a memorial plaque honoring Frank Ferrire, a former mayor of West Blocton (1964-1984).

t. Francis of Assisi Catholic Church, Blocton. (Courtesy of Alabama Historic onworks Commission.)

St. Francis Catholic Church, Interior. (Courtesy of Ferrire Collection.)

The West Blocton Improvement Committee dedicated a historical marker in 1996 near the site of the Italian Catholic Cemetery. Following a program at Faith Baptist Church (originally built as St. Thomas Aquinas Catholic Church), a priest from Montevallo and nuns from Sacred Heart School, Cullman, re-dedicated the cemetery in a moving ceremony. The same prayers of consecration used in the 1903 rite were read by the presiding priest.

The Episcopalians

The Church of Our Savior Episcopal Mission was founded in 1887 in Blocton. Although references to services at the Episcopal Mission are made in some of the early weekly papers, its location is not identified. Possibly worship was in the company "church schoolhouse" named in the *Bibb Blade*.

In the August 1910 *Blocton Enterprise*, Mrs. Truman S. Aldrich is listed as one of the founders of the mission along with the Morrisons, Dunlaps, Douthits, Wilkins, "and other earnest adherents of this faith." The editor of the *Blocton Enterprise* and his wife were members of this church from their arrival in 1906. Even though the church never attained more than "mission" status, the small congregation received considerable notice in the weekly paper.

Our Savior Episcopal Church, West Blocton, Erected 1900. For several decades the West Blocton Apostolic United Pentecostal Church has occupied this building. (Courtesy of Cahaba Trace Commission.)

Through the efforts of the membership, $1,000 was raised toward a building fund. Half that amount came from the Aldrichs even though they were residents of Gurnee at the time. The first clergyman to serve the mission was James Van Hoose of Birmingham, followed by Benjamin Dennis from Virginia. In 1897 Reverend Thomas Beard, general missionary of the Birmingham convocation, began a monthly weekday service which, according to 1910 diocesan records, continued to be the arrangement. Service times varied over the history of the mission. Some records also show Sunday afternoon communion services.

In 1900, the building fund was used to construct a chapel on land leased for $1.00 for 99 years from the TCI Company. As was the custom in the Episcopal Church, when all payments for construction and furnishings were paid, the church was consecrated. That ceremony occurred on September 29, 1906, and was officiated by Rev. C. M. Beckwith, bishop of the Diocese, "in the presence of a large congregation." The *Enterprise* reported:

> The events of the day will long be a memory to those who attended. The musical part of the service was conducted by organist and quartet from the Church of the Advent, Birmingham, kindly assisted by Mrs. Fred Maxwell, organist of Christ Church Tuscaloosa, and the hearts of the faithful little band that had worked for the establishment of the church greatly rejoiced at this confirmation of their long cherished hopes.

Prominent in the Episcopal mission's leadership in 1910 were lay leader J. W. Stewart, Mrs. L. E. Peacock, treasurer; and Mrs. C. E. Allen. Sparse records show other family names associated with the church such as Tucker, Sheridan, Clements, Wynn, Hughes, Greene, Ball, Bloomer, Jones, Robinson, and Burnap. From very early records, (1899 through 1904), there were from six to fifteen members. In 1899, there were six services; and in 1904, five services. In those years there were a few infant baptisms and later records show over twenty baptisms through the mid-1920s. One of the better record sheets reflects that in 1922–1923, twenty-two members were listed, seventeen shown as "communicants" presumably meaning some of the members were non-residents. Services were held rarely in the 1930s but generally were well attended by friends of the few member families joining in with the congregation by invitation. Bessie Robinson, whose father was Anglican, remembered Sundays when her family attended three services: a morning and night service at the Methodist Church and an afternoon service at the Episcopal Church. A 1943 letter to Bishop Charles C. J. Carpenter in response to an inquiry about church records, Mrs. Peacock reported "many children [were] baptized here, [and there were] a few burials and several marriages." It was her understanding that the mission's records were kept in Birmingham.

Minutes of the Executive Council of the Diocese recount the end of the Church of Our Savior Mission. The building needed repairs in 1941, and had been used very little during the past decade, (there were only three members) so it was voted that it be sold. Shortly before the church was sold, Bishop Carpenter celebrated Holy Communion there and pronounced the "sentence of deconsecration." The Mission was thus declared extinct in 1942, and the building was sold to the Church of Christ, trustees of that congregation being David Pickett and Benjamin Polk Williams. The sale netted $700, which was given to the Montevallo Chapel fund. Furnishings from the church were stored and later transferred to St. Mary's Birmingham and All Saints' Homewood. The communion vessels were to be used at Camp McDowell, the church's retreat in Walker County. The church built by local Episcopalians in 1900 today is the property of the Apostolic United Pentecostal Church.

The Baptists

Reverend Robert Praytor's research on the Number 1 Blocton Baptist Church was published in a booklet and presented to the church's congregation in October 1998. According to Reverend Praytor, the Blocton Baptist Church first appeared in Alabama Baptist records in 1887 when it reported to the Tuscaloosa Baptist Association. The minutes read:

> Blocton a new church admitted reports: Organized with a membership of twenty. The field is very hopeful and inviting. Two delegates from Blocton church were present, Elder Adams and D. A. Abston.

Baptist Church, Number 1, Blocton. (Courtesy of Will Reese.)

Early church records formalized the creation of the new church.

> On the 21st day of October A. D. 1888 the following named persons: D. A. Abston, C. T. Farlee, E. J. Ivey, Geo. Dunn, J. Campton Crawford, E. L. Nicholson, J. E. Barringer, J. E. Glover, W. B. Rachel, A. Abston, Fannie Farley, Jennie Ivie, Annie Duncan, E. D. Lolly, N. C. Nicholson, Alice Barringer, Martha Crawford, Rufa Crawford, Cora Perry, Mary Paul and ? Woods bearing letters of dismission from Missionary Baptist Churches, met at the school house in Blocton for the purpose of constituting a Baptist Church. Rev. J. T. Yearby and J. G. Lowery having been appointed by the presbytery to perfect the organization.

The 1910 anniversary edition of the *Blocton Enterprise* printed the histories of several religious groups. A column on the Baptist Church gave pertinent information about that congregation. It started with 27 members in 1888 and had increased to 252 by 1910, becoming one of the largest congregations in "this section of the state." The first minister of the church was E. L. Nicholson, who was selected at the first conference, and the following were the first deacons: S. A. Abston, George Dunn, and J. E. Barringer. The first church edifice was built in 1888 at a cost of nearly $2,000. The 1910 pastor of the church was Rev. John L. Ray. The deacons were W. W. Perry, Chairman, J. J. Lucas, H. G. Caldwell, W. F. Jones, D. W. Moss, D. A. Johnson, and J. H. Blake. The church's trustees were J. B. Davie and W. Jones. J. B. Davie was secretary and treasurer and served as superintendent of the Sunday School for 16 years.

The Baptists built the first church in Number 1 Blocton, the Methodists following in 1900 when they moved from the Number 2 community. The Baptists, like the Methodists, leased land from Aldrich's mining company. Praytor's research revealed that the pastorium was built in 1893 at the cost of approximately $600. In 1893, the church name was changed from Blocton Baptist Church to Blocton First Baptist Church. By August 1893, it had a membership of 132 and a Sunday school of 75 with five teachers.

The double steeples on the Number 1 Blocton church rivaled the bell tower on the TCI School building in height. One of the steeples contained the church bell that rang on Sunday mornings and other appropriate times calling members to worship. It joined the Catholic and Methodist bells and others in a Sunday concert that could be heard for miles. A remodeling in 1910 resulted in the removal of the dual towers and the construction of a central steeple.

The original church building and its pastorium along with a neighboring house burned on August 28, 1922. Praytor reports that in 1923 the church exchanged its lots in Blocton for property on "water tank hill" near the eastern edge of West Blocton. There the congregation built its second church in 1925. The building was classical temple design with four majestic Grecian columns and attractive memorial stained glass windows.

On June 28, 1968, the West Blocton First Baptist Church dedicated a "ground level" sanctuary. The adjacent 1925 building is used for Sunday school rooms. Within the decade of the 1990s, an activities annex was constructed on the east side of the church. This impressive church complex at the eastern end of Main Street exemplifies the strength and dedication of the church's membership. While the West Blocton First Baptist membership can be justly proud of many individual members' achievements, they take particular pride in two members having served as state presidents of the Women's Missionary Society, Mary Curd Street and Irene Lawrence.

Other Early Congregations

The African-American community's churches were catalogued by Willie Jo Parker in 1983. Her research found one of the earliest churches to be constituted was the Hope Hill Baptist congregation that first met near Number 3 Town in a brush arbor shelter. Their first church was built in December 1898, near Pipe Springs (exact location uncertain but close to Number 3 mine), later removed to Stacks Hill. Early deacons were M. Smitherman, D. Bailey, and R. Spenser. The church was still active in 1983, the pastor, Reverend Samuel Hollis.

According to Parker, the Liberty Baptist Church was built in Bucktown in 1903 on company property near Shiny Rock Cemetery. This church replaced another on this site that had been destroyed by a spring tornado in 1903. The black community had been served by Baptist and Methodist churches built sometime in the mid-1880s by the Cahaba Coal Company. Reverend Witherspoon was the first known pastor of the Liberty Baptist congregation and Reverend Cannon the second. This church was moved to the Heights in 1944 under the leadership of Reverend E. G. Massey. The New Hope African Methodist Episcopal (AME) Church in Bucktown dated from the earliest days of Blocton, and its first preacher is said to have been Reverend Hayden. This church, according to Parker, was moved to Bogus Row in West Blocton in 1944.

Union Baptist Church on the Heights was so named after the unification in the late 1950s of Friendship Baptist, organization date uncertain, and Peace Baptist, founded in 1888. Leaders of Friendship Baptist were Reverend Rice and Reverend Mitchell. The first preacher of Peace Baptist was Reverend Wagstaff and I. S. Dooley was its clerk. Deacons were T. W. Trott, L. C. Catlin, S. C. Cutts, and K. Martin. Peace church was also damaged by a storm and both churches were showing the need of repair so the bold decision was made to gain strength by joining the two churches rather than paying for costly individual repairs, thus the name "Union." The new, handsome building with attractive stained glass windows was finished in 1961 and stands proudly on one of the highest points in West Blocton.

Liberty Baptist Church. (Courtesy of Cahaba Trace Commission.)

Parker identified another church on the Heights, New Zion AME Church, which faded in the 1950s. Little is known of its beginning. Associated with this congregation were the Mitchell Nalls, Willie Woods, and Mamie Williams families.

The Blocton Cumberland Presbyterian Church was active from 1892 until about 1928. This congregation was begun by members of the Greenpond Presbyterian Church, and its first preacher and chief organizer was Reverend Sabert Gavelston Oglesby, a member of a pioneer family of the Greenpond-Gray Hill area of northern Bibb County. Oglesby's grandfather, also named Sabert, immigrated from Scotland in his youth and fought in the American Revolution and War of 1812. There were several Presbyterian clergymen in the Oglesby family. Information is unavailable on the church's construction, but its pews were handmade, reportedly by the preacher Oglesby himself. After the building was abandoned, it was used occasionally by other congregations including the Methodists in the mid-1930s while their new church was being built. The Presbyterians shared the building with the Baptists when their 1925 building was under construction. Located at the sharp curve in the road now known as Walter Owens Drive, its last use was by the Black Diamond Coal Company Store as a furniture warehouse. Prominent Blocton families associated with this church were the Alexanders, Durettes, Lawrences,

New Hope AME Church. (Courtesy of Cahaba Trace Commission.)

Union Baptist Church. (Courtesy of Cahaba Trace Commission.)

Edmonds, Farrars, Gatlins, and Manrings. Mrs. Manring, local hotel proprietress until the 1927 town fire, was one of the church's devoted Sunday School teachers. The preacher in 1921 was Fred R. Rosboro. Services were held on the first and third Sundays but Sunday school was every Sunday. During the 1920s, W. T. Alexander was its superintendent and Emmett Durette was leader of the Christian Endeavor Society. Reverend Harris was officially installed as pastor in October 1925 and served both the Blocton and Brent Presbyterian congregations. Harris, who attracted a sizable youth following, also took a great interest in high school sports and wrote some of the first high school sports columns in the local paper under the name, "Jerry." The last preacher's name appearing in the 1928 *Enterprise* was Reverend Thomas F. Campbell.

Otto Miller and Jack Lawrence, members of the church in their teen years, reported in the early 1990s that when the Presbyterian congregation was declared extinct, the preacher, records, and membership were transferred to the Greenpond Presbyterian Church. Today, that church has no evidence of these late 1920s transactions.

There is little known of the Disciples of Christ, whose church was located in Number 3 Town, or of an Independent Methodist group that has been reported in Campbell Town. The Welsh Congregationalists, made up mostly of immigrant Welsh citizens, held their services in the company building erected in the town's early days for school and church use. The Congregationalists were known for their musical talents and were a major part of the Philharmonic Society mentioned in early local papers. No evidence of the Congregationalist Church is found after the 1890s.

By 1914, a church had been built in West End for the Wesleyan Methodists. Their original building is today the sanctuary of the West Blocton Bible Methodist Church. It has been reported that a preacher named French helped organize the church and for years J. D. Muse was the Sunday School superintendent. Serving as preacher in the 1920s was Reverend E. M. Stamps. Other early members were the John Gilbert family, and the Cassidy and Bracknell families. The Arthur Finnen family is among the church's membership today. The John and Will Finnen families, the Fayettes, and the G. C. Frederick family were active in the church for several decades.

Holding services for around twenty years was a congregation of Free Will Baptists that built a Carpenter Gothic, steepled church *circa* 1913 about a block east of the Wesleyan Church, reportedly dating a year or two before the Wesleyan Church. The building was torn down in the 1940s and some of its materials used to construct a residence for the Jack Harmon family. Prominent in this church's founding was John Richard Weaver. Other families in the congregation were the Leige Stamps, Pat Boleys, Sparks, Gilberts, McCulleys, and Wootens. The building was used for elementary school classes during the mid-1930s. Adult literacy classes, a Depression era WPA program, were taught there by Mrs. Percy Burton.

The League of Brothers Synagogue

The earliest written record of a Jewish religious rite in West Blocton was the marriage of Israel Kronenberg of Blossberg and Annie Israel whose wedding was held in the Pratt Building. The Jewish community organized its congregation and built its synagogue in 1905. It is named in the incorporation records as Ah-Goodies Ah-Chem, an anglicized spelling of *Agudas Achim*, translated from Hebrew as "League of Brothers." The *Blocton Courier* of July 30, 1896, reported "a real Jewish wedding" with Rabbi Roseberg of Birmingham officiating. According to former members of the community, Jewish faith services were held in private residences and local meeting halls for more than a decade before the building of the synagogue in 1905. This was the first house of worship of any kind in the corporate limits of West Blocton. Ellison's history of Bibb County has a succinct account of the history of the congregation:

> There were as many as twenty families but that number accrued over a period of time; it is unlikely that total resided in any given year. Most of these families had been drawn to the boom town by its commercial opportunities. Some were refugees from prosecution in Russian-dominated countries, chiefly Poland, though in the case of the Krentzmans, Lithuania. Some of these immigrants, like William "Wolf" Israel, arrived as early as 1892. Many began as peddlers but eventually set up stores on West Blocton's Main Street. Israel is considered to be the patriarch of the Jewish community. His was the first store which he opened in early 1892 in partnership with H. Weinstein.

The congregation was formally organized on June 5, 1905, accepting the bylaws prepared by a committee composed of Morris M. Odess, Samuel B. Israel, and John B. Krentzman. The bylaws were largely the work of Israel, then only twenty years old. The building committee, whose members were William "Wolf" Israel, John Krentzman, and Samuel Baer, purchased lot number twenty-five as shown on the Uriah Smith map of West Blocton from A. D. Belcher for $100 on July 6, 1905. The synagogue was built on this lot on Main Street in September 1905 at a cost of $1,500. The frame building consisted of a sanctuary and a classroom for the children in which they were taught, among other subjects, Hebrew.

Ah Goodys Ah Chem Synagogue, about 1914. To the rear of Frank and Libby Jean Israel is the only known image of the synagogue in West Blocton. (Courtesy of Elmo Israel Ellis, Atlanta, Georgia.)

In 1997, the West Blocton Improvement Committee placed a historical marker at the former synagogue site. In addition to a sizeable audience of local citizens, the event was attended by more than thirty former Jewish residents and their offspring. The session was covered by the Hebrew newspaper *Shofar* printed in Birmingham. At the ceremony, the featured speaker was Dr. Joe Morgan Farrar of North Carolina, son of Jean Israel Farrar, former resident of West Blocton. Joe, who grew up in West Blocton, had been a star athlete in high school where his athletic ability and intellect won him a college scholarship.

The synagogue was described by Elmo Israel Ellis of Atlanta:

> Its exterior was painted white. The interior was painted in an unusual folk-art design that was popular at the time it was built. The walls were painted a combination of colors—yellow, brown, green, smoky gray—blended together by the artist sponging on the colors, or daubing them on with the tip of his brush. The building was constructed with a bannistered outside porch. One stepped into a small hall, and straight ahead was the Sunday School or religious education room. To the right was the sanctuary, approximately thirty by forty feet. It was a simple rectangular room, with a raised platform at the end, on which stood the Bema, the wooden case housing the Torah scrolls [Old Testament Books]. The platform held a small lectern and a couple of chairs. There were about six rows of pews in the sanctuary, and in the back on the right hand side was a table at which some of the older women

> used to sit during services. This was a holdover of Orthodox tradition that provided a separate place of worship for the women in the congregation. In the West Blocton synagogue that was never a rigid custom, and the younger women sat with the men.

Ellis believes that regular Sabbath services were held until about World War I. High Holy Days services were held yearly until the mid-1930s. Older West Blocton residents remember the large number of worshipers that gathered for those occasions and specifically remember the blowing of the ram's horn, the shofar, at the end of the service celebrating Yom Kippur, the day of atonement, the most solemn day of the Jewish year.

During the last years of the synagogue the only Jewish families left in the community were the Sam Israels, Joe Israels, and Klotzmans. To have enough men for the High Holy Days services, Sam Israel's nephew, Melvin Israel, brought a carload of his Kappa Nu fraternity brothers from the University of Alabama to augment the congregation, which required a minyan (ten adult males) to hold a proper service. "Wolf" Israel was Melvin's grandfather; his father, Julius, had grown up in West Blocton. Melvin is remembered as Mel Allen, renowned sports broadcaster of radio and television. He is often identified as "The Voice of the Yankees."

The synagogue was damaged in a mid-1930s windstorm and later removed. Family home services during the fall High Holy Days and at Passover were continued by Samuel B. Israel until his death. The last Jewish resident, Dave Israel, moved to North Carolina in 1992 to be near his nephew, ending a century of Jewish residents in the town.

Later Congregations

Episcopal diocese records from early 1942 show that David Pickett and Benjamin Polk Williams contracted to purchase for $700 the 1900 Episcopal Church for use by the new congregation of the Church of Christ. According to Jacqueline Mosley, this church group met in local homes before 1942. This congregation held services for a time in a Main Street storefront, later built a church on old Highway 5 and in recent years erected an attractive, modern building on new Highway 5. J. D. Mosley was the church's preacher for more than two decades.

The West Blocton Church of God of Prophecy was organized in 1950 according to members Margaret and Annie Jo Lawrence. The first meeting place was a Main Street store building, and the first pastor was Brother Lewis Houton of Greenpond. The first members were Houton and his wife, Marie; Sam and Pearl Miller; David and Mildred Elloff; Verta Lee White; Ellen Sewell; Elsie McCulley; and Alice Holsamback. The church now located on Honeysucle Lane was built in 1952 on property donated by Ellen Sewell.

Bible Methodist Church. (Courtesy of Cahaba Trace Commission.)

Reverend Sabert Galveston Oglesby. (Courtesy of Ken Penhale.)

West Blocton Church of Christ. (Courtesy of Cahaba Trace Commission.)

For several decades, the West Blocton Apostolic United Pentecostal Church has occupied the church building erected by the Episcopalians in 1900. It stands as the oldest church building in town. The sanctuary has been updated with the installation of air-conditioning and central heat. The church has a new roof and is kept neatly painted. Prior to this church's acquisition of the structure, the Assembly of God met in the building. In March 1982, the United Pentecostal Church purchased the church and its parsonage from that congregation. The Pentecostal Church then held services three times weekly.

There were eleven charter members and their families: Bobbie Jo Finnen (pastor), Dorothy Monk, Donise Cox, Marilyn Belcher, Frank Hardemon, Shirley Hardemon, Woodrow Shows, Annie Mae Jones, Ann Patterson, Pat Patterson, and Debbie Lightsey. In August 1983, the church became an official member of the United Pentecostal organization.

Church of God of Prophecy. (Courtesy of Cahaba Trace Commission.)

CHAPTER 3

Tennessee Coal and Iron Company Era

In January 1893, Truman Aldrich left the town he had founded in 1883–1884 to become general manager of Tennessee Coal and Iron Company based in Birmingham. Peter Bellis Thomas, who was serving as superintendent under Aldrich, was retained as the company's local manager in Blocton and remained until the summer of 1894. During his tenure, there was a brief strike over a "deferred payday" problem. The local weekly of November 3, 1893, reported the strike in a headlined column with bold face letters: "MASS MEETING! A Contract Was Violated. Miners and Laborers in the Employ of the T.C.I. & R.R. Co., Cahaba Division, Meet and Pass Resolutions. They Demand Pay, But Do Not Get It. As a Result all the Mines at Blocton are Idle."

In January 1893, Truman Aldrich left the town he had founded in 1883-1884 to become general manager of Tennessee Coal and Iron Company based in Birmingham.

The strike Thomas had to deal with was settled after a week-long work stoppage. The *Blocton Courier* reported the return to work in a story reflecting mixed results. It was a time of economic recession and the miners had struck because the company had deferred wage payments, as its officials said, "owing to the great depression in money matters." It appeared the company would reduce work to two days a week in order to meet their payday promise. It was an old story, the miners gained nothing in the strike.

The *Blocton Journal* of July 26, 1894, announced Thomas' resignation effective August 1. It praised him saying "his manly, conservative way of handling men has endeared him to each and every one in the place." Thomas had been in charge of company matters for nine years. During his stay, the weekly paper stated, Blocton had grown from a small mining camp of 800 to a town of more than 5,000 (including West Blocton and other nearby settlements). Coal tonnage had increased from 700 tons a day to more than 5,000. The *Journal* wrote, "the best wishes of the people of Blocton go with him for happiness and prosperity." In July, Aldrich also left the company office in Birmingham to run for Congress on the Republican ticket.

J. G. Moore, sometimes identified as Jones G. Moore, succeeded Thomas as superintendent of the Blocton operations. Moore, then in his early thirties, had been in mining since his early twenties when he was associated with the Pratt enterprises in Jefferson County as superintendent of the coke oven operations. Armes reports that he had worked with the New Castle Coal and Iron Company, the Coalburg Coal and Coke Company, and for a time, with Aldrich and DeBardeleben. He was a country boy, born and raised on a farm in Macon County near Tuskegee. After his tenure in Blocton, he became manager of the Sloss-Sheffield Steel and Iron Company. Moore's time in Blocton spanned an era of organized labor development. The first major problem confronting his administration was the strike of 1894, a milestone in the history of unionization of the coal industry. He must have been a remarkable personality. Though the strikes of the era were hard fought, Moore came out unscathed. In his last years in Blocton he challenged the established hold on politics in the county, running for state senator from Bibb and Perry counties and winning the post. He could not have been successful without a majority of miners supporting his candidacy. The local voters endorsed him by a sizeable vote of confidence.

Payday in Blocton, 1910. (Courtesy of Will Reese.)

Following the strike of 1894, Moore announced that for Christmas every widow in town would be given a turkey and a sack of flour, an act the local paper labeled a "charitable deed." Though widows or others of their gender could not vote at this time, Moore may have been looking ahead to a future in politics. Moore made a lasting impression on the local paper editor. Years later the *Blocton Messenger* of June 2, 1905, offered a tribute to Jones G. Moore.

> Although now a very prominent citizen of Birmingham and president and general manager of the Birmingham Southern Railroad, Mr. Moore's heart, sympathies and best feelings are with the people of Bibb. For six years as superintendent of the Tennessee Coal, Iron and Railroad Company at Blocton, he was one of us, and truly so. So much did he become one of us, so greatly did he win the affection and confidence of all the people of the county that we elected him by a very handsome majority to represent our district in the State Senate of Alabama. As a member of that body for the one session he became not only the pride of his district, but became a power for good in the politics of the state. The hand and heart of this distinguished Alabamian have never been withheld from any cause the purpose of which was the upbuilding of Alabama or the betterment of any man or community's condition.

Additional Organized Labor Developments

The subtitle for Ward and Rogers' *Labor Revolt in Alabama* is "The Great Strike of 1894," and it reveals important information relating to Blocton's role in the organized labor movement. There had been significant strikes in 1888 and 1890, but this work's focus is on the Alabama coal miner strike of 1894. In the words of the authors:

> The strike of 1894 was complex in that it pitted a union movement with its urban and industrial ideas against an agrarian society largely rural and unacquainted with the aspirations of labor. Attempts at unionization came at a highly unfortunate time; the economic depression that hit the nation in 1893 was felt with full force in 1894. The distress in the South was particularly acute. Alabama was still recovering from the difficulties of Reconstruction when the panic struck.
>
> In addition to the one clear-cut conflict between management and labor, the issue of wages, the

> Alabama strike involved the employment of Negroes as strike breakers, the question of convict labor, and the calling out of the state militia. To compound the situation, the state was engaged in a bitter political battle between the Bourbon Democrats and the Populists. Alabama's organized labor movement was strongly Populist, so the often cited lack of co-operation between union men and Populists did not hold true in Alabama.
>
> A violent and protracted strike began in Alabama coal mines on April 14, 1894. The strike involved more than 8,000 miners seeking to protect themselves against wage reductions in a time of depression, and hoping to gain peripheral improvements in living and working conditions. This strike, the largest demonstration by organized labor in the state during the nineteenth century, lasted through four violent months.

Unionization of the Blocton miners dated from about 1888 when the Knights of Labor were active in the district. This organization was reported to have had its meetings upstairs in the "Old Hall" in Number 2 Town. A local had been re-organized in 1890 with the coming of the United Mine Workers of America (UMWA). The *Alabama Sentinel* of Birmingham was a contemporary pro-labor newspaper, and in the summer of 1890 its pages carried frequent references to the organized labor scene. Prominent in UMWA District 20 affairs was its secretary-treasurer, John Harkin of Blocton, whose name appeared in reports on union activities. An organizational meeting of "white and colored" miners met in Blocton on May 18, 1890, and heard an executive board member, P. McBryde, explain the aims and objects of the national organization. Later, a series of resolutions relating to grievances of the Blocton miners was signed by Harkin as a result of a "large and enthusiastic" mass meeting of miners on June 7. The resolutions concerned pay for specific inside jobs, the need for organization, the need for bosses to be aware of the need for better air, and the recommendation for better access to the general manager of the company. The resolutions were presented to the General Manager (Peter Thomas) and he promised "to give them a fair and careful consideration." District President J. L. Conley of Adger was present and gave a rousing speech on the necessity of organization, and at the conclusion of his speech, eighty-one new members joined the union. The report concluded with the hope that Blocton would be organized to a man in the near future. Other names prominent in the Blocton organization were S. Lynch, Allan Pow, Charles Everleigh, Joseph Hooter, William Kaars, W. N. Pitts, and S. McNatt. Pow was listed as chairman and Samuel Lynch was on the district executive board.

Jones G. Moore, Blocton Mine Superintendent and State Senator. (Courtesy of Alabama Department of Archives and History, Montgomery, Alabama.)

The *Sentinel* reported a "little combat" between Blocton members and their employers over "radical changes in our mining rules" resulting in Harkin being asked to leave the camp. "By dint of presenting some hard facts and by cool reasoning we succeeded in having the changes desired considerably modified." Harkin remained in camp, and the union agreed to some of the company's changes. Subsequent encounters were not so peaceful. In December, members of the executive board including Harkin and Pow of Blocton, assumed to be promoting union organization, were arrested at Blue Creek and taken to the Birmingham jail to make bond for trespassing on DeBardleben Company property during a called strike. On Christmas morning in Blocton, union executive McBride was met at the depot by company officer Cornelius Cadle and presented with a letter stating he could not hold a meeting or address the miners on company property.

The flurry of union activity was met by strong company opposition and by mid-January 1891, the union was broken. Ward and Rogers, recorded: "No increase in wages had been won, and some miners were fired for union activity. In its initial trial of strength, the Alabama UMW had failed. It did not get a second chance." The names of those Blocton union officials disappeared from subsequent news items. They were not found in extant newspapers starting with the *Blocton Courier* of November 1891. It is assumed they were fired by the Cahaba Coal Company and obliged to leave the vicinity for lack of future employment opportunity.

The next union organizational activity came a few years later. A special new union council had been organized in Blocton by UMWA district president Frank Fournier in May 1894. During the 1894 strike an incident occurred in Blocton that brought on the wrath of Alabama's Governor Thomas G. Jones. During the first two months of the strike there had been a state of relative tranquillity in the Blocton district. Though the company officials had evicted strikers from company housing in some communities including the TCI's Blue Creek and Pratt mines, there had been no effort as yet to evict strikers from their homes in Blocton (the company owned over 400 structures, most of which were miners' cottages), or to bring in "blacklegs" (non-union Negro workers) to work the mines. Blocton had remained in a state of truce. But in July, the Tennessee Company decided to increase pressure on the Blocton strikers. On July 8, Governor Jones sent a detachment of troops into Blocton after receiving a report of a demonstration there. The alarm proved false, and the presence of troops brought a spirited rejoinder from influential citizens of the town:

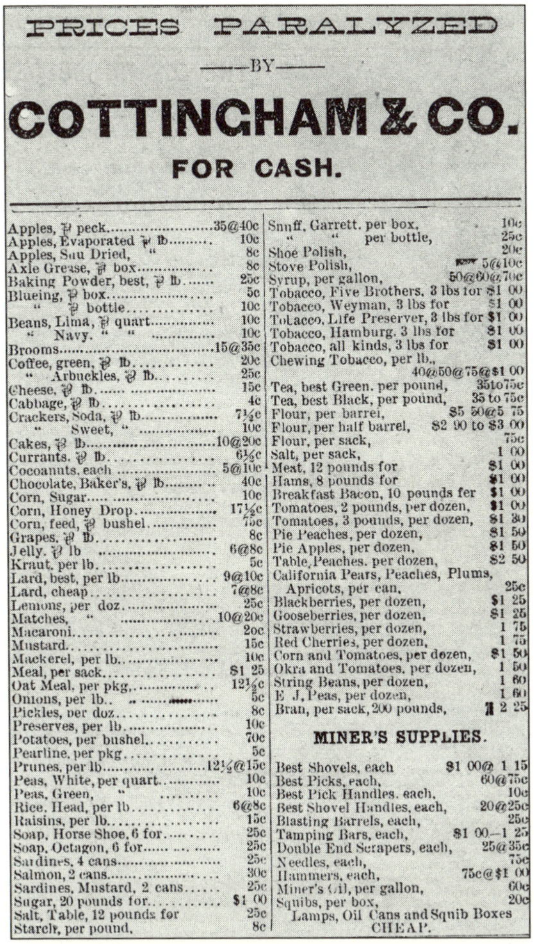

Cottingham & Company Advertisement, *Blocton Courier*, 1892. Miners were required to furnish their own tools and supplies, and many local stores attempted to provide for miners' needs.

Details on the "incident" were reported in the *Blocton Journal* of July 12. The item was titled MASS MEETING:

> Sunday evening, July 8, the L & N mail train came into Blocton with a Negro fireman and a Negro porter, who had taken the places of the white fireman and porter, who had refused to come out on the run owing to the strike. Parties approached the Negroes and questioned them in regard to taking the places of the white workers and accused them of being scabs. The Negroes concluded that matters were getting too warm for them to stay in Blocton that night so they decamped. The next day Governor Jones telegraphed N. L. Wilson, Esq. [local Justice of the Peace], inquiring into the matter. Mr. Wilson telegraphed the Governor there was no trouble in Blocton. The Governor also sent a telegram to Sheriff Latham [of Bibb County] at Centreville, to go to Blocton. The telegram sent to the Governor might not have reached him in time; however, Lt. Erwin [of the state militia] with sixty-five men came down Monday evening to protect the Negroes. The people of Blocton were surprised to see the soldiers, and were incensed at the Governor for sending them. Yesterday the citizens called a meeting to protest against soldiers being sent here and adopted the following resolution:
>
> Whereas, on the 9th day of July 1894, in regard to the alleged occurrence therefore be it resolved by the citizens of Blocton in mass meeting Thomas G. Jones, Governor of Alabama, did, through false representation and after protests against the same by a civil officer of Blocton, did send a detachment

Blocton Depot, 1910. (Courtesy of Will Reese.)

of troops into said town on the pretext of preserving the peace and protecting property, thereby causing the impression to be sent abroad that the people of Blocton were a lawless set of people and Whereas, the press of Alabama has been asking for information from here in regard to a demonstration against a Negro fireman and a dead beat [*sic*] porter said to have occurred here Sunday, July 8th and to correct the evil reports maliciously broadcast over assembled:

1. That no such demonstration ever occurred here as Governor Jones well knew after he made the inquiry.

2. That the citizens of Blocton denounce the unmeasured terms as an outrage on the Blocton people the above action of Governor Jones.

3. We denounce the action of the L. & N. Railroad and persons involved for their malicious misrepresentations of our people. We regard the actions of these parties as particularly deserving of condemnation, after their long residence here, and knowing the people to be opposed to any and all lawlessness, but we guarantee them our protection.

4. That we call upon the conservative and liberty loving citizens of Alabama to unite with us in some action to restrain this military mad governor in his unwarranted usurpation of authority in thus throwing troops into the midst of a peaceful and law-abiding people for no other purpose than to harass and annoy its citizens.

5. We hereby announce to the people of Alabama that we have implicit confidence in the ability of Sheriff Latham to preserve the peace in Bibb County and the attention to the fact that our mining population has been on a strike since April 14th without a solitary break of the peace, and that the people of Bibb County are ready to a man at any time to answer the call of Sheriff Latham to continue his efforts to keep the peace and good order inviolate.

6. That the thanks of this community are due to Lt. Erwin and the soldiers of his command

while on duty in our midst, for their soldierly and gentlemanly conduct towards all.

7. That we thereby instruct the chairman and secretary of this meeting to request the newspapers which have asked for information as to the alleged demonstration mentioned above, to publish these resolutions, and we ask the local press [to support] us in counteracting these vile and malicious misrepresentations.

[Signed by] Rev. F. M. Woods, Chairman [pastor of the Blocton Baptist Church] J. U. Ray, Jr., Secty., Thomas Wilson; Andrew Bryson; R. B. Wright; C. S. Smith, Committee.

Wesley McClendon. With a miner's lamp hanging from his cap, Wesley McClendon posed in miner's attire around 1900. In 1908, he was killed in a mine accident. (Courtesy of Billy Mack Roland.)

When a reporter from the *Birmingham Age-Herald* interviewed Governor Jones in regard to the matter, his response was, "I have not much time to devote to such matters now. The resolutions are signed by a reverend gentleman who, I am informed, forgets his duty to his divine Master in his zeal to misrepresent political opponents. There is not a word of truth in the statement that any civil officer in Blocton sent any protest against ordering troops there. No one there knew of such a purpose. The sheriff was miles away, and the only other civil officer I could learn of was Justice Wilson."

Additional pressure was soon placed on the Blocton miners. On July 21, the Tennessee Company ordered the strikers to vacate all company houses within ten days or suffer eviction. On July 23, Governor Jones conferred with the strikers at Blocton. When the miners explained why they could not accept the company's offer, the governor admitted he had no power to decide the merits of the controversy. The *Age-Herald* commented, "In view of the hopelessness of the strike and the fact that its continuance meant lawlessness he (Gov. Jones) had hoped the strikers would return to work. Although the miners might decline to call off the strike, he would be forced to protect other laborers who desired to work."

The Blocton miners had the situation made clear to them. If they refused to return to work it seemed probable that the company, supported by the state, would reopen the mines with non-union labor. Blocton miners held out to the last. Other miners in the state had given in to the governor's militia and to the company's use of non-union laborers as well as the employment of Pinkerton detectives.

As a result of the strike, mining families moved to other parts of the country hoping to find more favorable work opportunities. Evidence of the results of the 1894 strike appeared in the July 19 *Blocton Journal*: "Tuesday morning quite a crowd left Blocton for Thurber, Texas, to work in the coal mines. About twenty-five comprised the party. We wish them a pleasant trip, and hope that their brightest anticipation may be realized."

An estimated 1,000 miners in the Blocton area met to consider the company's offer on August 14. They were addressed by ex-superintendent Thomas, who informed them the strike was lost. Thomas, then working out of TCI's Birmingham headquarters, stood in for Superintendent Moore. It is not clear why but speculation is that the transition back to work would be easier if Moore were not the "messenger" bringing unwanted news. There was nothing to do but to go back to work and make the best of the situation. With only three dissenting votes, the Blocton miners accepted TCI's offer and resumed work. Again, the miners had lost the battle to better their wage level and working conditions.

The length of the 1894 strike caused considerable suffering in the community. The miners had raised funds in 1892 to send monetary support to the Homestead, Pennsylvania strikers. Now a plea was made locally to support needy miners' families. The *Blocton Courier* stated on May 11, "Next week the Miners Relief Committee will send out wagons to solicit contributions from their friends in the country. Friends, give something, even if it is but a chicken, a dozen eggs, or a pound of butter. The miners have always patronized you, and now you should do something in return in their time of trouble." The same issue reported the success of a miners' relief fund. Amounts ranging from $10 given by the Blocton Savings Bank to fifty cents from a local citizen had added to a total of $116.20. The lengthy list of contributors pledged to continue to help as long as the strike continued.

The miners were particularly interested in the election of Democrat Grover Cleveland in 1892 because of his perceived sympathy with the "working-man," asking for them the same "Fair Deal" he proclaimed for ownership. In a 1964 letter to the county paper, a former resident wrote, "I well remember the election of Cleveland in 1892, when hundreds of miners with lights burning on their miner's caps, paraded on the road from Blocton to West Blocton. Leading the procession was a wagon, drawn by two mules, with members of the band playing "'Dixie' and other familiar tunes."

However, though Cleveland supported better wages and shorter hours, his position was said by historians to have been ambiguous since he had said he resented the unions' "arrogant and domineering attitude" and that in fact he supported the open shop.

The Blocton District miners remained active in the national labor movement, and the area became known as a stronghold in organizational efforts. In May 1896, the *Blocton Times* reported that Eugene V. Debs, called "the greatest labor leader on the face of the earth" by the editor, visited Blocton and spoke twice in support of the American Railway Union to large audiences at 11:00 a.m. and 6:30 p.m. on Tuesday, May 19. The site was the "Tabernacle" which was probably another name given the Odd Fellows Hall, or a building used as a school and for religious groups that did not have their own sanctuaries. (The second floor of this early building was used by the Knights of Labor in the early days of Blocton.) That Debs considered the Blocton area for his spring tour of the South is remarkable. The following August, he was in Los Angeles at the city's largest arena speaking to a crowd of more than 5,000. In his active and radical political efforts, Debs had many confrontations with the era's political and business establishments, one incident in 1894 resulting in his being jailed for six months. While serving this sentence, Debs became a Socialist and subsequently formed the Socialist Democratic Party of America. As candidate for the Socialist Party, he received 96,116 votes for president in 1900, also running as that party's candidate in 1904, 1908, and 1912. During World War I, Debs was sentenced to ten years in jail for his pacifistic beliefs. His sentence was commuted in 1921, but while still in prison in 1920, he ran for president on the Socialist ticket, receiving almost one million votes. Whether it was an influence of Debs' visit, or from the sizable European immigrant population or the organized labor element generally, Socialism was a factor in Blocton history. While researching a thesis on Socialism in Alabama, Billie Faulk found voting patterns affirmed this notion. The Socialist Party garnered 8.4% of the Bibb County vote in 1908, 9% in 1912, 6.7% in 1916, and 6.3% in 1920. This vote undoubtedly was centered in the coal mining districts where the matter of worker unionization had been a part of the area's history since the formation of a chapter of the Knights of Labor in 1888. Debs was one of America's most colorful and interesting political figures. His visit to Blocton must have been a conversational topic for years.

When the Tennessee Company acquired the Cahaba Coal Mining Company holdings in December 1892, it became the largest coal and iron company in Alabama and the third largest in the country. Once the merger appeared in the press, several things occurred in the Blocton-West Blocton community that indicated new and more permanent business opportunities might exist in this Bibb County mining center. In November 1892, the Blocton Savings Bank, Bibb County's first bank was chartered in West Blocton. It was a historical landmark in Bibb County commercial history. The bank's opening was announced in the September 16 issue of the *Blocton Courier*:

> BANK
>
> BLOCTON TO HAVE ONE IN THE NEAR FUTURE
>
> Messrs. Irby Morgan of Nashville and Jesse Wadsworth of Birmingham were in town last Friday and perfected arrangements looking to the establishment of a bank. They leased the house [business place] now occupied by Mr. W. H. Douthit and at once have fitted it up with a nice set of bank fixtures, and expect to be ready for commerce by October 1st. This is something that Blocton has long needed, and we are glad to see so enterprising gentleman as the two above mentioned willing to take hold of it We predict great success for the enterprise.

The first official advertisement took front page prominence in subsequent local paper issues. The November 18 notice read:

> Irby Morgan, Jr., President J. B Wadsworth, Cashier.
>
> THE BLOCTON SAVINGS BANK, CAPITAL, $25,000. INDIVIDUAL RESPONSIBILITY, $50,000. We beg leave to inform the public that we have organized and established THE BLOCTON SAVINGS BANK, With a capital of $25,000, and are now open for business. We will transact a regular banking business, sell Foreign and Domestic Exchange on all parts of the world, make collections and remit promptly on day of payment. We will pay special attention to our SAVINGS DEPARTMENT, receive deposits from 25 cents upward, and pay interest on same. On time certificates we pay a special rate of interest. We solicit your business and promise you prompt and polite attention.
>
> We invite you to call and see us, and it will give us pleasure to show you our Fire and Burglar Proof Vault and Time Lock Safe.
>
> Yours very truly,
> J. B WADSWORTH, Cashier

Another indicator of change in the community was the arrival of Polish-Russian Jewish immigrant William "Wolf" Israel, mentioned earlier, and, later, his extended family, to open a department store in West Blocton. Based on newspaper listings of local businesses, there were no Jewish merchants in the town in December 1891. By March 1892, the firm of Weinstein and Israel had opened a department store offering "all dry goods cheaper than can be bought in New York." These astute businessmen believed in the possibilities of commercial development and stability in this mining center. That firm, continued by William's son, Samuel B. Israel, was to stay in business on Main Street, West Blocton for almost eight decades. The Israel family members would contribute to the civic development of the town for close to a century. Evidence that the store was in place by early 1892 is based on an item in the *Blocton Courier* dated March 11:

> A Negro woman stole a piece of cloth from the store of Weinstein and Israel on last Tuesday, and walked out of the house, and almost immediately the theft was discovered, and Mr. Weinstein started in hot pursuit. For a little while there was considerable excitement, but Weinstein was successful and captured his game. He carried her back to the store and sent for the constable, and in a short while the prisoner was convened before his honor, the Squire, and released on bond."

Confirmation of the founding of the Israel firm appeared in the August 1910 special edition of the *Blocton Enterprise*. The date given for its beginning was February 18, 1891 (The year should read 1892 because of other evidence found in the local papers).

Blocton Savings Bank Check. (Courtesy of Alabama Historic Ironworks Commission.)

The earliest Jewish resident was a coal miner named L. Levin, who Wolf Israel "rescued" from the coal mine once he was made known to the new town merchant. A transcription of an interview with Oscar Israel, Wolf's youngest son, reflected this fact. Levin's signature is found on the petition to incorporate West Blocton in 1893.

Another sign of change was the announcement in the December 16 *Courier* that application would be made to the next General Assembly of Alabama (the legislature) to establish a criminal court in Blocton Beat, Bibb County, Alabama.

While the official incorporation did not occur until 1901, in mid-summer 1893, the *Courier* announced in a brief item that West Blocton would probably be incorporated in the near future. "A movement is now on foot to secure the services of a special policeman, and it will probably be extended to incorporation." The official election notice on incorporation appeared in the *Courier* of August 25, 1893:

> THE STATE OF ALABAMA
> BIBB COUNTY
> IN PROBATE COURT OF SAID COUNTY.
> TO W. B. BRACKNELL, S. R. CAFFEE AND OTHERS.
>
> You will take notice that an election will be held on SATURDAY, AUGUST 26th, 1893, by the qualified voters in the following territory, to wit - Beginning at the southeast corner of the Uriah Smith land and run north one-half mile, thence west one-half mile to the section line, thence south one-half mile to the beginning, for the purpose of voting on incorporating the foregoing territory to be known as the town of West Blocton.
>
> Given under my hand this July 31st, 1893.
>
> M. Y. Hayes, Judge of Probate

The slate of candidates against incorporation was the winner. The paper announced: "The election for Intendent and Councilors went off quietly and the following officers were elected: A. D. Belcher, Intendent (Mayor), councilors, Jno. Seals, R. C. Lagrone, James Farrar, Uriah Smith and Jas. F. Jones. As a majority of them were in favor of no incorporation, it is supposed they will suspend operation during their term of office." It was a strange development. The next mention of incorporation is found in the *Blocton Journal*, successor to the *Courier*, in the following year in connection with a growing whisky problem in the town.

In an article on August 23, 1894, the new paper announced:

> We are requested by several good and substantial citizens of West Blocton to state there will be a meeting of the citizens of West Blocton on tomorrow [Friday] night for the purpose of taking steps to

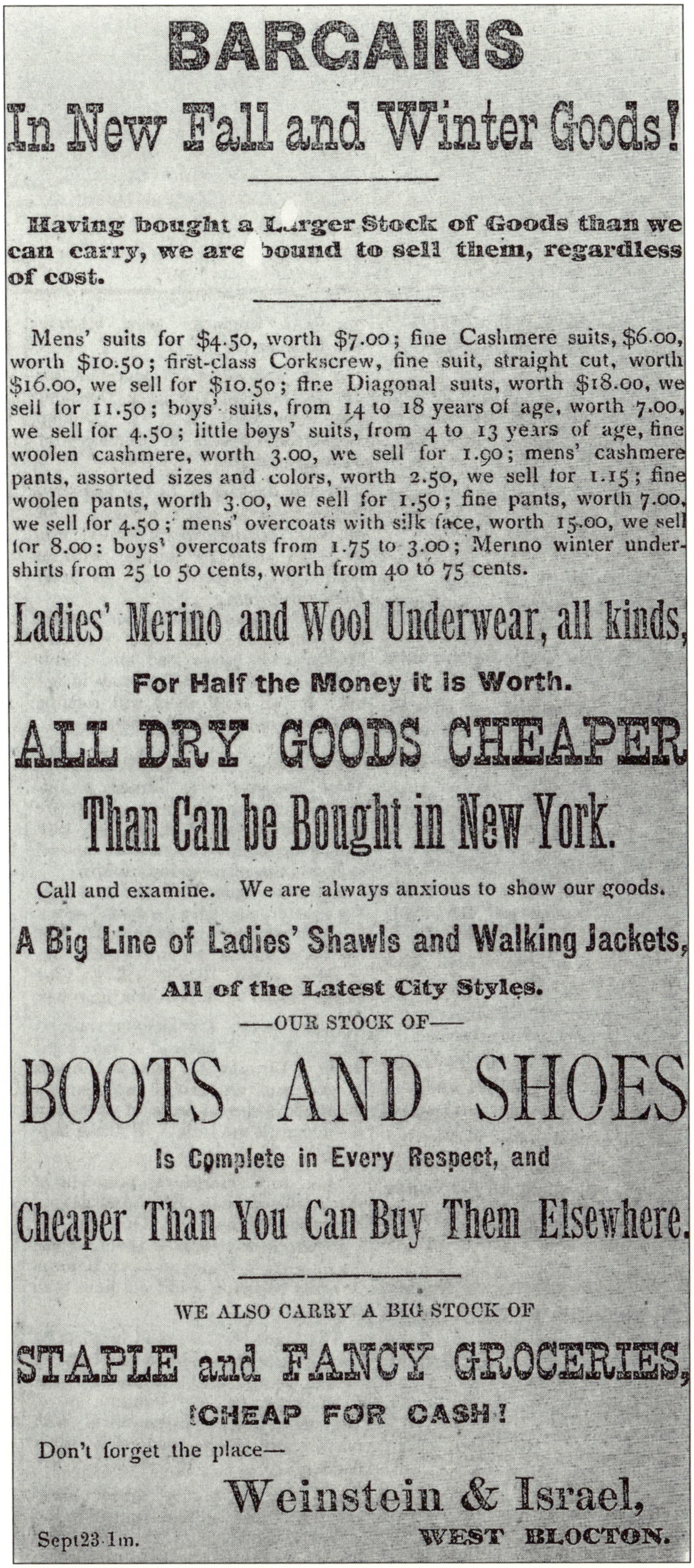

BARGAINS

In New Fall and Winter Goods!

Having bought a Larger Stock of Goods than we can carry, we are bound to sell them, regardless of cost.

Mens' suits for $4.50, worth $7.00; fine Cashmere suits, $6.00, worth $10.50; first-class Corkscrew, fine suit, straight cut, worth $16.00, we sell for $10.50; fine Diagonal suits, worth $18.00, we sell for 11.50; boys' suits, from 14 to 18 years of age, worth 7.00, we sell for 4.50; little boys' suits, from 4 to 13 years of age, fine woolen cashmere, worth 3.00, we sell for 1.90; mens' cashmere pants, assorted sizes and colors, worth 2.50, we sell for 1.15; fine woolen pants, worth 3.00, we sell for 1.50; fine pants, worth 7.00, we sell for 4.50; mens' overcoats with silk face, worth 15.00, we sell for 8.00: boys' overcoats from 1.75 to 3.00; Merino winter under-shirts from 25 to 50 cents, worth from 40 to 75 cents.

Ladies' Merino and Wool Underwear, all kinds,

For Half the Money it is Worth.

ALL DRY GOODS CHEAPER

Than Can be Bought in New York.

Call and examine. We are always anxious to show our goods.

A Big Line of Ladies' Shawls and Walking Jackets,

All of the Latest City Styles.

—OUR STOCK OF—

BOOTS AND SHOES

Is Complete in Every Respect, and

Cheaper Than You Can Buy Them Elsewhere.

WE ALSO CARRY A BIG STOCK OF

STAPLE and FANCY GROCERIES,

!CHEAP FOR CASH!

Don't forget the place—

Weinstein & Israel,

Sept23-1m. WEST BLOCTON.

Weinstein & Israel Advertisement, *Blocton Courier*, 1892.

> put down the sale of "blind tiger" liquor in this part of the moral vineyard. All citizens who are concerned in this matter should turn out and see the law is not violated. The meeting will be held at the office of T. C. Wallace on Friday Night at 8 o'clock.

The *Blocton Journal* crusaded against the "Blind Tigers," a name given unlawful saloons. It wrote on November 9:

> The Blind Tigers are in full blast. We are told there are more than a dozen in operation. From the number of people we see drunk, we judge there must be forty in operation. If the officers would do their duty, we would have none of that sort of business going on. The men who are engaged in the business are making no secret of it. The good citizens ought to make it their business to see that the last one of [them] are reported to the grand jury.

And further on November 15:

> The blind tigers that are now running in Blocton, the preachers are as much responsible for as the officers of the law. They are both neglecting their solemn duty. The officers are paid to arrest and punish all violators of the law, and the preachers are always howling against sin, and don't raise a finger to put a stop to the vilest of sin, when it is right under their nose. If these people are violating the United States revenue law, it would not be twenty-four hours before a United States Marshall would be down upon them. It would be much better to have legal rooms and let the state and county get a revenue and let those who are selling the stock sell.

The *Journal* continued to say its piece on November 22:

> We are told that it is useless to call the attention of the public to the blind tigers that are being run openly in our town, and that our readers are in sympathy with them, and will perjure themselves when called upon by the court to give evidence against these vipers of society. If this is true, a most fearful stigma is rising upon our people. We say this nefarious business can and will be broken up, but it will necessitate some work by the good people of Blocton. We suggest the preachers call their congregations in a mass meeting and formulate some line of action. There is strength in union. The officers of the law complain of not getting the support of the good citizens, and make an excuse for not taking the proper steps to oust them. It has gotten so it is dangerous for ladies to walk the streets. We saw on Wednesday evening not less than six drunken men on the street at once. Our grand jury meets on the 10th of December. Let's make this the last place the blind tiger man will reach.

Editors Glover and Mitchell pled with their readers in December 1894 to consider town incorporation. The reasoning was based on the need to control the sale of whiskey in the corporate limits, the subject of the citizens' meeting held the previous August. The prominent article stated:

> We understand there is a movement on foot to have West Blocton incorporated, and to ask the legislature to allow whiskey to be sold within the corporate limits. We are not advocates of whiskey selling, but as matters now stand, we are having all the evils of whiskey selling without receiving any protection or getting any revenue from the traffic. There is not a day but what we see two or three men drunk on the streets, and the conduct of some of them is most shameful - their oaths can be heard almost from one end of town to the other. It is true we have state laws covering these offenses, but the trouble is getting some one to make affidavits against them. "What is everybody's business is nobody's business." If the town is incorporated, then we will have policemen whose duty it will be to keep the peace, and to bring lawbreakers to court. With the revenue derived from whiskey license, we could have a nine months' free school and some decent streets. There is but one way to get the town incorporated, and that is to put our shoulders to the wheel and work.

The *Journal* took time out from their crusade against blind tigers in the December 27, 1894 issue to reflect on the season. However, the editor was unable to refrain from mentioning a drunk or two. It provides a look back on what it was like to spend Christmas in the locality in the 1890s:

> CHRISTMAS, 1894
>
> Tuesday morning, December 25th, 1894 dawned bright and beautiful, with not a cloud to obscure the rays of the rising sun. Long before the light of day had dawned, the citizens were awakened from their slumbers by the shooting of cannon crackers, and the children began to move around, each one anxious to get to his or her stocking to see what old Santa Claus had brought them while they were asleep. How their little hearts leaped with joy when they would empty the contents of their stockings into their laps and behold what presents they had!
>
> On Monday night there was a Christmas tree at the Baptist Church in Campbell Town, and a much nicer arranged tree we have not seen in a long time. Several beautiful songs were sung by the choir, and after a very appropriate address by the Rev. Mr. Selman, the pastor, the arrival of Santa Claus was announced. Mr. A. H. Crawford and Mr. Mel Hayes acted Mr. and Mrs. Santa Claus and they performed their parts to perfection. Walter Carlton delivered the presents as the names were called out. All of the children were remembered, and quite a number of presents were given out.
>
> Tuesday was the quietest day we have seen in Blocton on a similar occasion in several years, not but a very few drunken men being seen. It looked more like Sunday in Suttle Town than it did Christmas Day. In the early part of the day about fifteen young men of Suttle Town masked themselves, and, mounting horses rode through all the principal streets of Blocton, attracting a great deal of attention and causing much laughter.
>
> Tuesday night there were entertainments at the Methodist, Baptist and Presbyterian churches but as yet we have had no report of them.
>
> Thus ended one of the quietest and most enjoyable Christmases Blocton has ever experienced.

The argument against illegal alcohol was continued a month later in a January 1895 issue of the paper:

> The people of West Blocton are awakened to the necessity of incorporating our little city. It has grown to be a place of importance in the commercial world, and the people are realizing they can't keep pace with other cities without placing themselves upon an equal footing with other incorporated cities. We are sadly in need of sidewalks, good streets and police protection of property and the good morals of our town. Now is time to take action upon the matter, while the legislature is in session, so we can get a charter that will be in accord with our wishes. All who have this matter at heart are requested to meet at the Journal office Saturday night, 12th inst., at half past seven to discuss the matter and formulate some plan of action.

It is unclear just what the problem was that prevented action on incorporation within a reasonable time. It was not to be accomplished until 1901. The community depended on the county sheriff, local justices of the peace, and county appointed constables for maintenance of law and order. Blocton, the company town, had its own privately employed two-man police force. Their jurisdiction may have extended into the West Blocton vicinity.

Crime and Punishment

The number of "blind tigers" on West Blocton's Main Street had been the subject of the town weekly's drive for controlled sales of liquor. The easy availability of alcoholic beverages was considered a cause for uncivil, illegal acts. The *Centreville Press* of May 4, 1899, carried a strange report of a "shooting." It seems that J. H. Wallace had been selling whiskey from his store to J. V. Hosmer who had been summoned before the Grand Jury to testify to the effect that Wallace was the source of his purchases. He had been heard to state that "he would lie for no man" meaning that he was going to put the finger on Wallace. To confirm his story he approached the owner and asked if he had any whisky. The store owner said, "No, but wait and I'll get you some." He was gone for a few minutes and returned with the whisky. He gave it to Hosmer who went off. It is said he tasted it, and found that all was not right, and took it to a drug store and had it analyzed. The druggist said it contained enough strychnine to kill two or three people. Hosmer deposited the whiskey in the bank, got his pistol, returned to Wallace's store and opened fire on him. Two shots took effect in his arm and one in his side. Hosmer was placed under a $500 bond to await Grand Jury action. Incredibly, the news item said Wallace was not seriously wounded.

There were other reports of occasional rowdiness but in November 1899 there was more to report than misdemeanors. The *Centreville Press* of November 30 carried a bold-lettered story heading:

> KILLING AT BLOCTON
>
> One Man Sends a Bullet Through Another's Heart
>
> A killing occurred in Blocton last Saturday. This is an unusual occurrence and the town was excited over the news that Pat Finnen had been shot through the heart and instantly killed by a gambler named J. W. Maxwell. They were all in the Bell Top, a notorious whisky dive in Blocton, drinking and scuffling. Maxwell was drunk and Finnin took his pistol from him, then afterwards gave it back to him. Nothing was thought of this until a few minutes or rather some time after, Maxwell caught Finnin by the shoulder, and placed his pistol, a 45 Colts [*sic*], next [to] his heart and said, "boys watch me kill a ——————-," with that he pulled the trigger and the leaden missile went crashing through the young man's heart killing him instantly. After the shooting Maxwell made his escape and took to the woods and several posses went in pursuit, but failed to locate him. Sheriff Vance and Deputy Crawford [of Centreville] went up and aided in the search, and had given out catching him and had started home. When near Shultz Creek, they saw a man jump behind a tree and Vance and Crawford spurred up and had a chase through the wood, that soon overhauled the fugitive. It was merely by chance they happened to see him. He was covered by the sheriff while Deputy Crawford relieved him of a huge Colts [*sic*] pistol, he was then placed behind one of them [on horseback] and brought to jail. This will be on the case of temporary insanity, as from what we can learn it was a cold blooded murder.

Perhaps the *Centreville Press* writer left something to be desired in his grammar, but his story captured the excitement and drama of the moment.

The most sensational murder case of the period in Blocton was the result of an age-old motive, a love triangle. Lilly Gardner had arrived in early 1900 from Jacksonville, Florida, for an extended visit with her friend, Mamie Caddell. Lilly's marriage was on the rocks, and she took the occasion to get away from it all and put her life back together. Mamie's husband of a year was Festus Caddell, an employee of H. C. and W. B. Reynolds store in West Blocton. Apparently afflicted with a roving eye, Festus was soon smitten with this visiting *femme fatale* from Florida. A liaison between the visitor and host soon developed. As reported by a local author, the couple's relationship became "the talk of the town." Two marriage separations resulted, but each time Mamie's strong feeling for Festus brought her back. One thing led to another until Mamie asked the courts to intervene resulting in a charge of adultery being made against Festus and Lilly. It was May 29, 1900, and according to Elam's *Bloody Bibb* version, on that date Mamie had obtained a search warrant so she could go to Festus' house to retrieve her personal belongings. She was accompanied by deputies Jones and Peyton. Festus was outraged when the three entered his house. He cursed his wife and ordered her out of the house. Mamie told Festus she would kill him if he cursed her again. The deputies served the warrant. Lilly began laughing and talking to Mamie. Unnoticed by the lawmen, Mamie was carrying a pistol under her belt. In a bold move the observant Lilly suddenly grabbed the weapon from Mamie's belt and proceeded to fire repeatedly at her. Deputy Jones' attempt to divert the hand wielding the firearm was not in time, and Lilly inflicted a mortal shot to the right side of Mamie's head. Elam's account says Festus stood idly by while his lawful wife was being murdered. Dr. L. E. Peacock, who had been in the town since early 1898, was called. After thirty minutes he pronounced the pistol-packing Mamie dead. At that point the deputies

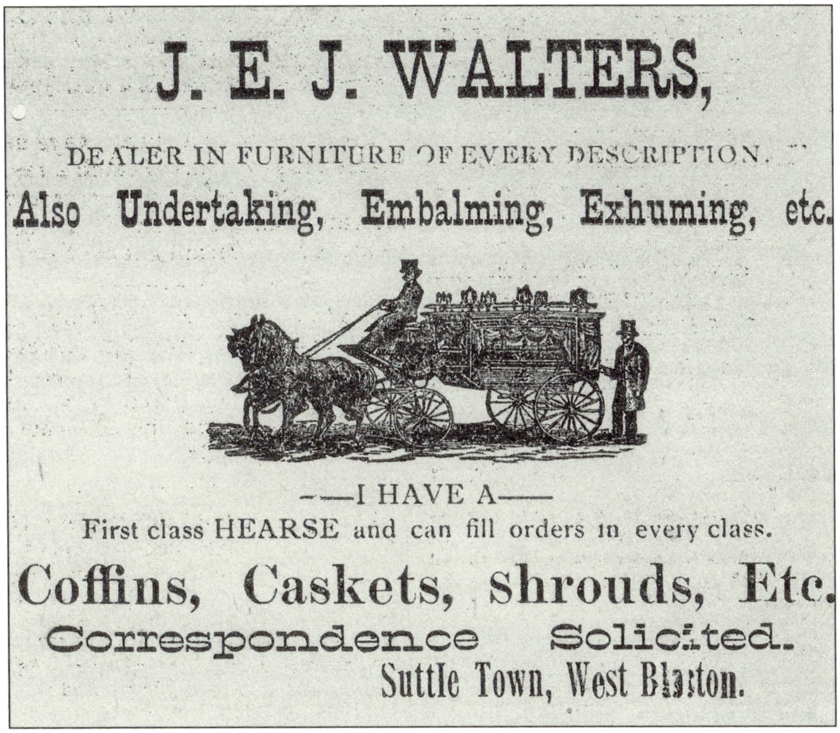

J. E. J. Walters Advertisement, *Blocton Courier*, 1892. Mining accidents and murders provided ample business for an enterprising undertaker.

arrested Festus and Lilly and escorted them to the town jail. A threatening crowd appeared and the lawmen decided to remove the prisoners to the county jail in Centreville. Appropriate measures were taken to prevent a mob from following for a possible ambush. Three buggies in convoy set out for the county seat, Festus in one, Lilly in the next and armed deputies in the third to guard the rear.

Elam's more lurid and colorful account goes on for nine pages. Ellison's county history shortens the case to two paragraphs:

> The case dragged on for three years. The *Blocton Messenger* on March 27, 1902 questioned the justice of the courts. The county's justice system was called to account. Why was not the case against Mrs. Lilly Gardner called? . . . She has never yet been tried. She is claimed to be insane. Festus Caddell, who is alleged to be an accessory to the killing of his wife Mrs. Gardner, has been tried twice. His case has been appealed to the [state] Supreme Court for the second time. . . . If the deed was committed by Mrs. Gardner during a moment of frenzy and insanity [as she says, unaided by Caddell] why is the case not disposed of legally?
>
> When Mrs. Gardner, pleading insanity, was finally tried in the fall term of circuit court the following year, in one of the first murder cases in the new county courthouse in Centreville, she was represented by a battery of legal talent: Col. Sam Will John of Birmingham, Dan Collier of Tuscaloosa, and the firm of Ellison and Thompson of Centreville. The county seat was crowded with witnesses and spectators for this long awaited event. The *Messenger* commented that the case had already cost the county thousands of dollars. On September 25, 1903, it reported Mrs. Gardner's acquittal. Festus Caddell, meanwhile, had been meted out a life sentence in the state penitentiary.

Elam's account says Lilly spent the rest of her life in the state insane asylum. While responsible citizens of Blocton probably did not like to admit it, the shootings in the area were a likely reason for the identification of the county as "Bloody Bibb." The Centreville weeklies often headlined stories as "Killing in Blocton" or "Shooting in Blocton."

Domestic violence was not confined to this small Alabama town. What may be a first, however, is a report of a Mrs. Wilson of Blocton whose personal notebook, dating from about 1900, outlines in 50 pages the abuse she suffered at the hands of her husband with names of witnesses to that treatment prior to her leaving him. Owner of the notebook, James Pat Cather of Birmingham, characterized by Joyce H. Lamont, University of Alabama Special Collections librarian, as a "true bibliophile," says this is "presumably a very early documented account of domestic violence in Alabama."

Crime in the new Bibb County town was reported as early as mid-1887. Frequently race was a factor in the crimes as well as in the reporting style of the local papers' editors, reflecting the treatment of the African-American in the era of Jim Crow. The following crime report from Blocton appeared in the July 21, 1887, *Bibb Blade*:

> AGAIN THE DARING FIEND DEFILES PURITY BY HIS POLLUTED TOUCH
> AND PAYS THE PENALTY AT THE MUZZLE OF ANGRY GUNS
>
> Last Friday, Blocton, a quiet mining town in Western Bibb, was thrown into a state of intense excitement by a horrible deed committed in open daylight. Mrs. Crawford, a highly respected white lady, was out with her little son gathering blackberries when she was attacked by a burley negro [*sic*] by the name of Johnston. His intent was plainly shown, and the lady made a break for home, hoping to save herself in that way. Nothing daunted the negro [*sic*] [who] pursued her, and even ran into her house and locked the door after him. Here he overpowered her and treated her in the most brutal manner. In the scuffle the lady's arm was very badly hurt. Her little boy ran out and gave the alarm. At once about a hundred men gathered and went to the scene. They found the lady unconscious and unable to tell the extent of her injuries, Ere long they caught the black devil and at once tied him up and riddled his body with bullets. Seven holes were put through his vile carcass and he was left dead. Saturday morning the body was still where they left it, the white men refusing to bury it, the negroes [*sic*] afraid to go near.
>
> He was a negro [*sic*] who the lady had known for a long time, and there could have been no mistake. He paid for his deviltry in a horrible way, but none too bad for the crime he committed. This thing is getting quite too frequent in its occurrence and if it keeps on this county, always so peaceable and safe, will be a place of danger. It will probably reflect on our name to send out the news that a negro [*sic*] was killed by a mob; but there is another name that must be kept sacred—the name of our fair women—and this is the way all such criminals may expect to be handled. Our people are law-abiding and orderly, but the law is insufficient to meet the demands of such occasions, and they do not care to trouble the courts with it.

In the 1890s, gangs roamed Bibb and adjacent counties stealing cattle and cotton and robbing victims of their valuables; associated murders also resulted. Notorious were the Miller Gang, the Morrison Boys, and the Thrasher Boys of whom Bart was the most famous. Bart roamed freely between Blocton, Woodstock, Big Springs, and Horse Creek. He was reported to have killed a man in a Blocton drug store in 1896 because the man had been a member of a posse that had killed a member of Bart's outlaw family. In a letter from Frank E. Carlton of Birmingham printed in the September 10, 1964, issue of the *Centreville Press*, the writer said he was a youth in Bibb during the outlaw era. "I was in dear old Bibb when Bart Thrasher walked into Harvey's Drug Store [in West Blocton] and shot Griff Bass to death." The bold and frequent treachery of these gangs contributed to the county being called "Bloody Bibb." The excessive violence and lawlessness led to the organization of vigilance groups called "beat clubs" in each beat of the county. In August 1891 when Miller and an accomplice named Cook were arrested for murder, the Blocton Vigilance Committee, around sixty armed citizens, escorted witnesses on their way to the trials.

In addition to Elam's *Bloody Bibb*, the outlaw era in Bibb is chronicled by Ellison's Spring 1988 article in the *Alabama Heritage* magazine titled, "Outlaws, Cat's Paws and Spotters." Outlaws, posses, brave lawmen, and vigilantes were a part of both Blocton's and West Blocton's past. It is not without historical basis that the town celebrates Wild West Blocton

Castelli Tombstone, Italian-Catholic Cemetery. (Author's collection.)

Days each summer. Re-enactments of street "shoot-outs" are frequently a feature.

One of the most gruesome crimes was headlined in the December 18, 1902, *Centreville Press*: "Murder Near Blocton" (Typically, the murdered Italian woman's name was misspelled and the location was identified as Dago Town.) This was the case of the hatchet murder of Elizabetta Castelli. Her tombstone in the Italian Catholic Cemetery near West Blocton tells the story in Italian. Translated, it says: "Here lie the ashes of poor deceased Elizabetta Castelli born in 1864, murdered at the young age of 36 years, robbed and killed by a murderer of the black race in the morning at 8 o'clock on December 15, 1902, leaving behind her husband and son, weeping and sorrowing."

Two Negroes were reported to have entered the Castelli store in Little Italy and brutally killed the victim. After their furtive entry, the front door had been closed and locked. According to the newspaper accounts, Elizabetta Castelli's head was cut in several places with a hatchet and crushed almost into a jelly, with part of her brains scattered over the floor. The hatchet was found nearby. The suspects, Ed Walker and Monroe Jackson, stole six hundred dollars from the store. Walker was reported to have paid-up standing debts and ordered a $23 tailored suit. Walker, apprehended with considerable money on his person, was charged and taken to the county jail at Centreville, tried and convicted in circuit court the following April. He claimed innocence, saying that a white man by the name of Bud Red had killed Mrs. Castelli. The jury was out less than one hour. The gallows was prepared for an April 24, 1903, hanging, only the second in the county since 1865. The *Centreville Press* editor described the final minutes:

> On the morning of the hanging, the accused had two colored preachers with him at the jail. Reverend Witherspoon of Blocton read a Bible chapter at his request and accompanied him to the gallows. Walker shook hands with the sheriff and his deputies, and Sheriff Crawford placed the rope over his head pulling it up snug, while one of the deputies tied his limbs. The black cap was placed over his head, and the sheriff waved his handkerchief, the trigger [on the gallows' trap door] was pulled and his body dropped with a sickening thud. His shoulders drew up several times and in ten minutes doctors pronounced him dead, and in twelve minutes his body was cut down and placed in a coffin and shipped to Blocton.

It has been reported that the rope was chopped into small lengths for souvenirs and handed out to the execution witnesses. Monroe Jackson, the other suspect, was not mentioned in the paper's subsequent reports.

The *Blocton Messenger* of January 23, 1903, carried an editorial supporting a proposal for the establishment of a branch of the circuit court in West Blocton with both civil and criminal jurisdiction for trial of non-capital felonies. A bill to establish the court and provide for grand juries at Blocton was reported to be in the works for the next session of the legislature.

> With no officers to act as conservators of the peace, enforce the majesty of the law and execute mandates, the Blocton Mineral District, Bibb's greatest center of population and industry, is rapidly verging into the wicked, lawless state that total absence of law and restraint will give rise to anywhere.
>
> Let us lift up our hearts and voices to our Christian Governor Jelks for succor, relief and protection.
>
> The fair name of Blocton, Bibb County and her thousands of peace loving, law abiding men, women and children, demand that their lives be protected from the vicious and that diabolical crime and lawlessness shall be stamped out.

Editor George D. Rice of the *Blocton Messenger* made this further observation on the subject of law and order in the March 20, 1903 edition:

> There is no doubt that some shocking crimes have been committed here, and while there have been judicial trials it seems the juries have been remiss as the people are. This we say with deliberation. At the last term of the court every Negro tried was convicted. No white man was convicted. Crimes in Bibb County, and especially in Blocton, have been numerous. While our natural sympathies are with the white people of the country, we think it rather peculiar. The juries of this country are the safeguard of American liberty.

Although the editor partially covered himself with his line about his "natural sympathies" he, nevertheless, made a bold statement for the time. Any stated sympathy for a black citizen had to be a brave move indeed for a small southern town weekly newspaper editor in 1903.

It was not prudent to get out of line in the days of editor Rice's tenure with the *Blocton Messenger*. The January 26, 1903, issue carried this account of an "affray" in Beat 11:

> DEGRADED WOMAN CAUSES TROUBLE IN BEAT 11
>
> Last Saturday night's shooting affray [took place] near the Crawf Reach Spring about seven o'clock. I— S—a woman who has caused this locality much trouble and has cost the county much money. A number of young men were visiting the S—— resort, two of them, A—— P—— of Belle Ellen, and W. T. B—— of Smith Hill, were there. P——— seeing a young man walking up in his shirt sleeves

> with a revolver in his pocket, walked up, jerked the gun out of his pocket and shot him, the ball taking effect in the young man's arm. Then P———— and B—— opened up a general fusticade [*sic*] against the crowd and succeeded in scattering them. B—— it will be remembered got on a razzle-dazzle in West Blocton last summer and was lodged in jail by marshal Gilbert, and in his frenzied condition of intoxication he tried to tare [*sic*] up the jail and fired the bedding in the jail, causing a terrible fight among the other prisoners. The woman, I— S——, recently stabbed her neighbor, Mrs. S—— C———, whose wound came very near proving fatal. I— was arrested and carried to jail where she remained until her bond was recently reduced from $5,000 to $500 and she succeeded in making bond and is again at liberty to prove a disturbing element in the otherwise quiet community of Reachtown. Neither B—— nor P—— have as yet been apprehended. I— is still at large.

The *Blocton Messenger* of early January 1905 carried a story that indicated official lawmen were not always called on to right the wrongs in West Blocton. Sometimes it was simply neighborhood citizens. The *Centreville Press* reprinted the item on January 26:

> WHITE CAPING IN BLOCTON
>
> Last Saturday night several persons went to the home of Mrs.——— in McGrawtown and entering her house whipped her severely with switches. The persons who administered the punishment are thought to have been women disguised as men. Mrs. ———'s husband left her recently. Since her husband left she has been a thorn in the side of many good wives in that community because of her fair face, subtle charm and easy virtue. Mrs. ——- has left the community and stood not upon the order of her going.

A "whitecap" is an archaic word describing a member or members of a self-appointed vigilance committee attempting by lynch-law methods to drive away or coerce persons obnoxious to it. They were called "white caps" because some early ones wore white hoods or masks, notably the Ku Klux Klan.

The town marshal had to remain alert to the unexpected. The local paper reported frequent "affrays" on the town's streets as it did in the *Enterprise* issue of November 17, 1906:

> A shooting affray occurred last Tuesday between Jim Prather and Columbus Hartley on Main Street, in which the latter was seriously wounded and had to be taken to the hospital at Birmingham where an operation was performed which we learn was successful and Mr. Hartley is in a fair way to recover. One of the shots fired struck Sylvester Morrison who was a block away from the scene of the shooting inflicting a slight flesh wound. Mr. Prather had a thumb shot off. Of the origin of the difficulty there are conflicting accounts, but we hope steps will be taken to prevent a repetition of affairs of this kind which jeopardize the lives and limbs of people who have no interest in the disagreements which lead up to them.

Prather family members say the report of a digit loss by Jim Prather was in error. No one remembers any such dismemberment. It appears the news media of the day was as prone to error as is today's. Even so, if a flesh wound was all that Prather suffered, the report is still proof that Main Street was not always a safe place

CHAPTER 4

The Bloctons: 1901 Until World War I

Eight years after the local paper editor first reported "a movement afoot to incorporate West Blocton," the event materialized. The charters creating West Blocton as an official municipality and establishing a dispensary resulted in the first local election and plans for selling liquor. The official record is in the 1900–1901 volume of *General Acts of Alabama* No. 976, and is titled: "An Act to incorporate the Town of West Blocton, in the County of Bibb." The charter defines the corporate limits, lists rights, defines the town government, states the town name, identifies mayor and councilmen's qualifications, gives election requirements, elected officials' duties, relates taxes and licenses responsibilities, and gives guidelines on street work and the street tax. The mayor's compensation was set at $25 annually plus $2 for each case tried before him. The act was approved March 4, 1901. The first election was to be as soon after April 1901 as practicable. J. B. Davie (misspelled "Davis"), G. W. Randall (misspelled "Randle"), and W. W. Perry were authorized to hold the election. The named citizens were community leaders. Davie owned one of the largest department stores; Randall was one of the town's dentists; and Perry was also a local merchant.

Eight years after the local paper editor first reported "a movement afoot to incorporate West Blocton," the event materialized.

A companion act was No. 933 that established the dispensary for the sale of spirituous, vinous, and malt liquors. Members of the board of commissioners were J. B. Wadsworth, J. B. Davie (again misspelled as "Davis"), W. W. Perry, A. D. Belcher, and Jonathan Krentzman. This act was also approved on the same date as the incorporation. Citizens listed in the incorporation act were Wadsworth, the local banker; Belcher, a merchant-landowner; and Krentzman, a department store owner. Over the years of the town's history, it was largely the merchant class that provided local municipal leadership. Medical professionals, such as general practitioner Dr. Peacock, and a dentist, Dr. Randall, also contributed their leadership abilities.

The *Centreville Press* of May 16, 1901, reported the West Blocton election:

> BLOCTON ELECTION
>
> At an election held in Blocton on last Monday for city officers of West Blocton, the following were elected. For mayor, E. D. Reynolds, Aldermen T. A. McCool, A. S. Sachs, L. J. Hays and W. A. Harvey were elected. Rufe Hubbard was elected Chief of Police. The newly elected officers will qualify and take charge of the affairs of the town at once. The dispensary will begin operation on Monday in the Douthit building, and this method of dispensing spirituous liquors in this manner will commence. Two policemen will be elected by the council and they will have police jurisdiction within a mile of West Blocton. No dispenser has been appointed but it is thought that Harvey Yeager will be offered the position.

The mayor and councilmen were all merchants. W. A. Harvey was the local druggist; Eugene D. Reynolds was a local merchant, and later coal mine investor, who managed a general merchandise store and in 1905 was listed in *Young and Co.'s Directory* as president of the West Blocton Savings Bank. H. C. and W. B. Reynolds had stores in West Blocton, Centreville, and Montevallo. Eugene Reynolds was an 1888 honor graduate of the University of Alabama. He is

West Blocton, *circa* 1900. This T. C. Tuggle photograph is believed to be an early view of Main Street, West Blocton. Scribbled on the reverse is "Willie Smith & Miss S. Hudson, Blocton, Ala. (Courtesy of Jimmy Lovejoy, Randolph, Alabama.)

remembered in a memorial window in the 1925 West Blocton First Baptist Church sanctuary. His father, H. C. Reynolds, had been the first president of the Girls' Industrial School at Montevallo, now the University of Montevallo. He left that post under a cloud having been fired by Governor Joseph S. Johnston because of an alleged fraudulent land deal. He was later exonerated in a jury trial and awarded compensation. That award was later overruled. Reynolds Hall on the Montevallo campus is named for him. Eugene D. Reynold's brother, H. E. Reynolds, who managed the Centreville business interests of the family, served as the mayor of Centreville at the turn of the century and later as state senator from 1907–1911. His legislative act creating county high schools will be examined later. The decision on the location of Bibb County's high school was a defining moment in local history.

In December 1902, the adjacent community of Smith Hill announced plans to incorporate and to also operate a dispensary. By May 22, 1903, legislation had been passed for incorporation and the dispensary was in operation. The *Messenger* editor, after touring the facility, reported an up-to-date operation with the latest in corkscrews and other equipment. But the most compliments were bestowed on the drinking rooms, "Three compartments are marked off for drinking; one for whites, another for colored people and a third for dagoes." He suggested the West Blocton Dispensary should take note. It is reported there was also a bottling works on Smith Hill and some claim it was really a brewery. No written confirmation of this establishment has been found.

Blocton's Worst Storm

Alabama communities have always had to deal with brutal spring storms because of their proximity to the Gulf of Mexico and the Appalachian Mountain chain. Bibb County has had its share of this severe weather. In the days before "tornado" was adopted to identify this most intense and destructive of natural weather phenomena, these storms were called "cyclones."

What was called the worst storm West Blocton had experienced in its first ten years occurred at 3:00 a.m., December 10, 1894. Identified by the *Blocton Journal* as a cyclone, it did widespread damage. The store house of W. Israel, a building owned by Uriah Smith, was all but destroyed. (Israel's goods were removed to the Douthit building where he reopened.)

The front was blown from the H. C. and W. B. Reynolds store, and it took workmen a week to repair it. T. J. Vickery's blacksmith and wood shop was badly damaged by a tree falling on it. The sign of the Blocton Savings Bank blew down smashing one of the bank's front windows. But the TCI Company sustained the heaviest loss. Twenty-eight company tenement houses for Negroes were blown down and otherwise damaged in what was known as Number 4 Quarters. There was no loss of life or injuries reported.

Another severe storm came nine years later when the area was struck by cyclonic winds and suffered some property loss. The main damage was in Blocton, the company town. The storm struck early Monday morning, April 13, 1903, between five and six o'clock. No lives were lost. The *Birmingham Age-Herald* listed the property damage and loss: Rufus Bearden's kitchen was blown away, house unroofed and furniture broken; Col. J. N. Campbell's kitchen was lost and part of his dwelling unroofed; Sam McCartney lost a roof and his kitchen was moved four feet; Martha Aldrich lost a chimney; Mr. Robinson lost a roof; Patrick Shannon's house was completely destroyed; Mrs. Pickney's summer dwelling had tree damage; stables and buggy houses of Col. J. N. Campbell and A. B. Roycroft were completely destroyed, and a dozen goats sheltered under the Roycroft warehouse were killed. At Primitive Ridge there was damage to houses. Ben Wallace's house lost its roof as did the home of Mrs. Hunter Allen. John Nabors lost his kitchen and the pillars of his store building gave way. Milt Cundiff's house was completely destroyed. At Bucktown, the Negro Baptist Church was swept away and one dwelling destroyed. The Number 2 mine tipple was blown down, the boiler house unroofed, and the scale house destroyed. The mine was not to resume work for several days. The cyclone damaged some West Blocton property. A tree fell on the Dan Belcher residence and many houses were blown down. Estimated total damage was around $7,500. The paper added "there was heavy rain and the heavens were aglow with lightning."

By November, rebuilding of the Negro Baptist Church had begun. The *Messenger* reported: "The colored Baptist Church at Bucktown [Blocton] are [*sic*] building a $2,700 church under the leadership of Rev. J. W. Witherspoon, their worthy pastor, who has few equals among his race. The building is now towering toward the heavens. All the people of Blocton should help him in his worthy undertaking." In July 1949, S. L. Johnson's *Centreville Press* column gave a brief description of this community.

> Directly west of the company store in Blocton and less than a mile away on some abandoned farm hill sides was located Bucktown, the exclusive Negro quarter. All the Negroes who worked in Blocton, except those who lived in Number Four town, had their homes in Bucktown. There were perhaps as many as forty houses and these were always fully occupied, and this was a lively and noisy place in the Nineties when I knew it. All is gone and the place where the town stood is gullies and briers."

Another community was named Campbelltown, and Johnson remembered it as follows:

> Campbelltown was located about two miles [this may be a distance recollection error] from the company store and to the southeast near the entrance of Number 7 mine. The town took its name from a man named [J. N.] Campbell who lived there and the town grew up around him when the mines were opened. Campbell owned some of the houses and some were owned by men who worked in Number 7 mine. There were about twenty houses and the town was partly in the woods and partly on what had been a farm. Campbell had a son named Charles who became a physician [who] married Lela Jones of the Rocky Church Community. She was a sister to J. A. Jones of Rockey and W. A. Jones of West Blocton.

Not mentioned about Campbelltown by Johnson was the location there of the Independent Methodist Church, a Baptist Church, and a school.

Most of the early development in the company town had been close to the mining operations, especially near the Numbers 1, 2 and 3 slopes. Businesses and miners' cottages had spread along both sides of Caffee Creek and on adjacent hillsides. Just to the west of Blocton, in a relatively level area, a local landowner, Uriah Smith, whose family had owned

Jasper Norbert Campbell was a pioneer landowner and coal prospector. He built several residences to the southwest of Number 7 Blocton mine, and the community there was known as Campbelltown. Among his children were Ruby Virginia (J. T. Bearden's grandmother); Adewin, an attorney who practiced locally in the firm of Campbell and Thomas; and Charles, a physician who practiced locally and later moved to Tuscaloosa County where he was the attending doctor at the birth of Governor Lurleen Burns Wallace. (Courtesy of J. T. Bearden, Tuscaloosa, Alabama.)

considerable acreage in the area since 1839, drew up a map and engaged himself in the real estate business by selling business and residential lots. Documents in the Bureau of Land Management show that Smith purchased 265 acres of government lands in the area between 1839 and 1886. The earliest evidence of the map's use was in a document deeding property in 1887 to J. B. Davie, a local merchant. Smith had already sold mineral rights to the coal interests but had retained surface ownership. Obviously an astute businessman, Smith sold property until his accidental death in 1898. He is buried in a residential section of West Blocton. The West Blocton Improvement Committee erected a historical marker at the site in 1996 with a number of Smith descendants present. Because he labeled his street layout "Map of West Blocton," this pioneer resident can be given credit for having named the present municipality.

In addition to the early suggested name of Smith Town for the developing community to the west of mining operations, there was also the often used name of Suttletown. Julien LaFayette Suttle was the son of a Centreville area planter, John W. Suttle, who had served as a sheriff and probate judge. Julien was raised on his father's plantation and educated in Selma. He had acquired storekeeping knowledge while working as a clerk in his uncle's store in Selma. He was in his early twenties when he opened a general merchandise and miners' supplies store on what was to become Main Street in West Blocton. He soon formed a partnership with E. N. Cottingham. Theirs was the first prominent private business operation in the community's developing commercial district. Suttle was well known and took part in community affairs. When fraternal lodges were formed, he was among the first on the rolls. He was a member of the Knights of Pythias, Red Men, Knights and Ladies of Honor, and Odd Fellows. His store, probably opened in 1884, was the earliest major center of business activity on the new street accounting for the town being known briefly as Suttletown.

Dr. Albert E. Meadow was one of the earliest medical professionals on record, arriving in 1885 to open practice in Blocton. He was joined in 1892 by his brother, Dr. John M. Meadow, both native Tennesseeans and graduates of Cincinnati's Pulte Medical College. The former had practiced briefly in Birmingham. Like other early residents, Dr. A. E. Meadow was active in the fraternal lodges. His name is on the Elks charter, and he also was a charter member of the Blocton Methodist Episcopal Church, South. By 1906, he had relocated to Birmingham. Though hired by the Cahaba Coal Company management, he stayed on to serve the TCI operation after 1892.

The *Bibb Blade* of July 1885 announced that a Dr. Rosser of Birmingham had been chosen by the company and citizens as physician and surgeon for the ensuing year. No further reference to Dr. Rosser's tenure is found, but this is evidence of the presence of a "company" doctor at an early point in the community's history. Another early company

Julien Lafayette Suttle, Sr., *circa* 1932. (Courtesy of Julien Lafayette Suttle, Jr., Winston-Salem, North Carolina.)

physician was Dr. Joseph Curtis who is reported to have followed Dr. Meadow. He was in Blocton for several years and according to historian Ken Penhale of Shelby County, three of Dr. Curtis' children were born in Blocton.

Another early resident was one of the first teachers in the town. Thaddeus C. Wallace, a native of nearby Six Mile, was a Confederate veteran and former county tax assessor. He settled in Blocton in 1885 where he remained for a decade as a teacher. He also served as the chief court official in town for four years as justice of the peace. He is reported to have disposed of over three thousand cases. The record is unclear, but he may have been employed as a teacher by the Cahaba Coal Company. His classes would have been held in the firm's "union church and school house," a facility identified in the *Bibb Blade*. Wallace was clerk and secretary of the Blocton Cumberland Presbyterian Church. He and his wife, Margaret Lightsey Wallace, were the parents of ten children.

Wallace was the organizer of Camp Clayton, No. 960, of the United Confederate Veterans. The earlier organization, Camp Pratt, dated from its 1884 founding in Randolph. That camp group was almost gone in September 1904 when the unit was re-organized in Blocton. Wallace had been a charter member of the Randolph group. His war record was recounted in the August 4, 1910, *Blocton Enterprise*. He had been a member of Captain R. H. Pratt's command, Company D of the 20th Alabama Infantry, serving all through the war except for a short time out for illness. Upon recovery, he returned to his unit. He participated in the battle of Franklin, Tennessee, and the siege of Atlanta, after which his unit marched to Nashville where he was taken prisoner and sent to Fort Douglas, Chicago, until the war's end. Upon the organization of Camp Clayton, Wallace was elected its commander. The weekly paper identifies him as Capt. Wallace, likely an honorary title, since the county's official muster roll lists him as a sergeant. While Blocton and West Blocton were not on the map during the Civil War, the work opportunities provided there two decades after the war's end were attractive to Confederate veterans. Camp Clayton initially enrolled eighty veterans as members from all over the county but most were local residents. By 1910, that number had dwindled to forty in good standing. The members of Camp Clayton sponsored an annual Confederate Veterans' Reunion at Scottsville until the end of World War I. Elderly residents of the area often remembered those gatherings. Harry James Adams and Dorothy Adams Caffee were among those who remembered traveling by train in their childhood days with their parents to the campgrounds for the festivities.

A prominent businessman, William H. Douthit, was one of those Civil War veterans. Reared in Tuscaloosa, his father had been an attorney and his mother the daughter of a riverboat captain. He was the proprietor of the Douthit House, a hotel, the first such facility erected in West Blocton, dating from the earliest days of the community.

The hotel was sold after he operated it for eight years, but the name was retained. He later opened a general store

J. B. Lowrey Tombstone, Confederate Soldier. J. B. Lowrey served in the 44th Alabama Infantry, one of the few Bibb County regiments sent to Virginia to fight under Gen. Robert E. Lee. (Courtesy of Cahaba Trace Commission.)

and other businesses. Douthit is also credited with opening the town's first bakery. It was in one of Douthit's buildings that the Blocton Savings Bank opened. He was aware that the town's youth would patronize entertainment halls. In July of 1892, he announced the arrival of new skates for his rink. The facility was open on Tuesday, Thursday, and Saturday nights and days the mines were not running. In an advertisement headed "Amusement Hall-The Skating Rink," prospective patrons were warned that "Hereafter all skates broken must be paid for by the parties breaking them, unless purely accidental." A prize was offered to the lady skater that improved the fastest. Earlier in the year the paper announced that Douthit would begin the erection of two stores with an amusement hall above. It was to be 44 x 60 feet with iron weatherboarding and iron roof, and with fire escapes. The hall was to be "fitted with opera chairs and have a convenient stage." There was no equal to Douthit's business acumen in early West Blocton history.

More information about Suttle, Meadow, Wallace, and Douthit is recorded in *The Memorial Record of Alabama* published in 1893 by Brant and Fuller.

The spring of 1892 was a busy expansion time in Suttletown-West Blocton. The grapevine must have carried the news far and wide of the coming sale of Aldrich's company to TCI. Business was "on-the-build" in anticipation of bigger and better times in the new Bibb town. The official announcement of the Cahaba Coal Mining Company sale to TCI appeared in the June 24 *Blocton Courier*. Another business was announced in April. W. B. Lowrey was to open both a shooting gallery and bowling alley in West Blocton. The editor said, "We may expect some lively sports since Mr. Lowrey knows how to please his customers. He will leave no stone unturned to make his alley a place of pleasure and amusement."

As the turn of the century neared, West Blocton's Main Street developed and expanded. Already, the western reaches of the thoroughfare was called West End. The town's hotels were the most prominent structures along the street. The Douthit House and the Manring Hotel provided lodging for business travelers. In the winter of 1899, W. C. Coleman, then a student at the University of Kansas School of Law, made a stop in one of the hostelries calling on a number of businesses and professional offices to sell typewriters and other wares. One of his stops was the drug store on the northwest corner of Main and Florida Streets. His story appeared in *The Coleman Company, Inc. 1987 Annual Report.* (The publication identified the town visited as New Brockton, the error made so often by the Post Office department workers.) The following is extracted:

COLEMAN MEANS QUALITY

For three generations of the founding family the Coleman Company has stood for quality... A symbolic new logo states it aptly: Coleman means quality. Always has. Always will. The story of how this came to be began on a wintry night in 1899. W. C. Coleman, a young Kansan who gave up

> teaching to study law, was in the mining town of [West Blocton], Alabama selling typewriters to pay his way through his final year in law school. Quality was important to him. For instance, the reason he selected the model of typewriter he sold was that he considered it "the best of its kind." Lighting also concerned him because he had very poor eyesight. The yellow flame from wick-fed kerosene lamps and the dim carbon filament in electric bulbs of the time did little to ease his handicap. On this cold, rainy night when he saw a bright, steady light beaming from a drugstore window, he quickly sloshed across the muddy street to examine the source of the rays. The lamp he found hanging there burned gasoline under pressure. Instead of a wick, it used a mantle. Instead of a smoky glow, it gave off brilliance. It was so bright, that for the first time, he could really see well at night. He could even read the fine print from labels on the store's apothecary jars and medicine bottles. Here, he thought, was a quality product. It could help not just him and others with poor vision. It could provide better light needed between sunset and sunrise by millions of Americans. He determined he would try to serve them with good lamps. The one in the window was called the "Efficient." He traced it to its distributor, a crockery wholesaler in Memphis, and arranged to handle the line in the Oklahoma Territory. The enterprise that would become the Coleman Company went into business on January 1, 1900, when W. C. Coleman arrived in the town of Kingfisher. It was soon the best-lighted town in Oklahoma.

The Coleman Company is no longer a family held corporation, and the drug store where Coleman observed the "Efficient" lamp in 1899 on the north corner of Main and Florida in West Blocton was superseded by two other drug stores. Today the seventy-year-old building on the site where the Coleman Company got its inspired start is in a dilapidated condition and unoccupied.

West Blocton School

When West Blocton was incorporated its town limits did not include a school or a church. For almost two decades the residents of the new town depended on the company schools, and occasionally private institutions, to educate their children. It was not until the fall of 1903 that the West Blocton High School opened its doors with eight grades. Among the reasons to incorporate the town was the need to have a legal means by which to raise revenue for supporting a public school. September 1901 came and went without action on the school presumably because the new official municipal entity had to get organized. Potential revenue sources, mainly the dispensary, had to be evaluated. How much and how often the town could expect working capital was being tested. Yet, the school year began in September 1902 also without action. Finally, the *Blocton Messenger* editorialized for a "free school" in its February 13, 1903, edition. By March, plans for a street fair were announced with proceeds to be used for a school fund. In mid-March, with revenue now available, the *Messenger* announced the creation of West Blocton High School by resolution of the Town Council. The dispensary had cleared $14,000 in March, with one-half of that amount retained by the town government.

On May 1, the *Messenger* reported that the mayor and town council had purchased from John Harmon of Eoline a pretty, shady lot next to Mr. Vickery for a public school site. A large, comfortable school building was to be erected right away so that school could be opened in the fall

The council asked for an election to endorse the issuance of a $2,000 municipal bond to be used in the erection of a public school building. There were only two dissenting votes on the subject of a bond issuance. The July 10 *Messenger* editorialized on the subject. Editor Rice wrote eloquently of his philosophy of education:

> The town will now go to work to sell its bonds and establish the finest school in the county. We will educate our children. Education means civilization, broad mindedness wider conception, intellectual expansion, regard for the rights, privileges and feelings of others. It means a nobler, better type of manhood and womanhood, and progressive citizenship. It is gratifying to know that West Blocton people are practically on a unit on this line, as they have so expressed themselves at the ballot box.

West Blocton School. (Courtesy of Bessie Robinson Collection.)

Further July issues of the paper recorded the school building's progress. "Trustees of West Blocton High School are now ready to receive bids for the school building to be completed by September 15, for the opening session of nine months. [It] will employ three teachers." The bid was awarded to Vickery and Deason, a local construction company that started work immediately. "Soon we will see a two-story, eight room building loom up in our hustling village." The school opened on September 11, 1903. A. E. Handley was elected principal, Miss Almon from Moulton was in charge of the intermediate department. The school enrolled 132 pupils on opening day.

The December 11, *Blocton Messenger* reported:

> The Blocton Public School [the editor was not consistent with the school's name] now numbers about 175 pupils. It is the only absolutely "free school" in Bibb County. Under the leadership of Professor Handley it is doing a really marvelous work in educational advancement.

The one area of continual progress and community support was embodied in the company schools. The Tennessee Company had inherited a somewhat tenuous company school in 1892, but developed a first class school program, and by the century's end began to have a respectable school for company employees' children. Other children in the community attended for a modest fee.

The Tragic 1904 Strike

While there was progress on the education front, the community experienced a setback during 1904. The destructive storms of 1896 and 1903 were short-lived natural phenomena over which man had no control. Recovery was fast and scars erased. The more long-lasting community tragedies were difficult to overcome. There had been an element of perceived impermanence in the Bloctons almost from the beginning. West Blocton's civic leadership frequently came from the merchant class. When business was slow, they tended to be less likely to vote for improvements that would require the town's going into debt. When civic improvements were advocated there was a tendency to find reasons to wait for better times. Inevitably, the continuing labor unrest contributed to this pervasive, negative posture

In Blocton, the company town, the leadership was vested in the mine superintendent and his mine bosses. They made decisions based on company policy formulated in Birmingham and after 1907, from as far away as Pittsburgh, Pennsylvania, headquarters of United States Steel.

The coal industry strikes in Blocton followed the national trend. The major work stoppages for improved wages and conditions for miners generally were nationwide. Until the 1904-1905 strikes, the Blocton scene had been relatively quiet; some pushing and shoving and the occasional intervention of government to effect civil control and property protection had been the extent of the labor troubles. No property had been damaged nor lives lost. The local papers had used the phrase "Blocton is quiet" in reporting the strikes that occurred from 1888 through the turn of the century. The very size of mining operations in the Blocton District, and the number of miners in the Blocton area (1,200 to 2,000) caused the area to be mentioned in state papers.

The 1904-1905 strike was a different matter. Genuine and serious trouble was on the horizon for the Blocton strikers in this first decade of the twentieth century. The earlier strikes had been broken by management and the politicians that the ruling class had helped elect. But organized labor had started to rebuild and by 1902, organization efforts had begun to bear fruit. The picture can be better focused by noting the observation of the authors of *Alabama: The History of a Deep South State*:

> Once again a slow rebuilding was in order, this time under the aegis of the national union. In 1902 the UMW claimed to have organized 65 percent of Alabama's miners, with blacks making up about half of the membership. The companies struck back with the open shop, [the optional employment of non-union labor] and in 1904 the UMW went on strike. The walkout lacked the focus of earlier strikes, and the standoff dragged on for two years.

There had been problems in the summer of 1903, but the Blocton news corespondent, attorney E. M. Bishop, wrote to the *Centreville Press* in August 1903 that work had resumed. However, he bemoaned the fact that business had been at a low point with the workers out on strike. He emphasized that the town depended almost entirely on the mining industry. Work had resumed the previous week, but he stated that "the woeful expression of uncertainty and uneasiness has given place to a buoyant look of cheerfulness and confidence." That outlook was to change drastically the following spring.

The UMW Blocton local had called a strike on May 3, 1904, because the Tennessee Coal, Iron and Railroad Company refused to take action on their request to remove Superintendent W. S. Lang. The company president, D. H. Bacon, responded through J. M. Franklin, manager of the coal mines. The statement to the miners was: "Your message of this morning received. I hold Mr. Lang in high esteem. Nothing has reached my ears to show his unfitness for the position he holds, and it is not my intention to make a change. You are at liberty to make this public."

The major conflict between Lang and organized labor was likely the importation of non-union labor. Lang was a native of Pennsylvania and had been employed by TCI for three years. Although he was said by the company to be a just man as well as an exceptionally efficient mine manager, there were reported threats on his life by the striking miners. On Saturday night, November 12, 1904, Lang and deputy sheriff John F. Golson were investigating the trespass of three union miners onto company property. The *Birmingham Age-Herald* said the site was Dagotown. Two of the three trespassers were arrested after being pursued some distance up the railroad tracks near the mine (likely Number 3). While the party was returning to the compound entrance, Lang was shot, supposedly by the missing third intruder. The shot, coming from the crest of a nearby rock dump, entered his hip and ranged downward coming out several inches lower and in the front portion of his leg. He was helped to the company office and attended by the company doctor. Contact by telegraph was made with the TCI office in Birmingham, and a special train was dispatched to Blocton bearing J. M. Franklin, general superintendent of the company, and Dr. Cunningham Wilson. Lang was taken to St. Vincent's Hospital where initially it was believed the wound, though painful, was not life threatening and early improvement of Lang's condition was indicated. On the following Monday, however, inflammation set in and Lang developed blood poisoning. His condition steadily worsened and he died the following Thursday. The *Birmingham Age-Herald* carried daily updates on the event during the week. When Lang died, the paper's November 18th editorial page printed a long column titled "Death of Superintendent Lang" recounting the whole story. For some understanding of the union miners' problem with the superintendent, there is a subtle line that suggests Lang had personality traits not making for the best management-employee relationships. The editorial follows:

> W. S. Lang, who was shot at Blocton Saturday night, died yesterday at St. Vincent's Hospital, in this city, from the effects of the wound. The death by violence of any reputable citizen is deplorable, but in the fatal shooting of Mr. Lang the community is especially stirred, regardless of the personality of the deceased. Mr. Lang enjoyed the reputation of being a just man as well as an exceptionally efficient mine manager. But he incurred the enmity of some of the union miners working under him prior to the strike and the Tennessee Company was asked to discharge him. The company made an investigation and then publicly announced that the miners only had no grounds for a grievance, but that Mr. Lang was giving eminent satisfaction as superintendent. Since the strike the Tennessee company's mines have been worked by non-union men.
>
> The trouble Saturday night followed the arrest of two union miners who were said to be trespassing on the company's property at Blocton and the escape of a third union miner who was alleged to belong to the trespassing party. While one of the theories of the shooting is that the union miner who escaped fired the bullet, there are other theories, and in the absence of developments the public is at a loss to know what to believe concerning the identity of the assassin.
>
> But it behooves every right-minded citizen in this district, regardless of union or non-union sympathies, to give expression condemnatory of the outrage which resulted in Mr. Lang's ultimate end. Lawlessness was in evidence in the mining camps soon after the strike was declared, and more than one man was shot. But Mr. Lang was the first high official to meet his death, and this tragedy therefore rises to a more serious level and becomes a burning public matter. The Tennessee Company has offered a large reward for information that will lead to the arrest of the man who shot Mr. Lang, and the murderer is almost sure to be brought to justice eventually. While a large part of this community feels friendly to the union miners and while every man in Birmingham feels sure that the officials of the United Mine Workers deplore the occurrence at Blocton as much as any persons in the district, it will be especially in order at this juncture for the miner's union to join actively with the citizens at large in every movement to maintain law and order. The union miners have a perfect right to strike. No man questions this right. But the Tennessee Company and every other mining concern has a right to operate their mines with non-union miners if they see fit. There is no place in this district for the murderer, be he union or non-union.

Boldfaced advertisements appeared in Birmingham papers and the *Centreville Press*:

> NOTICE—$5000 REWARD
>
> By authority of the Board of Directors of the Tennessee Coal Iron Railroad Company, a reward of Five Thousand Dollars is hereby offered for information that will lead to the arrest and conviction of the person or persons who shot Superintendent W. S. Lang at Blocton, Saturday, Nov. 12.
>
> Tennessee Coal, Iron & Railroad Company.
>
> By D. H. Bacon, President

The amount of the reward was a small fortune at the time; a miner might work years, even a lifetime, and never see that amount. But despite the huge reward, research has not turned up record of a conviction. The *Centreville Press* of January 24, 1907, had a terse report in its local news column: "Lee McGuire, formerly of Blocton, is reported to have been hung [*sic*] in Arkansas. McGuire was a suspect in the shooting of Superintendent Lang several years ago."

With tempers obviously on a short fuse, there was a rash of violence during the strike. In November 1904, Will Thompson, a guard for TCI, was shot in a Negro's home. George McCool, marshal of Smith Hill, shot a Negro in the arm when a gun was drawn on him, and a Negro was accused of shooting a non-union white man in the leg purportedly as a result of a drunken brawl.

Evidence that the 1904 strike caused severe economic conditions in the community was reflected in the quarterly report of the departing Methodist preacher, Reverend Robert Echols. Echols wrote that "in view of the disorganized condition of affairs in the towns we hope for new life soon . . . while our people have sustained heavy financial losses they have supported the church . . . our loss in membership has been great but we are hopeful of better things."

Among the more active members of Echol's church to depart was banker J. B. Wadsworth. He and his family had been on the church rolls since the founding of the Blocton Savings Bank in late 1892. There is evidence that his family had moved to Birmingham several years before his departure, likely as early as 1899. His collected letters on file at Samford University suggest he lived in West Blocton during the week and commuted to Birmingham on weekends. In a letter to his wife he complained about the illness of his cashier saying, "There is a great deal of fever here. It is a wonder to me they don't all die, the place is so filthy. The Jews and low white trash won't clean up." The banker, a leader in the Methodist church, let frustration bring his prejudices to the surface. It would not have served the bank business in the community not to have the Jewish merchant class as clients. Wadsworth had been associated with other contemporary financial institutions, including the first bank in Centreville, the Bibb County Banking and Trust Company, which opened in July 1897. There is evidence that besides the residence in Birmingham, he found it necessary to have a home in Centreville as well. In any event, he apparently moved away from West Blocton for good in 1904 but kept an absentee management assignment with the bank for several years.

The death of Superintendent Lang and the unrest of 1904 spilled over into the following year. There were continued complaints in the *Blocton Messenger* about the strikebreakers brought in by the company. An April 21 editorial stated the local attitude:

> The non-union people of Blocton are giving considerable work to the deputy sheriffs and justice court of Blocton. Arrests are made every day. Some of them are taking berths in the Blocton and Centreville jails daily. They are the wrong class of people to add to the moral, social and industrial prosperity of any community. They are really such a low grade, inferior quality of citizenship and manhood that they can make no effort to command the respect of their fellow man because they are so grossly lacking in selfrespect. The man who respects not himself and his calling certainly cannot look for the respect of others.

In a most remarkable episode in 1905, the union element, claiming special rights in the community's main cemetery, Mt. Carmel, refused to let a non-union worker be interred there. Their authority was based on the claim the union had contributed to the purchase of several acres of the cemetery. The story implies that the confrontation took place at the time of grave site preparation.

A June 30, 1905, *Messenger* entry further indicates how ridiculous the situation had become:

> When eighteen Germans were standing at the station waiting for the train to take them away from a place to which they had been fooled into coming as strike-breakers, last Monday morning the constable appeared and arrested one of them who was charged with having milked his neighbor's cow without the knowledge or consent of the owner. The Germans protested with the man himself that he was the wrong man, that the man who milked the cow was still up at No. 3 mines, Mr. W. B. Williams

of Smith Hill, had pointed him out as the guilty milk maid and Messers Golson and Williams took the alleged lover of sweet milk away while his sympathizing fellow countrymen shook the non-union dust from their feet and boarded the train sadder and wiser men from their unfortunate experience.

The cow that was thusly milked was a union man's cow and she doubtlessly kicked when she found there was a scab on her tete [*sic*].

The same newspaper issue editorialized as follows:

IMMIGRATION BAIT

About all the recent T. C. I. R. R. Co.'s Blocton immigrants emigrated from Blocton this week. They were fooled into coming here. But they needed no misrepresentation to get them away, not a bit. So [as] soon as they realized their unfortunate and shameful predicament they kicked and kicked and kicked out."

In Brown and Davis, *It is Union and Liberty*, it is reported that regardless of the lack of work and problems related to the strikes, the UMWA attempted to keep up workers' morale by spending considerable sums sponsoring entertainment. "In July 1905 more than six thousand people (the number is exaggerated) attended a union barbecue at Blocton, featuring ball games, dancing, a parade, music, and speeches."

The community's leaders, as best they could, dealt with strikes and storms. The early mayors of West Blocton (Reynolds for two terms followed by Dr. Peacock) were educated men capable of exercising their duties and responsibilities. One pressing civic need was a water system for West Blocton. The earliest water source was at the Reach Spring near the old, recently revitalized Hill Creek ball-park. An August 1910 special edition of the *Blocton Enterprise* reported the system had been placed in operation in 1902. It consisted of a ground-level cement collection tank and a steam operated pump. At appointed times an operator would come on duty to keep the pump running in order to move water to tanks on West Blocton Heights at a hilltop location on Florida Street (earlier named Belcher Street). During several of the years this system was in use, Leige Stamps was employed as pump operator by the town. Adequate water pressure was maintained by the elevation at that site. The wooden reservoirs were constructed in the fashion of steam locomotive water station tanks. Perpendicular, notched, treated wood slats were held in place by steel bands to form the huge circular holding tanks. The Reach Spring installation was a prominent landmark. "Pump House" became a place name and is listed in Ellison's *Place Names of Bibb County*. According to that publication it was a voting precinct later identified as McGraws. It is labeled Beat Eleven or McGrawtown in 1997. Years after Reach Spring's abandonment for a greater water flow spring, Arthur Finnen recalled in 1997 that he and boyhood friends swam in the ground level tank. Others remember the spring as a drinking-water source during ball games at the nearby Hill Creek park. In the days before electric motors and automatic switches, the early pumps, as noted, were operated by a steam engine requiring the employment of a pump operator. Fires quickly depleted the water supply in the tanks, and the pump man had to be ready to stoke the engine's firebox to keep the water supply at a safe, sufficient level.

Another story from Arthur Finnen gives an account of the town's waterworks maintenance man, John Gilbert, who had to climb the wooden structures from time to time to tighten bolts known as "screw jacks" on the steel retaining bands. Gilbert told Finnen of an incident that occurred while he worked 40 feet above ground straddling the edge of one of the tanks. While working on a top band's adjustment, his huge monkey wrench slipped from his grasp into the full tank. Looking in all directions for possible witnesses, he removed his outer clothing and dove in to retrieve the wrench. West Blocton households unknowingly were supplied with "bath" water for a few days.

In 1911, Dr. L. E. Peacock, serving as county health officer, investigated a far more serious water contamination problem. After complaints made to him by local citizens, he carefully examined the town's Gillespie Springs reservoir and found it contaminated by human and animal waste. He brought the problem to the attention of town officials in an eleven-page, handwritten 1911 annual report to the State Board of Health. An added demand for water, said the report, had led the town to take its needed additional water from a branch and it was this move that led to the use of contaminated water. The mayor at the time, M. C. Davie, Peacock's neighbor and a prominent businessman, is assumed to have acted promptly to resolve the problem.

In a December 1906 issue of the *Blocton Enterprise*, a list of town improvements during that year included the addition of a new water tank, improvement of the old tank, and the addition of a new pumping station at the Gillespie Springs (this was the old water works site near the Frog Level community). Another water project improvement was about three miles of 3 1/2 inch pipe extending the main, and the addition of four fireplugs to improve fire protection. While the installation date is uncertain, eventually a second spring was tapped and a small retaining wall was built a short distance from the original Gillespie Spring site. In 1915, an up-to-date mechanism was installed so pressure in the tanks would automatically stop and start the pumps.

Among other improvements listed for 1906 was the grading and graveling of all streets and alleys, including the splendid walkway to the L&N Depot. These were accomplishments during the term of the town's third mayor, W. H. Logan. The "walkway" was a pedestrian shortcut from Main Street, West Blocton that traversed the small rock bluff edge north of Hill's Creek, passed in front of the mine superintendent's home, and ended at the bottom of the hill near the Methodist Church in Number 1 Blocton. For many years, pedestrian shoppers, church-goers, and Sunday social groups going to meet the train, walked this path. Enterprising West Blocton merchants painted advertisements on flat-faced rock outcroppings that were still visible during the the 1930s. The path was scraped away when the Cahaba River Drive bypass was constructed in the 1960s.

Intra-county Rivalry

A new industrial community superimposed on the agrarian society existing in Bibb County during the last decade and a half of the nineteenth century was destined to experience conflict over political power. When disagreements occur within county government, political scientists and historians term the condition "intra-county sectionalism." Developments in the 1890s and during the first decade of the new century brought the conflicts within the county into sharp focus. The population in the coal mining district and the amount of county revenue collected there were arguments for more influence by the mining area in county affairs.

There was early evidence of dissatisfaction in Blocton with the court system based in the county seat. In late 1892, a division of the county court dealing with local misdemeanor cases had been established in West Blocton by the legislature. After the court had been in place only a matter of months, the editor of the *Blocton Courier* made an editorial comment in the March 24, 1893, issue that must have raised the eyebrows of a few elected officials.

After reporting that the first term of the Blocton division of the county court had been held successfully in Wright's Amusement Hall and that it pleased the attorneys who were in attendance, the editor said the majority of those attorneys were in favor of moving the court house to Blocton and voiced the further opinion that an effort would be made in the next session of the legislature to do so. He added that "if this is done it is very probable that Blocton will be the choice of the people by a large majority."

The *Courier* continued this theme on May 19:

> The Blocton division of the County Court will never be a success without a jury term. As it is, it is simply a useless expense to the county. If Blocton is to have a court at all, let's have a court in reality. The citizens of the county who favor the removal of the courthouse to this place should at once begin to look about and make preparations to this end. Blocton furnishes the greater part of the revenue to the county and should be recognized as the seat of justice.

In January 1903, the *Blocton Messenger* editor wrote that application would be made to the next session of the legislature to establish a circuit court branch at West Blocton with both civil and criminal jurisdiction. Trial in the criminal jurisdiction was to be for non-capital felonies. There was to be a provision for grand juries. The record does not indicate this was accomplished. In July 1907, the *Blocton Enterprise* reported that the Blocton Division of the county court had been extended by legislative act. The towns of Coleanor, Garnsey, and Piper now were in this court's jurisdiction. The court was to continue handling non-jury, misdemeanor cases. The higher court would not be coming to West Blocton. The local court sessions were most often held in Wright's Hall, which for a time the locals jokingly called the "Court House," and on occasion court was held at the Opera House. There is no record of this court after about 1914. Perhaps the completion of better roads to the county seat was a factor in the demise of the Blocton Division of the county court. Too, as early as 1905, the M&O passenger train schedule had been arranged to make it possible to travel to and from Centreville in a day. The *Centreville Press* said, "Arrangements have been made whereby the people from Blocton can spend the day in Centreville and return to their homes in Blocton in the evening. The train out of Blocton will leave in the morning at seven o'clock, reaching Centreville at eight o'clock, and returning in the evening on the later train and arrive at home in time for supper. This is going to be a great convenience to the people of these two cities who have business in the two places."

There was occasional rivalry between the towns revealed in the local weekly newspapers. In October 1903, the *Centreville Press* editor expressed his dissatisfaction with the contents of a recent issue of the *Blocton Messenger* featuring sketches of prominent men in the county as well as a half-page illustration, by West Blocton photographer T. C. Tuggle, of the new courthouse. H. C. Nunnally, the *Press* editor, wrote: "The paper is well illustrated and we congratulate our neighbor on his enterprise. We suggest that Brother Rice [the *Messenger* editor] pray long and without ceasing, to get forgiveness for the things he said and what he left unsaid in his Industrial Edition." In Nunnally's opinion, the articles had been grossly slanted in Blocton's favor.

The Blocton editor's claim about the Blocton District's contribution to the county's revenue had been greatly enhanced with the coming of the dispensaries (town operated liquor stores) in West Blocton and Smith Hill. Centreville had legislative permission to establish a similar city-controlled liquor store but declined to do so, the local voters having rejected it. The *Centreville Press* of January 17, 1901, reported a mass meeting in Centreville to debate the dispensary question. The resolution against won 68 to 47. The delegation from Blocton had previously declared for the dispensary. They argued that illegal whiskey and beer in the amount of $80,000 worth had been shipped into Blocton alone in 1900.

When West Blocton was incorporated in early 1901, the dispensary system was installed. Half the profits from the dispensaries went to the county. The division of this revenue on an equal basis was challenged in late 1902. Legislative action was explored to have two-thirds of the profits go to the county. The plan to do so was advocated by J. F. Thompson, the county's legislator, a Centreville attorney. The idea was dropped. Evidence of local feeling on the matter was voiced by *Blocton Messenger* editor George D. Rice in January 1903: "The people of all Blocton and all this section will say to him with hardly a dissenting voice, give the dispensary towns half the profits."

Advertisements for the dispensary were rare. A simple item appeared in the May 15, 1903, *Messenger*:

AVAILABLE AT THE WEST BLOCTON DISPENSARY:
Golden Creme Ryes
King Edward Rye
Royal Name and Royal Whiskey
For a pure article, for tonic or beverage,
ask for 1880 Rye Malt.

Adequate revenue support had long eluded the county government and now, almost as a windfall, here was real money available and arriving in the till monthly. But it was relatively short-lived. The county option law (each county could vote whether it chose to be "wet" or "dry") was passed in 1908 and Bibb County voted for prohibition. The chairman of the prohibition movement in the Blocton area was J. B. Davie, prominent West Blocton merchant and Baptist deacon.

Profits varied for the two dispensaries over the period of their operation. There is evidence that the Smith Hill facility had a shaky start and did not produce profits in its first few months. The reports were not regularly printed but occasional accountings were given. In an early report in the *Centreville Press*, the gross sales of the West Blocton store were listed as $4,100 for February 1903. Profit was something over $1,200. The paper reported that the Smith Hill store had shown only a $765 profit by February 1904 after being in operation nine and a half months. Sales totaled $29,000. There was an implication that profit was insufficient in the newer facility. The *Blocton Messenger* in April 1903 reported a profit of $14,000 from the West Blocton dispensary, "one-half to the city, one-half to the county for roads and schools." The *Centreville Press* of April 16, 1903, claimed there was dissatisfaction with the West Blocton Dispensary because of its kinship to the open bar. The paper said, in essence, that the dispensary system was the lesser of two evils. The impact on the county school system was evidenced by a report in the *Centreville Press* of January 31, 1907. The County Board of Education had met in regular session to accomplish only one thing, to deal with the appropriation of the dispensary funds for school purposes. The board's share for the year was $6,827.25.

Toward the end of the era of city-managed liquor sales in West Blocton and Smith Hill, a story appeared in the *Birmingham Age-Herald* originating from its Centreville correspondent suggesting the merger of the two stores as a cost-cutting move. It quoted county treasurer Fuller as saying for several months the Smith Hill facility had paid three times as much into the county treasury as the West Blocton store. There was immediate response by E. M. Bishop, a West Blocton attorney, representing the dispensary. Bishop wrote to the *Centreville Press*:

> At the request of the managers of the West Blocton Dispensary I inspected the books of the county treasurer, Mr. Fuller, from January 1st 1903, to February 1st 1907 As to the amount each of these dispensaries had paid over to him during that time. The Smith Hill Dispensary was established and commenced business sometime in 1903, I think about May 1st. During the remainder of that year nothing was paid by it to the county treasurer. That year West Blocton Dispensary paid $4,484.30. In 1904, Smith Hill Dispensary paid $1,258.75; West Blocton, $2,247.85. In 1905, Smith Hill paid $1,107.62; West Blocton $3,192.57. In 1906, and up to February 1st 1907, Smith Hill paid $3,707.45; West Blocton $3,839.48. These are the figures as they appear on the books of County Treasurer, Mr. Fuller.
>
> While it may be true that for the last few months the Smith Hill Dispensary has paid more than the one at West Blocton, taking the year as a whole, or any preceding year, it falls far short of yielding the revenue to the county that is derived from the dispensary at West Blocton. A further sum of $1,030 is now due from the West Blocton Dispensary, payable on demand of the treasurer.
>
> I simply wish to say this injustice to the managers of the West Blocton Dispensary, who are competent businessmen, of the highest honor and integrity. I have heard the remark by persons not citizens of the town that the dispensary here was one of the best managed institutions of its kind in the state.

Attorney W. W. Lavender of Centreville had written the Birmingham newspaper article that caused the detailed response by Attorney Bishop of West Blocton. Lavender's reaction to the report appeared in the February 24 issue of the *Centreville Press*. After taking issue with the figures, Lavender added sardonically:

> If the difference in the amounts of revenue yielded by these two dispensaries is due to the fact that the people of West Blocton are more temperate than the people of Smith Hill I do not see why some of the people of West Blocton should complain because the falling off of revenue realized should be made public. As to whether two dispensaries located in the same town within one mile of each other will do more to advance and promote the cause of temperance than one dispensary or no dispensary at all I leave it for the President of the Alabama Anti-saloon League and his co-workers to answer.

In the August 22, 1907, *Centreville Press* the examiner, G. F. Sedberry, reported the books of the West Blocton facility in good condition and properly kept. A similar report appeared in the same issue giving the Smith Hill store good grades. The latter report added: "While the dispensary system is condemned by some, it is considered by many to be in the interest of good municipal government and appears to be the solution of the whiskey question for small towns."

During the period covered by Bishop's report on the financial matters of the dispensaries the county realized $20,775 in revenue, equivalent to many more thousands today. Earmarked for roads and schools, additional amounts were paid the county until the dispensaries were closed because of the prohibitionist vote. There was an element in the county that frowned on the system but those counting the revenue dollars at the county seat were anxious to be sure the county's share was placed on the ledgers promptly.

The liquor stores' end as a revenue source could be foreseen in a September 19, 1908, *Centreville Press* item, reporting on a mass meeting in Blocton the week before to discuss the possibility of calling a prohibition election for the county.

A series of public meetings were held by the prohibition element. One bizarre event was reported in the October 1, 1908, *Blocton Enterprise* that resulted from an illustrated lecture on the evils of drink. Dr. J. F. Alexander, the company physician at Lucile, "while assisting in the chemical illustration at the Methodist Church . . . was injured by an explosion." It seems a piece of glass from a broken bottle pierced the doctor's right eyeball causing him to lose his sight in that eye. Churches throughout the county hosted meetings pleading with the populace to vote for prohibition.

The November 6, 1908, *Blocton Enterprise* reported that the prohibition law would take effect on January 1, 1909. Thus dispensaries were out of business, so town and county were without one of their most lucrative sources of revenue. Later in the month, the *Enterprise* announced a mass meeting to find revenue to finance the town government. Subsequent reports are unclear as to what action was taken. The school operated with reserve funds for a short period and then began charging fees. Within a short time a new business and professional office license schedule was adopted by the town council, perhaps a partial remedy for raising school and other municipal service revenue. Visiting relatives in Centreville, West Blocton businessman and former mayor E. D. Reynolds was quoted in the January 2, 1909, *Centreville Press* as saying he had no regret at the closing of the dispensaries in that business would improve because better moral conditions would prevail.

Dissension concerning the governor's appointment of Nathaniel Edmond Stewart, a resident of the Scottsville area, as road supervisor on the west side of the Cahaba River, in August 1907, caused the *Enterprise* editor to observe, "It looks like the north end of Bibb is doomed as far as getting office is concerned, either by appointment or otherwise." The writer had overlooked the election of Jones G. Moore to the state senate (1898–1901). Moore had been TCI superintendent of the Blocton mine operations at the time. Perhaps Moore was perceived as a temporary Blocton District resident and his relevance as a true local representative was thus considerably lessened.

The intra-county rivalry episodes by comparison amounted to mere skirmishes until the question of the location of the county high school arose in 1907. The school issue caused a major rift in feelings between leadership in the north and south ends of the county. In the legislative session of 1907, Herbert E. Reynolds, Bibb County's representative in the state legislature and former mayor of Centreville (as well as brother of West Blocton businessman Eugene D. Reynolds, the town's first mayor) introduced a bill that would provide state support for ten high schools, one in each congressional district in the state. Reynolds, a law school graduate of the University of Alabama, who also had studied at the University of Virginia, had been a resident of Centreville since 1898.

In the *Centreville Press* of February 28, 1907, the editor's comment on the high school location read: "If the bill passes I think one of them will be placed at Centreville and we will have a first-class high school for the county. The towns where the schools are located will have to furnish the building and it must be a substantial one." Further information in the article said the initial appropriation would be $1,500 per year but would doubtless be increased in a short time. At a subsequent legislative session the number of schools was increased. There was to be a high school partially financed by the state in each county, and it was definite that a high school would be located in Bibb. The civic leaders of Centreville launched a successful money-raising campaign that funded the erection of a "substantial" school building. This building still exists facing the intersection at Four Points and now houses junior high school grades.

Evidence came early that West Blocton was also interested in obtaining the high school. The competition in Bibb was noted in the October 3, 1907, *Centreville Press*: "Senator Joseph F. Johnson spoke in Blocton last week in the interest of the high school that the citizens of that town are attempting to have the state locate there. Blocton has raised $5,285 and they are confident of raising the five thousand [additional] dollars, required by the state."

The high school topic appeared often in the pages of the Centreville weekly newspaper during the spring and summer as Centreville was in the midst of a drive to get donors to give money for a two-story, brick school building. A substantial amount had been raised toward a goal of $7,000 according to the *Press* of April 4, 1907, and plans were announced for a meeting at which a Professor DuBose of the University would speak on the subject of education. It is assumed that the legislative act proposing state-supported high schools precipitated this building fund drive. Historically, previous attempts to establish and sustain a public school in Centreville had not met with continued success. Local private institutions had provided instruction to those able to afford the fees.

After county high school promotional committee meetings in West Blocton, a note of apprehension was voiced in late fall. The *Centreville Press* editor wrote on November 28, "Are the people of Centreville going to let [indifference] keep them from receiving a continual feast by the failure to locate the high school in Centreville[?] Are we going to sit here and see this school located at some other point, when by bestirring ourselves we can get it?"

In the September 12, 1907, *Blocton Enterprise,* a meeting had been called on the subject of locating the high school in West Blocton. Mayor W. H. Logan of West Blocton and Mayor Wood of Smith Hill, J. B. Davie, and others endorsed the concept. There was a subsequent meeting on October 10 at Wright's Hall and another at the Piper schoolhouse. The announcement was made that $4,000 had been raised.

Meanwhile, the *Centreville Press* of October 31, 1907, reported on a barbecue rally that took place on October 23. The occasion was the visit to Centreville of the high school commission made up of Auditor W. W. Brandon, State School Superintendent Gunnells, and Governor B. B. Comer. Comer and Brandon spoke to a crowd of 2,000 promising, said *Press* editor Nunnally, to "give our merits careful consideration."

In the October 24 *Blocton Enterprise*, a report on the "Rally at Centreville" said that a delegation from Blocton [while there] conferred with the commission in reference to getting the county high school located in West Blocton. Among the delegation were Mayor Logan, J. B. Davie, Tandy Medlin, Pat Ginnett [likely Jennett], Pat Shannon, J. R. Golson, J. M. McMasters, C. H. Cole, Peter Shannon, M. C. McCool, E. M. Bishop, and L. J. Hayes. The commissioners agreed to come to Blocton, but Governor Comer said he had a problem with an early date because of a legislative special session. At subsequent meetings in November, a local committee made plans for the commission's visit. The *Enterprise* of December 26, 1907, stated, "Interest in the county high school is not abating, and the people of this end of the county are more determined than ever to have it located where it will do the most good." In mid-January 1908, the Blocton paper reported the governor had promised to visit Blocton. In the next issue it listed a committee of prominent Centreville citizens who had gone to Montgomery to present its case for their town as the high school site. Editor C. E. Allen offered the idea that "a committee from this end of the county be sent to represent the great majority who desire placement where it will do [the] most good." The February 6 issue devoted most of its editorial space to the high school question. The commission's visit was eagerly anticipated. Editor Allen of the Blocton weekly proclaimed:

> That West Blocton is the proper place for the County High School there can be no manner of doubt, but it rests with the citizens of the district to make such a showing to the gentlemen comprising the commission as will prove it to their entire satisfaction. We have within an easy radius three-fourths of the population of the county. The pupils who can attend the school if it is located here are in the ratio of seven out of ten of the eligible school children of the county.

On February 20, 1908, the paper discounted a report that Centreville had been chosen for the school. Editor Allen said it was not so, "the law plainly says it should be placed where it will reach the most pupils." This point was emphatically sounded over and over by the forces from Blocton and the north end of the county. However, there were other elements at work that helped thwart the merits of this argument, and over the next several months various reasons for Governor Comer and the commission's delay in visiting Blocton were given.

A confident *Centreville Press* editor reintroduced the school site topic in the February 13, 1908, issue: "When Centreville gets the high school there will be so many boarding pupils in our little village that some means will have to be devised to improve our sidewalks and streets, so as to make a better appearance when parents come to the city to see the boys and girls." In the June 30 issue, the editor again laid claim to the high school and referred to sharp competition throughout the state for site selection. "Centreville is putting in for the high school location. It seems to be a cinch that this town will get it and it will be a great thing for the people of the county."

Meanwhile in the Blocton area organized labor difficulties clouded, and quite possibly seriously affected, the school site decision. In the summer of 1908 Blocton miners became embroiled in a major coal strike, one of the largest and most violent the local industry had ever experienced. Union organization had been in progress with the local paper reporting on the movement's recent efforts in Belle Ellen where a large crowd had gathered on Nelson's field and on Smith Hill at the picnic grounds. Here UMW District 20's president Kennamer and an Italian interpreter, Joe Poginno, addressed the meeting. One gathering of miners from Lucile formed into a march to Blocton starting with 400 men and growing to 600 by the time it reached its destination. The *Blocton Enterprise* reported: "The new local starts out with a membership of over 500 and bids fair to be the banner local in the district."

The July 23 issue of the *Blocton Enterprise* gave an optimistic outlook because contracts had been signed with the Howell Coal Company, operators of Klondyke, as well as with Braehead mines. It also said, "Everything is quiet and orderly in the Blocton District." That did not last. The company sent in the state militia and strikebreakers, the latter dubbed "scabs" by the union workmen

On September 24, 1908, the *Centreville Press* announced that the strike was over and the affairs of state were running smoothly; therefore, the High School Commission should get together and locate the undesignated schools' sites. It further stated: "This high school proposition was not passed to see which town could raise the most money, but where the school would do the most good, to the children of the county. It should not be auctioned off like a piece of merchandise."

It was not specifically articulated locally during the strike or while efforts to obtain the school were in progress, but the messiness of the strike situation in Blocton doubtless had an effect on the commission's school location decision. Since Governor Comer was directly involved, the Blocton incident probably affected his willingness to visit Blocton with the school site selection committee and prejudiced his general attitude about the town.

The strike came about with the coal companies threatening to reduce wages by ten percent. As soon as the UMW struck, the companies countered by importing laborers to replace the striking miners. The miners attempted to keep them out by persuasion or violence. In Blocton violence erupted on Sunday morning, August 9 when a train bearing non-union workers was fired upon resulting in three deaths. The ambush of the train created chaos. The incident was reported in the *Blocton Enterprise* on August 13:

> A special train of three cars loaded with strikebreakers guarded by deputy sheriffs and soldiers left Birmingham Saturday night about 11:00 p.m. for Blocton. As the train was entering a cut about a mile north of the Blocton depot about 1:30 Sunday morning, the engineer saw an obstruction on the track several hundred yards ahead and slackened speed, when, according to accounts, a volley of shots was fired at the cars from the embankment on the left of the track. The fire was returned from the cars, but the train continued on its way until the depot was reached. It was then found that the conductor, Joseph T. Collins, O. Z. Dent, a deputy sheriff, and Willard Howell, an imported strikebreaker had been killed and nine others including three company officials had been wounded.
>
> Several hours later, a train bearing a company of militia arrived at the place where the attack was made. The train remained at Blocton until daybreak, the injured being attended to in the meantime by surgeons, when it returned to Birmingham, where the dead were removed to undertaking

Southern Railroad near Blocton. A location similar to the one pictured here was the scene of the ambush on Sunday morning, August 9, 1908. (Courtesy of Will Reese.)

> establishments and the badly wounded taken to St. Vincent's hospital. All but eleven of the strikebreakers returned to Birmingham on the same train, many of them refusing to leave the coaches. Several hours later a train bearing a company of militia [Alabama National Guardsmen] arrived at the place where the attack was made. Sheriff Oakley and a number of his deputies were awaiting them. Four bloodhounds were placed on the scent but failed in their efforts. However, about fifty persons, mostly foreigners and the rest negroes, were arrested Sunday for complicity, some of whom proved alibis and were let go. Some of the arrests were said to have been made on the strength of an alleged confession by a frightened Hungarian who had been put through the "third degree." A sock said to belong to him was found near the place where the track had been obstructed.
>
> Further arrests have been made during the week and there are now about forty prisoners in the [county] jail charged with complicity in the attack. They are closely guarded . . .the cases likely will be before the grand jury and Robert Hayes, president of the Blocton local, was arrested Monday morning and was taken with many others previously arrested to Centreville on the Monday noon train. Mr. Hayes is a young man well known and respected in this community and no one believes for a moment that he was in any way connected with the affair. In case of indictments [they] will be acted on in Circuit Court to be convened a week from next Monday.

The *Centreville Press* featured the news on its front page in inch-high, bold headlines. Next to the story on the ambush killing was a story laying the blame on the Tennessee Company's 1904 importation of strikebreakers who were characterized as "the scum of the foreign element." On September 17 and 24, the *Press* reported that Wasilije Winovick and Mipalo

Lazanovick were found guilty of the shootings and sentenced to life imprisonment. Others were allowed to plead guilty and received five years. Ellison's history says that a third defendant, Boresea Melita, was convicted and so reported in the *Centreville Press* of March 4, 1909. Throughout the trials, the *Press* referred to the defendants as the "foreigners" in keeping with the local editor's earlier identification of the prisoners.

Perhaps seen as a plus for the Blocton District securing the high school location was the acquisition of the Tennessee Company (TCI) by United States Steel. The November 7, 1907, *Enterprise* reported the event under the headline:

> THE TENNESSEE COMPANY CHANGES HANDS
>
> The United States Steel Company has purchased a controlling interest of the Tennessee Coal, Iron and Railroad Company and the transfers are now being completed. The consideration was $15,000,000 for about 70% of the stock, the holders of which were caught in late financial flurry and found it best to let the great steel corporation take over the properties. The deal is regarded in the Birmingham District and here with great favor, since it is known that the general policy of the great corporation is to develop on large lines, and that it is the only industrial institution to put up cash and push improvements in spite of adverse financial conditions. Great developments are looked for in the Blocton District as the result of the transfer.

The *Blocton Enterprise* of April 16, 1908, reported that highly positioned U. S. Steel officials had inspected the TCI Blocton holdings as follows:

> President W. E. Corey, Vice President W. D. Dickinson, George Crawford of Birmingham, President of T. C. I., Birmingham, and a number of high officials of the U. S. Steel Corporation, arrived in Blocton last Thursday on a tour of inspection of the T. C. I. Properties here. The party arrived in a special [railroad] car, and were met by Supt. Deloney, of this place, who escorted them over the mines and other places of interest. This is the first trip of President Corey to this district, and it may mean much for this section in the way of further improvements and developments. The Blocton District is considered one of the best mining districts in the State, and if any improvements or further developments are made, it is safe to predict she will get her share.

Labor troubles and related violence certainly did not help West Blocton's efforts in securing the high school and neither did the town's choice for state senator and governor. Senator Reynolds was endorsed by the Blocton *Messenger* in February 1906, and in the same issue the editor proclaimed B. B. Comer as the likely local voters' choice for governor. As expected, Reynolds, chairman of the Senate Committee on Education, sided with his home town for the high school location, and it was his bill that laid out the state high school support plan. Neither Comer nor Reynolds was destined to serve as friends of the Blocton District's civic wishes relating to the high school. The Chairman of the County Board of Education was Centreville attorney John T. Ellison, a bright, energetic, and enthusiastic civic-minded resident. For years Ellison was the TCI attorney in the resolution of labor conflicts in the Blocton District and a personal friend of Governor Comer. Ellison's daughter, Rhoda, remembered that as a small child she sat on the governor's lap when the governor had been a guest in their Centreville home. Attorney Ellison, a frequent visitor to Montgomery, undoubtedly had the governor's ear on the high school decision. That friendship, while valuable for Centreville, may not have been a major factor in the governor's position on the school site. As a Birmingham textile manufacturer, Governor Comer looked with disdain at the organized labor movement. In the coal industry strike of 1908, he had threatened to call the state legislature into session to ask for an amended vagrancy law by which he could have striking miners arrested. He could hardly have been expected to look with favor upon a community that had been the scene of one of his administration's major labor problems.

Another consideration the committee may have applied in making its selection was the good reputation of the company-financed school in Number 1 Blocton. Since the beginning of TCI's tenure in 1892, the school had developed an excellent, well-staffed, and well-equipped program for miners' children and others able to pay a modest tuition. The

TCI Teacher's House, 1917. (Courtesy of Alabama Historic Ironworks Commission.)

TCI Teacher's House Living Room, 1917. The company provided clean and comfortable housing for the school staff. (Courtesy of Alabama Historic Ironworks Commission.)

school had eleven grades (as did most state high schools) during the time of the school selection decision and while it was not identified as a true "high school," graduates with additional individual study were admitted to the state institutions of higher learning.

In an article listing the rules and regulations governing the company's Number 1 School in the *Blocton Enterprise* of September 8, 1910, the grading of the school was explained, "The high school department consists of four years work as follows: Eighth grade, Ninth grade, Junior Year and Senior Year."

So it could have been argued that Blocton already had adequate school opportunity for its youth and thus did not have as urgent a need for a school as did Centreville. West Blocton had opened its "West Blocton High School," a two-story, frame building, in September 1903. It had nine grades in the early years but was adequately financed, at that time, by the dispensary revenues. Meanwhile, in 1908, Centreville had built an attractive, "substantial" brick school building.

In June 1908, county school location hopes were still alive in West Blocton, but the reluctance of the selection committee to make a definite commitment to a visitation was looming large as an indicator that likelihood of success, if it ever existed, was fading. J. B. Davie, leader in the efforts to award the school to West Blocton, wrote in the June 4 *Enterprise*:

> So far as I am informed we have now no date fixed for the visit, however, assurance from each member of the commission written at various times, that the commission will surely visit West Blocton before fixing the place for the Bibb County High School. We feel confident that after the commission has visited our town and become fully acquainted with the solid facts upon which we have our claim to the high school, they will have no hesitancy in naming West Blocton as the place in Bibb County entitled to the county high school.

The main item in the paper on June 4 was a long story on plans for the commission's visit. It reviewed all the reasons why West Blocton should win the high school. Governor B. B. Comer, State Auditor Gen. W. W. Brandon, and State Superintendent H. C. Gunnels were expected at an early date. One hundred and five ladies, the wives of every prominent professional, businessman, and company official in the area, were listed as special hostesses to plan and present an appropriate reception. Mrs. J. B. Davie chaired the group. For the gathering an extra coach was to be run from Piper, Coleaner, Garnsey, Belle Ellen, and Hargrove. All were to be notified by the *Enterprise,* or in case of short notice, by poster, telegram, telephone, or other means of communication. There was to be no visit until the following March 11, a year and five months after the visit by the commission to Centreville. Finally a commitment from the governor was obtained and detailed plans were made. The official reception committee was made up of the town's leading merchants, five doctors, several local attorneys, and town government officials.

On the day of the long anticipated visit, the *Blocton Enterprise* printed a bold-faced invitation on its editorial page.

> WE WANT THE HIGH SCHOOL!
>
> Today, [Thursday, March 11] we will be given an opportunity of showing the State High School Commission that we have in Blocton the best location for a High School in Bibb County. We want every boy and girl of school age in the Blocton District in the parade that day. The Blocton District contains more school children than are in the public schools of Montgomery. "We are the people." We will feed and care for the children that day. We would be glad for everyone who can do so to send a picnic basket that morning, and make the occasion more enjoyable. Schools are asked to bring a banner with the name of the school on it, to meet at No. 1 School, not later than 9:00 A. M. Come out, parents, children and friends, and neighbors, and help us to have the High School put where it really belongs, at Blocton. By order of Committee of Arrangements of HIGH SCHOOL COMMITTEE

Jesse Bates Davie. (Courtesy of J. W. "Bill" Davie, Houston, Texas.)

The March 11 event was reported in detail by the *Enterprise.* The day started with the morning arrival of the High School Commission from Birmingham by train at 8:50. A special committee met the delegation at the depot led by J. B. Davie, I. B. Krentzman, E. D. Reynolds, Frank Gafford, Dr. G. W. Randall, Prof. D. W. Lewis, and a large number of other citizens. As the committee alighted from the train, the band began playing and the visitors were greeted with cheer after cheer from the children massed on the hill. After this friendly demonstration they were taken to Smith Hill and then proceeded to West Blocton where a stop was made while the procession of teachers and school children, led by Rev. H. E. Wheeler, passed in review. Congregated at Number 1 Blocton School were children from Piper, Belle Ellen, Hargrove, and Braehead as well as those from the immediate neighborhood. Their number was estimated to be

Dr. George Washington Randall, "Bibb's Silver Tongued Orator." (Courtesy of Diane Kerr, Tuscaloosa, Alabama.)

close to 2,000. The procession was a spectacle, reported *Enterprise* Editor Allen, and must have been gratifying to all interested in the cause of education. After counter marching, they were halted in front of the Opera House. Addresses to the assembled crowd were made by Superintendent Gunnells and General Brandon. The children then were taken to the pavilion (the old Four Orders Park, then a TCI Company maintained park), but rain interfered with the entertainment planned there. The stores along the line of march were decorated with flags and festoons of red, white, and blue giving the town a holiday appearance. The highly complimented music was furnished by a band under the direction of Prof. J. W. Wallwork, director of the old Cadle Silver Cornet Band.

The members of the commission were conducted into the Opera House and seated on the stage in company with prominent citizens and a committee of ladies. Rev. H. E. Wheeler offered prayer, after which Dr. G. W. Randall, a local dentist, presented the facts in regard to the advantage of locating the high school in West Blocton in a masterful and eloquent manner. Randall, who was called "Bibb's Silver Tongued Orator," delivered a speech filled with classic illustrations, appropriate local history, and a dash of hyperbole to stress his points extolling the town. Excerpts reflect the tone and tenor of the oration:

> Sirs, clad in armor of a cause which statistics tells me to be a just one, and in the language of the "great Commoner," believing pebbles of facts will be more effective than the javelin of fiction, we shall proceed to express our claims without challenging any undue regard from the flock of Philistines.
>
> Gentlemen, we are glad to greet you thus alone. We have long sought this opportunity, and after many disappointments, we offer to you on this, which to us is an auspicious occasion, a greeting as great as in our power lie. Many adverse rumors have caused some of us to despair. Among the many things we heard that Blocton had been reported to this board as being but the bad end of the once famous "Bloody Bibb," that it was but a rough, temporary town. But gentlemen, not so. We are a people of lawyers and of merchants and of doctors, and of laymen classes as workmen who need not be ashamed of their work. It's true that a goodly number of our people are miners; but, sirs, after having labored in their midst for ten years I have learned to love them . . . though his hands be callused he has a heart of gold . . . his face has furrows of character. It is also true we have the Italian, the Greek, the Slav, the Macedonian, the Hungarian and the Bulgarian, and, gentlemen, we have nothing to conceal from your inspection. That terrible tragedy that happened on that memorable Sunday morning when conductor Collins and another lost their lives has been lain at the door of some of those foreigners, but be that as it may, the memory of it still exists and gives us trouble by creating in the world of us a reputation earned by a very small minority.
>
> Our object in asking you to come and inspect our claim has not been one of idle fancy, but, sirs, if I have interpreted correctly the heart's desire, and caught the true inspiration of this great body of people whom I have the honor of representing today, it's but the reflection from the crystals of hope. It has been the custom heretofore for all the good things to go the other way, but when it came to our

> ears that this board construed the law in regard to the locating of high schools in these few words, couched as they are in wisdom, "THE GREATEST GOOD TO THE GREATEST NUMBER" — when these words came to our ears we took new hope, and upon them we have anchored our hope.

The speaker sharply focused his feelings about the rival town for the school by saying:

> Now, Gentlemen, a word about Centreville. She was born some hundred or so years ago upon the banks of the beautiful Cahaba, whose name in Indian vernacular we know not what it means but since the paleface has superseded the red man . . . they seem to have rendered its meaning as "Ask and it shall be given."
>
> But you know we like Centreville. When the county commissioners, without submitting it to a vote of the people, placed that magnificent new courthouse there, we didn't kick, while we knew it wasn't fair. But when Centreville began to show the spirit Atlanta did when Birmingham was trying to land the Sub-Treasury we hallooed halt, and we meant it.
>
> Centreville comes forward with the lame excuse that "we have scarcely no school now, so let us have the high school." We have been informed that at one time they sold the only school house they had, because from every appearance they didn't need one bad, but since this matter has come up they have built a new one.

Dr. George Washington Randall spoke for the better part of an hour using prepared illustrated maps showing the Blocton District's resources, citing the number of children in the district, listing the businesses and town structures, and identifying the schools and churches. He quoted the value of the mineral deposits, the worth of the timber in the area and of surrounding agricultural lands, and he pointed to the railroad facilities as a great asset. He estimated the value of the schools and churches, and he listed the number of fraternal lodges in the town. He added in sarcastic humor, "Why, we have nearly as many Odd Fellows as Centreville has citizens." He added, "State and county tax, in a radius of seven miles, equals three-fourths of the entire county." He ended the eloquent speech as follows:

> In closing, I say that if I were an orator and possessed the power of subtle persuasion possibly I could have been more effective, but this one thing we would have you take back home with you: Our argument may have been void of eloquence, but if I have failed to impart an honest earnestness, then I have failed to execute my commission for I can assure you that the hearts of this people, backed by these statistics, are pregnant with a pungent longing for this institution; and when you go back home we want you to take with you this blueprint, and we want you to take with you the anticipation of this people.
>
> Here we rest, but when you reach on the roll the county of Bibb, we trust that in the defense of justice, and in the defense of this great majority of boys and girls, and the still greater number that our underdeveloped wealth assures us, and in defense of the "King's business," that Minerva, the goddess of wisdom, will appear at the hour of decision with ravishing beauty and radiant power to plead for us who have never as yet had any good things from State or county to come our way.

The good doctor had sprinkled his address with Bible quotations, invoked the blessings of classical mythical gods, made historical and geographic comparisons, used sophisticated metaphors and brilliant descriptions of the human condition, all in support of his adopted hometown. The locals were spellbound; the commission listened stoically and patiently.

The *Enterprise* reported that Randall was followed by Governor Comer, who spoke on educational matters and was enthusiastically applauded all through his remarks. He said there was a deficit in the state treasury, but the money went to the schools, and that fact seemed to give him great satisfaction. Superintendent Gunnels also spoke and was listened to with great interest.

Garden of White School Children, Blocton, May 2, 1917. The well-rounded education provided by the company school at Blocton worked against the community's efforts to get state financed high school in West Blocton. An important element of the company school's curriculum was gardening. (Courtesy of Alabama Historic Ironworks Commission.)

After the speeches, the Governor and party retired to the Douthit House, where an elegant repast awaited them, prepared under the supervision of Mrs. J. W. Stewart. Covers were laid for fourteen and the menu in part was as follows:

Baked Chicken
Roast Beef
Chicken Salad garnished with Olives
Sliced Tomatoes
Creamed Irish Potatoes
Imported English Peas
Candied Sweet Potatoes
Macaroni
Cooked Tomatoes
Pickles
Preserves
Jellies
Cream Biscuit
Cocoa
Governor's Cup Coffee
Vanilla Ice Cream
Old Time Sponge Cake

The paper listed the guests, which included the Governor and other commission members, J. B. Davie, Dr. Randall, Prof. D. M. Lewis, Dr. Charles Martin of Woodstock, Dr. Alexander of Lucile, Mayor Y. S. Smith (Smith had succeeded Mayor Logan in 1908), I. B. Krentzman, and Supt. Frank Gifford of TCI in Blocton.

The Douthit House dining room was prettily decorated with flags, potted flowers, ferns, and palms. A pretty feature was a silver fern bowl in the center of the table containing blooming white and blue hyacinths and another flower of red. Mrs. Stewart, the genial hostess, left nothing undone that would add to the pleasure and comfort of her guests.

It could not be said that the host town failed to go all out, with 2,000 school children on display, a band-led parade over a mile-long route, decorated Main Street buildings, a gathering at the Opera House with a local eloquent speaker pleading for the commission to listen to the community's argument that it had the site where the most children would be served, and a sumptuous meal to close the day's events. Two weeks later the commission gave its verdict.

The *Enterprise* of April 1, 1909, carried a syndicated column on its front page of items under the heading "Alabama Happenings." It included the following:

> CENTERVILLE [*sic*] GETS THE SCHOOL
>
> After hearing from a delegation from Centerville, headed by Senator H. E. Reynolds, Friday, the high school commission located the Bibb county high school at Centerville. A hard fight was made by Blocton, but the other was too strong.

The *Enterprise* editor voiced the opinion of the paper and the community:

> That Blocton was done a great injustice by the High School Commission is an indisputable fact, but there seems to be nothing she can do now but take her medicine and profit by it, remembering to go and pay your poll tax and at the proper time vote for men who when elected will do as they are sworn to do, work for the people as a whole, giving justice where it is due and special privileges to none. Broad minded men, who do not belong to a political ring, and who are not afraid to cast their vote where it will do the most good to the greatest number. Take these few thoughts with you and think about them, decide upon your men and begin now to talk for them, give yourself the time, work hard, and victory will be yours in the end.
>
> The general opinion is that we will have a high school here as good as any in the county, with or without State support.

There were other reactions. Rev. A. A. Green wrote an open letter to the State High School Commission stating his astonishment at the decision. He said, "Blocton has nine-tenths more children to educate than Centreville. To them you committed a great injustice. Your decision may have been controlled by the political ring at the capital of Bibb County but let me tell you the day will come when 'little Blocton' will speak at the polls."

A mass meeting was called for the West Blocton Opera House "to consider matters pertaining to the good and welfare of north Bibb." A committee was formed to bring together different elements of citizenship to take part in a movement that had as its object the betterment of all.

The *Enterprise* editor reported on April 15 that it had been suggested the proposed new county jail would be situated in West Blocton and alternate sessions of the circuit court held in town. It remained a suggestion. The three-story jail was built in Centreville in 1910, and the court was not rearranged.

On October 7, 1909, the Bibb County High School opened with ceremonies at the courthouse. On December 9, the *Press* reported an enrollment of sixty-nine. In April 1910, the paper pled for the subscribers to the school building fund to pay up so an indebtedness of $1,500 could be paid off. It reported, "The committee [is] at row's end in meeting obligations." In December there was another plea issued in the *Press* for "pay up" on the two-year-old school note. The first class graduated in 1912; there were three members. When and how the note payments were completed is not clear. West Blocton was not to have an accredited high school until the fall of 1920. It graduated a class of three in May 1921.

The Company School in Blocton had been remodeled in 1907 into a two-story building. This is likely one of the local improvements within the TCI's welfare system instituted by George G. Crawford who ran the company after it was purchased by U. S. Steel.

Over the years of the school's operation, improvements were made and the local paper recorded those occasions. In October 1913 the following appeared in the *Enterprise*:

> The Blocton Public School property has been much improved. The TCI Company installed an up-to-date heating plant in the building and enclosed the grounds with a new fence. The school improvement committee had the yard graded and will set the lawn in grass. But the pride of the

Graduating Class of 1914, TCI School. Claude Howard (left) and Ben Huey (right) were the only male members of the class of 1914. The girls all made their graduation dresses in the school's domestic science class. Though two are not included in this photograph, the female graduates were Annie Merle Farrar, Willard Thomas, Jessie Winston, Eunice Thomas, Mary Wallace, Margaret Melvin, Mabel Wallace, Lydia Morse, Kate Morse, Louvella McIntosh and Annie Mae Morrison. (Courtesy of Alabama Historic Ironworks Commission.)

> association is the new drop curtain recently hung by the scenic artist, J. R. McNair of Bessemer. The center of the curtain is a painting, capture of Manila by [Admiral] Dewey's ships surrounded by artistic ads of local business firms. Come and see the curtain.

In March 1915, the Number 1 Blocton School received a superior rating by J. B. Hobdy, Supervisor of Rural Schools and Professor R. E. Tidwell, Superintendent of TCI Company schools. (Tidwell was later Dean of Extension at the University of Alabama. Tidwell Hall at UAB is named for him.) The *Blocton Enterprise* reported that only a few schools in the state were so rated.

In 1982, George M. Neal of Birmingham in responding to an oral history researcher, described one of the TCI schools in Jefferson County. Since all company schools were designed alike, his narrative offers insights into the Blocton School. In addition to basic courses, the girls had an opportunity to take "domestic science" courses, sewing and cooking. The boys were taught how to maintain a vegetable garden. The garden in Blocton was photographed and shown widely so it must have been a particularly successful one. It even appeared in an Alabama geography text. Neal spoke of a sports program. In addition to baseball, softball, and field events, there was also a speedball team. Each May, all the company schools, including Blocton, met in Birmingham for annual competitions where medals were awarded to teams and individuals.

Neal, once an ore mine superintendent for TCI, recalled that the company provided inexpensive housing, pure water, utilities, good health maintenance, and good schools. "Just about everything that you could want to carry forward the proper upbringing of a family." It was his recollection that the schools were discontinued under company maintenance around 1933. As will be seen later, the Blocton TCI School closed after the spring 1928 session.

In 1910, Blocton and West Blocton voters had their first opportunity to have their say in the county after losing out on the high school. Political rhetoric in that election year echoed the county school arguments of 1908. In the March 31, 1910, *Enterprise*, for example, an advertisement for sheriff echoed the old intra-county rivalry. Sheriff candidate John A. Wood had been a resident of West Blocton for twenty-two years and Justice of Peace for sixteen.

A letter of support, signed by eighty-four prominent citizens addressed the voters of Bibb. The letter said in part:

> JOHN A. WOOD FOR SHERIFF
>
> We feel that as the mineral district of the county has the largest population and pays the largest part of the taxes of the county, and in addition the fact that seldom any of the county officers are chosen from this end of the county, that Mr. Wood has a high claim to this office, we feel that the other portions of the county will be fair and just to concede this much to the Blocton section.

Wood's opponents were C. L. Cleveland and Charles L. Oakley of Centreville. Cleveland received 185 votes, Wood, 619 and Oakley, the winner, 1,058. Despite the larger population in the north end of the county, many of the residents were either transients or foreigners without the right to vote. However, a decade later, Huey Wood, a West Blocton candidate, was successful in the sheriff's race.

Social and Political Life

The 1904-1905 strikes, the 1908 strike and train ambush, the battle for the county high school, the temperance campaign that ended legal alcohol sales, and the acquisition of the Tennessee Company by U. S. Steel dominated the local news for almost five years. However, there were other noteworthy developments in the community during these years.

Rev. Dallas P. Lee was serving as a Baptist minister in Blocton in 1908 when his daughter Mildred was born. The family shortly moved to Opp where he continued his ministry, and later moved to Georgia. Mildred Lee was recognized

"Blocton Colored School Garden, May 16. 1914." (Courtesy of Alabama Historic Ironworks Commission.)

Blocton Speedball Team. (Courtesy of Alabama Historic Ironworks Commission.)

in a 1978 publication, *Twentieth-Century Children's Writers*. Among her numerous books is *The Skating Rink* (1969) that merited the Alabama Library Association's Author's Award in 1971. Mildred Lee (Scudder) lives in St. Petersburg, Florida.

Since its founding in 1884, the Blocton Post Office in Number 2 town had served the entire community as its only post office. In 1904, the West Blocton Post Office was established as a fourth class office. In April 1910, it was upgraded to third class.

The *Enterprise* reported on June 20, 1907, that a moving picture show would soon be among the attractions in West Blocton. A building had been erected facing Belcher Avenue (now Florida Avenue) just north of Main Street. T. C. Tuggle, the town's photographer, was the proprietor and the facility was named "The Theatorium" after a movie-vaudeville house in Birmingham. A gasoline motor driven generator provided the electrical power for the film projector. Tuggle's picture show is noted as the first in Bibb County. Centreville acquired a moving picture theatre in 1923 opened by Ike Nathews and named "The Frances" for his oldest daughter. Nathews was a hotel resident serving as manager of the Centreville branch of West Blocton's I&N Quality Store. It is reported there was an earlier theater in Centreville.

A notice concerning the Theatorium's competition appeared in the paper in September:

> Special Schedule for the coming season at the Opera House. Doors will open at 7:15. Curtain will rise promptly at 8:15 (except on Saturday nights, then at 8:30). Please leave your peanuts and tobacco on the outside, as the use of them is prohibited on the inside.

T. C. Tuggle was a local businessman for more than five decades. He is remembered mainly for his photography business and is given credit for being Bibb County's official photographer for many years. A common sight in the 1930s was his A-Model Ford sedan loaded with camera and lighting equipment en route to assignments. He maintained a home studio until his last illness in the late 1940s prevented him from working.

Tuggle had a variety of careers during his life, including, in the early days of Blocton, that of agent for International Harvester. In a *Birmingham News* story of January 21, 1949, he recounted how in this capacity he had come upon the career of the notorious swindler Charles A. Ponzi of Boston, who at the time had recently died. A Ponzi Scheme is an investment swindle in which early investors are paid dividends from the money of later investors. Ponzi, whose obituary recounted a rags-to-riches-to-rags-again career, had been a multi-millionaire, but died a pauper, and in his infamous career, had, according to Tuggle, come to Blocton.

Tuggle reported that Ponzi just "showed up" and found friendly quarters on Smith Hill close to Little Italy. Everyone who met him liked him. His stay lasted around six months, and during that period he proposed selling stock to fund a $10,000 waterworks system on Smith Hill. Essential to the project was the land sale approval by Rufus D. Smith, local landowner and mayor of the small municipality of Smith Hill. Tuggle stated that Ponzi approached him with the water system scheme thinking perhaps that through Tuggle's affiliation with International Harvester he could get the needed heavy construction machinery. Tuggle cautiously considered the idea. A meeting was called and Tuggle made a talk on the system's advantages. Ponzi translated the remarks for the 130 or so native Italians in the audience. Landowner Smith did not agree to the stock plan so failing that, the Italian citizens proposed a cooperative venture. Ponzi would have nothing of that since he wanted to be boss, Tuggle said, and the project fizzled out.

Ponzi then proposed a foreign currency exchange business to Tuggle centered in the then sizable Italian community. Ponzi remarked to Tuggle, "We'll clean up." Tuggle excused himself, citing the difficulty in giving up time from his other businesses. After sticking around for a short while, Tuggle remembered, Ponzi simply disappeared. Local Italians believed he was the same man as the Boston banker who was reported years later to have made millions in a foreign money exchange scam. They were sure, Tuggle said, that the Boston Ponzi had spent time in Blocton in 1904.

Ponzi amassed a fortune during the 1920s by making glib promises to widows, old maids, young men, and able businessmen. Many made profits but the majority were stuck. He served a jail sentence for fraud and was deported to his

Blocton Post Office. (Courtesy of Will Reese.)

Unidentified Family. T. C. Tuggle photographed the people of the Bloctons for over fifty years. (Courtesy of Bessie Robinson Collection.)

native Italy. He died in Brazil in a charity ward, semi-paralyzed and partially blind. Whether he was the same Ponzi that had visited Blocton we may never know, but it does make for an interesting story.

Historians, professional and amateur, like to find events that fit together. That is not the case on the Smith Hill water works project. No reference is found of a 1904 project to supply water to Smith Hill. There is, however, an October 31, 1912, story in the *Enterprise* indicating that plans were being made for the small town to have its own system. Could it be that Tuggle was off in his memory by eight years? No other evidence appears in the local papers to confirm or deny the Ponzi story. The December 1998 *Smithsonian Magazine* reported that Ponzi arrived in this country from Italy in November 1903. He served time in Canada for forging checks and served time in Federal prison in Atlanta for bringing five Italians into the country from Canada. The story says Ponzi was up and down the east coast from Boston to Miami during his "swindlcr" years. There is no hard evidence that he made a detour away from the coast for a stay in Blocton.

T. C. Tuggle was an active and productive citizen of West Blocton for a half century. He died in October 1949 and is buried, with other members of his family, in Hickory Hill Cemetery.

Dr. Lovic Edward Peacock served as West Blocton's second mayor from 1904-06. Centreville's *County Press* reported that he had moved his practice to West Blocton in 1897 from Tuscaloosa County. That year other physicians in Blocton were A. E. Meadow, J. U. Ray, Jr., J. F. Curtis, and C. M. Campbell. Peacock's tenure as one of the town's doctors was to span a half-century. He was born in Sao Paulo, Brazil, the son of a disillusioned Alabama planter who migrated to a colony in South America as a refugee from the Confederacy. The people who joined this movement have been identified as "Confederados." His father, Andrew Johnson Peacock, had married the daughter of E. S. Trigg, one of the Southern colony's founders.

Dr. Lovic Edward Peacock. Allen Cleveland (right) and Dr. L. E. Peacock posed on the porch of the Peacock and Dr. C. W. Jones office, a structure destroyed during the 1927 town fire. (Courtesy of George H. Young, Centreville, Alabama.)

As a youth in Brazil, Peacock stepped on a thorn resulting in an infection that required a partial foot amputation and as a result, he had a pronounced limp for the rest of his life. The Peacock family chose to return to Alabama because they found Brazil disappointingly unlike their native state. When they returned, their plantation in south Alabama had shrunk to a farm, and young Peacock had to walk seven miles to the nearest school. An uncle, a doctor himself, decided young Lovic, with a mangled foot, could never lead the life of a plowman, so he sent the boy to Alabama Medical College in Mobile.

Dr. Peacock leaving his home. (Courtesy of Jane Greene Collins, Birmingham, Alabama.)

Dr. Peacock served the town in other capacities and spent at least one term as Bibb County Health Officer, a post rotated between members of the county medical association. In June 1901, he and other local doctors had to deal with an outbreak of smallpox. Some forty cases were identified in the Black community of Blocton, enough to classify the episode as an epidemic. Again, in June 1903, there were six cases in West Blocton, however, the disease did not spread this time, and all smallpox flags had disappeared from residences by early July.

The following appeared in the May 17, 1906 *Centreville Press* and is typical of periodic announcements made by Dr. Peacock and his colleagues when they held the office of county health officer:

NOTICE

To All Householders and Citizens:

To you who are interested not only in your own health, but in the health of the community at large, let me urge upon you the necessity of removing all causes of diseases that originate in unhealthy and unhygienic surroundings. The warm weather will soon be upon us, and it is your duty to yourself, your neighbors and your country to see that all causes of diseases are removed from your premises. In order to do this, have your immediate premises cleaned of all trash, cans, rubbish, chips, brush, refuse matter, etc., by burning. Have all closets and sinks cleaned and limed every two weeks during warm weather. See that all ponds and mud holes are properly drained, for they give ride to untold troubles, as they are the breeding places for mosquitoes. Have your wells and springs cleaned before warm weather comes, so your water will be pure. Use lime freely around your premises. It is to your interests to do so, and see that your neighbor does likewise.

I shall gladly assist you in this work as far as it is in my power. All complaints of conditions that are, or likely to become a nuisance to the public health will receive prompt attention.

I call your attention to the fact that in the majority of instances the following diseases are either caused from filth, or result from conditions that can be removed by each householder: Malaria, Typhoid Fever, Dysentery, Diarrhea, and numerous others.

All communications addressed to me in regard to conditions stated above will receive my attention.

Respectfully,
L. E. Peacock, M. D.
County Health Officer

The Doctor, as well as Mrs. Peacock, were subjects of frequent social items in the local paper. They were much admired by the community. In 1947, on his 77th birthday, Dr. Peacock was feted by some 40 neighbors, friends, and relatives gathered to celebrate at his home. The local news column writer, Lyda Keys, wrote a poem to be read at the birthday celebration, expressing the sentiment of many of his patients and friends in West Blocton:

To Our Doctor

As a young doctor, he came to
Our town,
A long time ago,
Just why he chose this place,
We'll surely never know.
But he did and he stayed,
As the years rolled by,
He looked about the same
And was just about as spry.
He stayed informed
And up-to-date in his profession,
As other doctors came and left,
Each in succession;
He doctored our ails,
And delivered our babies,

And told us how to raise 'em
Without any "Maybe's."
He was personal friend,
As well as physician,
And to worthy causes
Gave his liberal contribution
Through war years and epidemics,
He's been our stay,
No soldier could have served,
In a more unselfish way.

In the summer of 1948, when he retired from active practice at age seventy-eight, the community turned out for an old-fashioned picnic and day of recognition. The *Birmingham News* reported the crowd in attendance to be more than eight hundred. The committee was chaired by Margaret Lane Edmonds Hartley, local correspondent for the *Birmingham News* and *Centreville Press*. The venerable doctor died the following December 23 and is buried in his wife's family cemetery plot in Marietta, Georgia. After her husband's death, Mrs. Peacock continued to live in West Blocton except during illnesses, and in the early 1950s moved to be near family in Georgia. The sad news reached the community in August 1956 that Mrs. Peacock, "Aunt Jerry" to many, had died in Marietta, Georgia.

There is no mention of special music for Dr. Peacock's recognition day. In the late 1940s company-sponsored bands were a thing of the past and the West Blocton High School Band could be assembled only while school was in session. But earlier in the community's history, bands were a large part of local celebrations.

Musical Blocton and West Blocton

Some of the earlier musical groups in the community have been identified, such as Cadle's Silver Cornet Band and the Philharmonic Society. The *Blocton Messenger* of May 22, 1903, boasted, "Blocton has fine musical talent, our local orchestra is equal to most any in Birmingham or Tuscaloosa and far superior to many of them, and yet this is like many other eccentricities about Blocton."

Dr. Peacock with loving cup presented at his retirement picnic. (Courtesy of Jane Greene Collins.)

In November 1906, the *Blocton Enterprise* reported the formation of a band in "Italy" made up of "gentlemen of different nationalities but principally Italian." The leader, J. W. Wallwork, announced that he hoped to be able to furnish Bibb County with a first-class band of "home talent." Conductor Wallwork had been a local bandleader for more than a decade. He was the manager of Cadle's Silver Cornet Band in the early 1890s. The name initially given the group was "Little Italy Band," and the paper's editor pronounced that the band would be a credit to its leader and all of Blocton and that "West Blocton would be given a little music soon." The members and their instruments were listed: Solo cornet, Ernest Mattioli; first cornet, Giacomo Malvolti; second cornet, Willy Werer; solo alto, Liugi Gardeugli; first alto, Master Oedo Biggi; second alto Oscar Morrison; Solo tenor, Frank Neicci; first tenor, Tom Come; second tenor, Ottouis Mornin; baritone, Tom Mattich; b. bass, Puff Puffie; e. bass, Clelio Mattioli; snare drum, Master Birt Briles; bass drum,

Blocton Home Band. During a Fourth of July celebration, T. C. Tuggle recorded the band's new uniforms. The band's sponsor, Rock Contri, is seated in the middle with a dog. Contri was a popular merchant who later moved to Birmingham but not before becoming the first member of the Blocton Elks Lodge of Italian descent. (Author's collection.)

Larcuzini Ricorolo. This musical aggregation followed the earlier Cadle's Silver Cornet Band that flourished in the 1890s under the auspices of the old Cahaba Coal Mining Company. The band members' names seem to be, in some cases, only stabs at correct spelling, the local weekly papers having a continued difficulty in dealing with foreign names. Too, it seems that some singers were included in the listing so perhaps this was a band with a quartet. In later advertisements, the band was identified as the Little Italy Operatic Band, suggesting the inclusion of singers among the membership. That the band was given such prominent notice and encouragement may have been evidence that the Italian community was becoming more socially accepted. The leader, Wallwork, and band member Oscar Morrison, represented the non-Italian element in the group. On January 5, 1907, the paper reported, "the Little Italy Band discoursed sweet strains of music on our streets on New Year's Day to the delight of everyone in hearing distance."

The band was included in the Columbus Day Parade sponsored by the Italian community in 1910. There was a picnic on the grounds at the Italian Hall. The band led the parade from the hall to Smith Hill and West Blocton. There were floats featuring impersonations of Christopher Columbus, Miss Italy, and Miss America. Part of the celebration was oratory in the new park at Number Three Town. Rock Contri and S. Lazzare were committeemen responsible for the special celebration.

The staying power of this band is evident. The local paper reported in October 1913 that the Italian Concert Band appeared at the Bibb County Fair in Centreville. The *Blocton Enterprise* of February 12, 1920, announced the Fifth Operatic Concert to be held in the Theatorium by the Italian Concert Band. Among the ambitious offerings were the *Thunderbolt March*, the *William Tell Overture*, the *Nebuchadnezzer Overture*, and the *Semper Fidelis March*.

While it lasted only seven years, the company sponsored Blocton Home Band was active from its formation in February 1913. The *Blocton Enterprise* reported a Blocton Band in early January that serenaded at midnight, New Year's

Eve and played pretty pieces all New Year's Day. It was composed of local talent and was predicted to make Blocton people proud. Later the paper said contributions for the band were to be collected by Dr. G. W. Randall and Charles Milam. Its first concert was held in the Opera House on February 25, with admission set at 25, 35, and 50 cents. Known members of the band were Harry and Jim Adams, Will, John, and Andy Wyper, Lester and Frank Nucci, Johnny Stewart, and Oscar Morrison. Perhaps Charles Milam was also a bandsman. Local businessman, Rock Contri, represented the company as band sponsor. This band played for a number of town celebrations, at numerous dances, won a prize in a Birmingham contest, saw soldiers off to World War I in a special program at the Bibb County Courthouse Square, and celebrated the end of the war in 1918. The Blocton Home Band was discontinued as a company-sponsored project as part of the strike breaking moves of TCI in 1919–1920.

Local Politics, Development of Utilities

The town election of 1907 was reported in the January 31 *Centreville Press*. W. H. Logan, a local attorney, was elected mayor. Councilmen were M. C. Davie, merchant; L. J. Hays, cafe owner; M. M. Odess, merchant; E. D. Reynolds, banker-mine investor; E. M. Stamps, and T. J. Vickery. Named as executive committeemen in the same report were E. M. Bishop, an attorney; R. B. Keith, barber; L. C. Lagrone, market owner; G. W. Randall, dentist; W. D. Sparks, mining executive; and J. W. Stewart, opera house owner. This was a period during which the tenure of office was two years.

Another milestone in the community's development was the coming of long distance telephone connections. There was evidence that telephones were used as early as the 1890s in mine operations. A limited number of residential and business telephones had been in place from as early as 1903, and intra-county connections with Centreville were mentioned in a 1903 newspaper item. The newspaper editor complained in July of that year that the "Telephone system experienced much difficulty in ringing up the right party. A central office is much needed to make this phone [system] a success."

The *Blocton Enterprise* reported on March 2, 1907:

> The Bell Telephone people are quite busy here now putting in a local exchange. When completed Blocton will be in a position to talk to her neighbor next door or with President Roosevelt at Washington, D. C. The manager of the work is here with a force of men putting up poles and making other necessary arrangements as fast as the weather will permit. The exchange will be in the brick building of W. A. Harvey [the drug store], so the Birmingham people may look out for the hello call from West Blocton in the near future.

The improved long distance phone system was in operation by April 20, 1907. In June the public was requested to sign up for electric lights but even though the cost was reasonable (50 cents each for the first four lights, 40 cents for each additional) the citizens were reluctant to sign up. The *Enterprise* commented, "electric lights were more of a luxury than a necessity so can't be compared to water." The promotion failed. However, many of the Main Street businesses had their own gasoline generator-powered light systems at the time so the convenience was not unknown. Alabama Power was not to arrive until almost a decade later.

Social Clubs, Fraternal Lodges and Societies

In July 1907, the *Centreville Press* announced that a chapter of the United Daughters of the Confederacy had been organized at Blocton. The President was Lily Pratt of River Bend; Vice President, Mary Mayday, Centreville; Secretary, Mrs. J. O. Sturdivant, Blocton; Treasurer, Mrs. Dr. Curtis (residence or given name not listed; the treasurer's title was a construction used for many years to give special status to doctors' wives); Historian, Mrs. Thompson (no given name listed), Blocton. This group was identified as the West Blocton Chapter of the United Daughters of the Confederacy. The same issue announced that a lawn party would be held at Schultz Creek Baptist Church for the chapter's benefit. The paper said the ladies would be given every encouragement in building up a strong organization of daughters of those who fought for the Confederacy.

Blocton Elks Lodge. On July 25, 1901, the Blocton Elks Lodge was formed. Dr. A. E. Meadow was the first Exalted Ruler. Other officers were G. B. Bainbridge, W. L. Rogers, John Wood, D. M. Lewis, W. D. McGinnis, E. E. Jones, W. O. Broadhead, Frank Downing, and Gonzlo Jackson. (Courtesy of Blocton Lodge 710, B.P.O.E.)

Company Waterworks and Filter Plant, 1914. (Courtesy of Alabama Historic Ironworks Commission.)

The UDC chapter joined other clubs and fraternal groups that flourished in West Blocton at the time. There was the Ladies Needle and Social Club; the Three Arts Club; the Book Club; the Blocton Dam No. 48, Independent Order of Beavers; the Order of the Owls, Blocton Nest; The Yellowhammers Nest, a girls' club sponsored by the Daughters of the Confederacy; and the Prospectors Club, a debating society promoted by Dr. Randall. (The dentist also promoted oratorical skills and scholarship in the local schools. For a number of years, "the Randall Medal" was an annual award given outstanding students). Other lodges were Masonic Lodge 474; Knights Templar; Improved Order of Red Men; Camp Clayton United Confederate Veterans Post No. 966; Knights of Pythias; Independent Order of Odd Fellows; Woodmen of the World, and the Benevolent and Protective Order of Elks, Lodge No. 710.

In February 1913, Apollo Lodge No. 209, Knights of Pythias, moved into their new hall on the corner next to Roycroft's merchandise building. The hall was described by the *Enterprise* editor as "handsomely fitted up and electric lights and fans installed." The editor further commented that the town should seek electric power for all, following the example of the K. of P. Lodge. In May 1913, a "new picture show' opened downstairs in the hall with the intriguing name of "The Pandora."

An up-beat edition of the *Blocton Enterprise* of August 4, 1910, was billed as an anniversary edition of the paper under the ownership of Mr. and Mrs. C. E. Allen. It provided information not found elsewhere on the contemporary status of the community as well as an overview of the town's merchants, professional men, schools, fraternal lodges, and churches at that time. It also reflected the "get-up and civic pride of the citizens residing in the community." The edition's "two fold view" was to show the people of the Blocton District that the paper was a growing business and, secondly, to show the world the commercial importance of the Blocton District.

The anniversary paper announced a homecoming barbecue, and "confidently expected that the city will be swarmed with visitors." The event was scheduled for the pavilion and was to feature two baseball games, and plenty of dancing with music provided by a splendid stringed orchestra. Senator Bankhead and Congressman Underwood were to be guest speakers. (Bankhead did not show.) The over-all theme of the gathering was dedicated to the Good Roads Movement. The chairman was E. D. Reynolds, and committee members were Martin Riley, Pat Shannon, I. B. Krentzman, and M. C. Davie.

The Blocton area depended on coal company services for much of its early history and thus can lay claim to a number of "firsts" because of the resources to which the coal company operators had access. The first filter plant water works, and the first company-sponsored band, are among those claims in old Blocton. In West Blocton, the first bank in Bibb County, the Blocton Savings Bank, was chartered in November 1892. The local paper said the public school, opened in 1903, was the first "free" school in the county and in 1907, Tuggle had opened the county's first moving picture theater. Later, Ben L. Edmonds was to open the county's first dry cleaning plant in July 1914.

Bibb County's First Hospital

The January 10, 1910, *Enterprise* reported, "The T. C. I. Company is preparing a hospital in connection with the doctor's office at Blocton and we are told that when everything is in readiness it will be an up-to-date hospital with all the necessary equipments, and patients that have heretofore had to be carried to Birmingham can be treated here with all the comfort that can be had in a larger city."

So, in May 1910 another "first" came to Bibb County. The May 5 *Blocton Enterprise* reported that "The TCI and RR Company opened their hospital last Monday with Dr. A. W. Trigg and Dr. M. C. Thomas, physicians and surgeons, in charge and Mrs. J. J. Lucas as matron." Incidentally, Dr. Trigg was employed by TCI as company doctor from 1894 until 1919. (Even though Dr. L. E. Peacock's mother's maiden name was "Trigg," there was no known relationship.) Dr. Trigg died in October 1928 still serving the community.

During the first year of operation, the statistics for the hospital as reported by Mrs. Lucas were modest. There were five major operations, thirteen minor ones, fifteen major accident cases, eleven private patients, 900 dressings, 1,200 yards of gauze used, three deaths by accident (this likely meant mine accidents), no deaths from operations, and forty-four cared

TCI Blocton Hospital. (Courtesy of J. E. "Ed" Lewis, Greenpond, Alabama.)

for (assumed to be those admitted to the hospital for stays). There were 202 minor accident cases. The report was dated April 18, 1911. Items concerning the hospital appeared in the local weekly paper from time to time. In January 1912, it was announced that visitation of relatives and friends would be allowed between 10 a.m. and 8 p.m. Compliance, the notice said, would save the visitors and nurses time and inconvenience.

In September 1912, 18–year-old Phillip Florine became the first patient to die from an operation at the hospital. The paper said he refused surgery for appendicitis until it was too late. Drs. Jordan and Alexander were the attending physicians. A successful blessed event was reported at the hospital in 1915. Mr. and Mrs. Henry Grady Murphy became the parents of a little son born November 6.

In October 1912, the *Enterprise* editor made a personal inspection of the facility and wrote the following:

> We found the hospital in the best of condition, the different wards being conveniently arranged and nicely kept with all modern equipments. The operating room was especially noticeable being white enameled with every requirement to correspond, and arranged just as it should be. The room was well ventilated and lighted, a skylight being arranged above the operating table, besides two windows and several electric lights which were for use should an operation be performed at night. The hospital, while not large, is well arranged, and one need not fear to go there when sick or in need of surgical treatment.

The Blocton hospital was closed in August 1917 when it was announced that Dr. M. C. Thomas, its director, had been appointed district physician for the TCI Fairfield Hospital (later named Lloyd Noland Hospital). The Blocton facility continued to be a dispensary of medicine as in former days.

A second "hospital" was placed in operation as reported by the *Blocton Enterprise* of April 13, 1912: "The upstairs of the Harvey Drug Store has been fitted up as a hospital for the patients of Dr. Bloomer." This accommodation was apparently not a success, no further information has been found about its operation.

The decade of the teens was an active and relatively prosperous time for the Blocton District. The population during the years of World War I was probably the highest in the area's history. It is not unreasonable to estimate that the immediate area comprising Beats Ten and Eleven had five thousand residents. In 1910, an indicator of the numbers was the enrollment in the Blocton School of 250 pupils and over 100 in the West Blocton School. The Catholic School enrolled 140 pupils in these years.

The teens were a time of building and remodeling. The Blocton Savings Bank showed deposits of over $250,000, making it the largest in the county. The bank was in a new building with up-to-date furnishings and appointments, including Italian marble decoration. One could buy buggies at J. B. Davie's department store (the best rubber tire model was $90 and the least expensive, $35). The Davie store had a unique device for making change or charges. The youngest grandson, Bill, described the device in a 1980s family memoir:

> An intriguing invention used in the store before the 1927 fire was a change transport system whereby payments or charge slips were sent by way of a screw-top cup over an ingenious trolley-like device operated by a system of wires under tension and set into motion by a lever mechanism. All sales transactions could be sent winging to a bookkeeper in a raised platform cage toward the rear of the store who would return change or charge-approvals zipping back to clerks' stations using the same procedure.

The Standard Bottling Works, established several years earlier, had become the Coca-Cola Bottling works by late 1909, remaining in business until 1972. The Coca-Cola plant in West Blocton not only qualifies as a "first" but also as "the one and only." In 1910, Birmingham soft drink magnate Crawford Johnson, Sr., an important personage in the Birmingham franchise of the Coca-Cola Company, became president of the West Blocton plant. The August 1910 anniversary edition of the local paper wrote, "Mr. Johnson is one of the largest stockholders of the company and uses every effort to keep the company in the lead in the business it has undertaken." There was an earlier bottling works on Smith Hill that may actually have been a brewery.

In October 1911, the "Blocton Circus Grounds" featured the World's Wonder Show, staring Annie Oakley, the famous sharpshooter. The circus company advertised "The Parade of the Century" on the public streets. Entertainment was found not only on the circus grounds. Sometimes the pages of the *Blocton Enterprise* were worth careful scanning for a smile or two. In the March 24, 1910, issue an article appeared as follows:

> FLOOR GAVE WAY
>
> During a reunion of the Baraca and Philathea class of the First Methodist Church at the home of Mr. and Mrs. R. C. Parker Friday night, the floor of the parlor collapsed with about 40 of the young

Blocton Baseball Team. The local baseball team was a source of great civic pride. Games provided an opportunity to visit other towns, renew friendships, and exchange information. (Courtesy of Robert Majors.)

> people. Fortunately no one was seriously hurt, although Miss Benjie Dexter had her shoulder and arm injured by a vase which fell from the top of the piano. She was saved from more serious injury by Mr. Crump, who caught and held the piano from overturning on her. She was seated at the piano when the floor gave way on the other side of the room causing a veritable "landslide." Miss Lottie Elliot was slightly burned by the lamp that upset. Lively times ensued for a few minutes, for all the lights were put out, some crying "earthquake" and some "cyclone." The guests were preparing to go home when the crash came, and they lost no time in following their original intention. The damage was repaired Saturday morning.

An achievement of some note was accomplished in 1911 by a local immigrant. A syndicated United Press item by Harman W. Nichols appeared in a September 1953 *Centreville Press* column reporting:

> Mike Lutenberger of Blocton, Alabama wasn't behind the barn door when brains were passed out. He invented in 1911—October 24, to be exact—what he called an "air cooled rocking chair." Not only did this wonderful invention have a clever mechanical fan attached for the user's comfort which was activated by a gentle rocking on a warm day; it also had a storage place under the bottom for fruits, "bottled goods" and other material for the convenience of the occupant. Would that the air-cooled rocking chair were on the market today, and I'm not trying to get commercial. Running to the icebox for bottled goods can wear a man to a frazzle.

In 1912, the Park of the Four Orders was improved by the Odd Fellows and the TCI Company in a joint project that included the painting of the pavilion, improving the grounds, and installing swings. A second ball park was opened in West End by the Hayes Roycroft family. (In the mid-1930s, the West Blocton High School was built on part of the site formerly occupied by the park.) Bleachers for eight hundred spectators were provided. In May 1912, a ladies baseball game was attended by 581 fans, and if this wasn't a "first" it was certainly an early involvement of local woman in a major sport. Mayor M. C. Davie pitched the first ball across the plate. In June, the West Blocton men's team played the University of Alabama's nine at the field. In August 1912, Blocton claimed the state amateur baseball championship by defeating Edgewater, 2–1 and 3–2. The *Enterprise* of July 18, 1912, listed the officers of the West Blocton Baseball Association as I. B. Krentzman, Manager; John Finnen, Assistant Manager; J. W. Steward, Treasurer; and J. F. Alexander, Secretary. John R. Smith also had officer status, though his position is unclear.

In March 1968, J. Leonard Carroll, a local sports fan and old baseball player, remembered this "professional team" in a *Centreville Press* story. At one time Brown Frederick had management duties. He knew of the pitching talent of one Herman Jones, an eighteen-year-old from Primitive Ridge, and prevailed on manager Krentzman to play him. In a game with the Birmingham Outlaws he pitched, according to Carroll, a perfect game, striking out twenty-seven batters and during the game, one ball was hit to the outfield and caught by Ed West, Carroll's future father-in-law. West Blocton won 1 to 0. Herman Jones became a legend as one of Bibb County's greatest pitchers.

The first automobile advertisements appeared in the pages of the local weekly in December 1911. Prospective buyers could contact a Birmingham agent to order a Studebaker "Flanders 20" for $800. Also, one could also contact Albert Lebourg in Marvel to purchase a Thor Motorcycle. Apparently, there were some automobile owners in West Blocton by December 1912. The town Council passed Ordinance #120 which required lights on autos, back and front, and they were to be driven no faster than eight miles per hour. The first advertisement for a local dealership appeared in the August 24, 1916, *Enterprise*. Shannon Hardware Company offered a Willys Overland 31 1/2 horse power for $635. In May 1916, auto owners could call Griffin and Company, operators of the newly opened West Blocton Garage, at phone 107, for any kind of automobile repairing, vulcanizing, storage, etc. They also offered taxi service and were salesmen for Ford, Saxon, and Chevrolet cars.

State records from 1912 show there were 3,385 automobiles registered in Alabama, so the Blocton area must have had its share since there were obviously enough to merit an ordinance governing their operation. Because autos were on the increase, the Alabama Good Roads Association was organized and in 1911, it successfully lobbied the state legislature to create the State Highway Department. In April 1914, West Blocton formed a local Good Roads Association chapter, the beginning of another intra-county spat between West Blocton and the county seat.

Roads were poor and only occasionally worked. S. L. Johnson's column related a word about their maintenance in the early days. "When there was any work done [on the roads, it was accomplished] by an overseer, and the citizens, who by law were due [to contribute] three days of labor per year to the public roads. These work crews worked only in summer and the winter rains came and washed out their work. Roads in lowlands were bogs in winter and the roads in the hills were gullies."

The town government also asked citizens to maintain city streets. William Israel of St. Petersburg, Florida has an original notice sent his father, Oscar Israel, in 1915; it reads:

> TOWN OF WEST BLOCTON TO OSKAR ISRAEL
>
> You are hereby notified to meet in front of Post office the 14 day of July 1915 7 o'clock A. M. , for the purpose of working the streets of West Blocton for five days. Bring a good shovel. Also bring the notice or pay Street Tax ($3.50) or make excuse. Five days after workdate, all who fail to work, pay or report excuses will be returned.
>
> This 8 day of July 1915.
>
> JOHN GILBERT
> Street Contractor

The roads that horsemen, buggies, and by 1914, automobiles had to take from West Blocton and Blocton to Centreville were over little more than primitive trails. From Blocton, one route took travelers through Primitive Ridge and by way of Pratt's Ferry to Centreville. Another route to Centreville was by way of Mars (near Johntown) and via Scottsville. Several of its miles were over an old stagecoach route between Tuscaloosa and Montgomery. Depending on the route taken, which in rainy seasons was dictated by whichever road was passable, the distance was either sixteen or twenty miles. In 1914, a new road was proposed to connect Centreville with West Blocton and extend on to Woodstock. Several routes were debated. As early as February that year the *Enterprise* reported a mass meeting at Wright's Hall to discuss the road issue. The main topic before the meeting was the matter of where the roadwork would begin, Centreville or West Blocton. Again, it seems, West Blocton leadership had been out-foxed, at that point the road already had extended four miles north of Centreville. The Blocton leaders supported the roadwork beginning on the Blocton end of the road putting forth the old argument about greater population and the majority of county revenue being in or around Blocton. The paper reported, "there is a "steam roller" in process in the south end [of the county] with the commissioners [intending] to start from Centreville." On February 5, 1914, the *Enterprise* wrote, "Blocton is certainly entitled to some consideration in the handling of funds of Bibb County, and it is hoped that the commissioners will recognize this fact and act accordingly." In July a report on the road matter indicated a local committee was to go to Centreville to petition for the road route through River Bend. The committee also carried plans on a bond issue to help finance the road. It also imparted the message that construction should begin at West Blocton near West End. The *Enterprise* of November 12 reported a "Meeting of Protest" held by the local chapter of the Good Roads Association. The leaders took issue with the contract saying it was different from that compiled by experts. It seems the route through River Bend had been recommended. A resolution of protest was called for by the chairman, Jessie B. Davie.

By February 1916, the argument over the road construction between Centreville and West Blocton was over. There appeared to have been no "winner" proclaimed in the route dispute, however, the road did not go by way of River Bend. The local paper turned its attention to the stretch of road between West Blocton and Woodstock. It reported, "the public road to Woodstock is almost complete giving a route from Blocton to Birmingham and from Blocton to Centreville." Most of the gravel-base road constructed between 1914-1916 exists today and is known locally as old Alabama Highway 5. Asphalt paving of the route occurred in the mid-1930s.

Long active in civic affairs, Jesse B. Davie excited the local folks in January 1914 by his entry into county politics. The *Enterprise* reported the move:

> Mr. J. B. Davie, the well known West Blocton Businessman, announces in this issue of the *Enterprise* as a candidate for Representative from Bibb County in the Alabama Legislature. We are informed that one of the features of his platform will be the doing away of the iniquitous and brutalizing convict system now in vogue, and using the inmates of our penal institutions in road making and other pursuits where they will not interfere with the right of free citizens to earn a living. The Blocton District, containing by far the largest portion of the population of the county, has never heretofore been represented in the legislative councils of the state and her claims to such representation being undoubtedly just, now is the time to secure it, and we know no man is more capable of looking after its material interests at the state capital than Mr. J. B. Davie of West Blocton.

The local paper's editor had forgotten the senate office won by Bloctonian J. G. Moore for the term 1898–1903. The April 9 *Enterprise* announced Davie's victory over C. D. Logan.

Convict Labor and the Lucile Prison Fire

One of candidate Davie's planks related to doing away with the convict labor system. His efforts to this end, though not found in the local press, were commendable.

Today, Lucile is a community on the West Blocton mail route, but in years past, mines were operated there and at nearby Moffat and Braehead. West Blocton served as their trading and service center. The Lucile mine was the scene of a convict prison fire tragedy in 1910.

The use of convicts in the mines by TCI in the Blocton District had not been a local issue, although it was one of the social issues in many of the regional coal industry strikes. The nearest mines that used convict labor were in Belle Ellen, four miles to the northeast, and in Lucile, four miles to the southwest of Blocton. Prisoners were worked in the Belle Ellen mines as early as 1910 and were still employed there as late as 1927.

Just after the Civil War, Governor Robert M. Patton began the lease of prisoners outside penitentiary walls. In an 1880 coal miners' strike at Coketon, convicts had been used to break the strike. But in December 1911, the *Blocton Enterprise* took notice that "the TCI Co. will work no more convicts in their mines. This is a commendable example, and it is hoped will be followed by other companies." However, the system was not abolished until the administration of Gov. Bibb Graves who asked the legislature to void the practice. The system ended officially on June 30, 1928.

Ellison writes of the problem in Bibb County as follows:

> By the turn of the century, however, in spite of the protests of other labor, convicts were shipped to two of the newest mines in Bibb County: Lucile and Belle Ellen. The procedure of turning convicted men over directly from the county jail soon became routine. The *Centreville Press* of September 9, 1900, noted: "Eighteen prisoners convicted at the recent term of the circuit and county courts were transferred to the mines on Tuesday."
>
> Prisoners were hurried off to this difficult and dangerous life for a variety of law violations. On June 12, 1913, the *Press* observed: "Two Negroes convicted in county court, one for petit larceny, the other for defamation of character, were taken at once to the mines to serve out their sentences." During World War I, the Lucile and Belle Ellen mines depended more than ever on this readily available labor. The conditions under which the inmates lived encouraged manslaughter. In reporting such an event, the *Press* commented: "This is the third murder of a convict by a fellow convict in Bibb in the past year."

The monthly rates agreed upon in 1888 between TCI and the president of the State Board of Inspectors of Convicts were $18 for first-class convicts, $13.50 for second-class convicts, $9 for third-class, and maintenance (clothing, food and

Belle Ellen Prison, similar to the Lucile prison. (Courtesy of Will Reese.)

lodging) for fourth-class convicts. How long these rates remained in effect is not certain. It is likely these were typical amounts contracted with other mining companies. One can understand why leaders in the union movement targeted this system in their numerous conflicts with the mine owners.

The Lucile mine tragedy, the most horrible convict prison tragedy in Bibb County, and perhaps in the state, took place on Monday, May 16, 1910, at 2 a.m. While it was not life lost in the mines, it was by far the largest one-day loss of life in the history of the mining industry of the county. A fire in the wooden stockade resulted in the death of at least twenty-six convicts. The *Blocton Enterprise* reported that thirty-three out of 145 convicts were missing. Later reports, probably accurate, say twenty-six died and none escaped. The *Centreville Press* of May 19, 1910, reported the incident. The item was a shortened version of the following story from the *Birmingham Age-Herald:*

> FIRE SET TO PRISON AT LUCILLE MINES
>
> Result of Attempt of Prisoners to Escape
>
> Twenty-eight state convicts suffered death when the prison and stockade located at Lucile Mines, operated by the Red Feather Coal Company of which W. H. Perry is president, were completely destroyed by fire at 2 Monday morning. Three negro convicts who were attempting to escape by setting fire to the prison are responsible for the terrible holocaust. Lucile is four miles south of Blocton in Bibb County.
>
> Twenty-seven negro convicts were burned to death.
>
> One of the negroes who set fire to the building was shot and instantly killed while attempting to escape.
>
> Twenty-two prisoners were more or less seriously burned and several of them may die.

> Two men were shot seriously while attempting to escape by climbing over the stockade fence. From the best information obtainable here the fire was started by three convicts who stole oil from their lamps which they used in the mines. It is said they spread oil promiscuously about one section of the prison and when they set it off the fire spread rapidly, and many of the sleeping men confined were caught by the angry flames before they could make their way to the door leading to the prison yard. The buildings and the stockade, being of dry pine and parts of them saturated with oil, burned with startling rapidity.
>
> All of the convicts at the Lucile mines are state time men who are leased by the state convict department to the coal company for the purpose of their mining coal. The financial loss will reach several thousand dollars.

By June the Lucile prison fire incident had made it into the judicial system. An inquest was ordered by Circuit Judge B. M. Miller, and at the request of J. F. Thompson, solicitor. D. W. Crawford, an attorney and justice of the peace in the Centreville precinct, convened a coroner's jury at Lucile. Summoned to serve were Nelson C. Fuller, former city attorney and legislator; H. H. Jones, teacher; D. L. Morrison, farmer; S. W. Price, liveryman; Cal Riley, bank boss for Red Feather Coal Company; and C. M. McCary, of Blocton (affiliation unidentified). D. W. Crawford was appointed acting coroner.

The *Centreville Press* of June 9 gave a detailed account of the inquest. The jury found the prison watchman, Mr. McFarland (no given name appears in the article), guilty of criminal negligence in leaving the cells locked, and the jury declared the guards did much reckless shooting. It laid the origin of the fire to Ed Porter and John McCoy, victims of the disaster and to George Porter, who survived.

One convict bravely rescued Warden A. O. Thompson who was overcome by smoke and had become unconscious while trying to open cells on the second floor of the prison. Munroe Harris, a lifer from Colbert County, had pulled Thompson to safety. All officials associated with the investigation recommended to the governor Harris's full pardon.

Joining the coroner's jury at Lucile was the entire state convict board. They were accompanied by Thomas W. Martin, Assistant Attorney General (by 1920, Martin had become president of the Alabama Power Company), who said he was in attendance under the direction of Governor B. B. Comer. Martin objected to the inquest, declaring the acting coroner was without lawful authority to hold the inquest. Solicitor Thompson argued that the acting coroner had full authority to proceed. What had developed, said the *Centreville Press*, "was the unique situation of two arms of the state's legal department being arrayed against each other—one arm acting under the circuit judge pushing the investigation, the other arm, under direction of the governor and the state convict board, resisting the investigation." The implication in the report was that the governor's office did not want the stockade officers charged; that the state was not responsible for the killing of prisoners.

Further reports on the outcome of the acting coroner's recommendation, that a charge of criminal negligence be declared against the night watchman, Mr. McFarland, have not been found. It appears no court punishment was meted out in the case.

The coroner's jury report, guarded in its tone, contained severe censure of some of the prison officials, declaring that the state maintained an insufficient number of guards, and some of these guards were not properly impressed with their duty toward their fellow men. Implicating the guards, the official report said, "We find that the guards did much shooting and by said reckless shooting four prisoners were wounded while making their escape from the fire." The jury also found that Ed Porter, one of the prisoners said to have set the fire, came to his death from a gunshot wound at the hands of guard Stewart.

The *Blocton Enterprise* editor on June 9 offered his comment about the fire and the jury's findings:

> The disaster is one of the most serious that has ever happened to Bibb County and this state, this being the first destruction by fire that has ever happened in a state prison in Alabama, and the jury

expressed a hope that the next Grand Jury of Bibb County would make a thorough investigation into the sad affair, and that such measures would be installed by the state as is necessary to prevent a repetition of this disaster.

Thus ended one of the worst injustices ever perpetrated by the convict system of the period. In the social and political atmosphere of 1910, the lack of more serious public concern for the victims may be expected. The local panel did as much as the contemporary system would allow. That acting coroner Crawford had the fortitude to stand his ground against the power of the Governor's office was unorthodox and commendable.

William J. Murphy of Tuscaloosa remembered hearing Dan Lamar, one of the prisoners, tell of the incident. Lamar related that he had to play dead to keep from being shot by the guards. Edward Perry, a life-long resident of Lucile, remembered hearing stories handed-down of the fire. He said the recollections always included the belief that many of the dead were shot when they attempted to escape the flames. The newspaper accounts in the *Centreville Press* of May 19 and June 9; and in the *Birmingham Post-Herald* of May 17 and 18, 1910, report that only two prisoners were shot because they were going over the fence to escape the stockade.

The *Post-Herald* stated the dead were buried in the stockade cemetery. The law permitted families to exhume the bodies at their own expense. Dr. J. F. Alexander, resident company doctor, was in charge of the treatment of the victims and also had responsibility of burying the dead and dealing with family requests. Dr. E. M. Fuqua was engaged to work with the severely burned, his place of practice was not given.

Edward Perry of Lucile has identified the location of the cemetery where there are no markers. Bill Collins of Woodstock, who worked as a miner at Youngblood, Blocton Number 10, and Hill Creek mines, remembered hearing one of the guards, Andrew Bright, tell of his experience on the day of the fire. He was ordered to hold his shotgun on those who escaped the flames to keep them from going over the wall. The newspaper accounts say the prisoners were moved into the mine manway that opened into the stockade to keep them from escaping.

Reconstruction of the stockade was begun immediately. During its rebuilding, the company commissary was converted to prisoners' quarters. The June 2 *Enterprise* reported in Braehead Briefs, "Work on the new prison at Lucile is being pushed rapidly. Mr. McFarland, the contractor, has a large force of hands at work, and the new building will be ready for occupancy in a few weeks. We are glad to learn that the convicts who were so badly burned and otherwise crippled in escaping from the recent fire, are getting along splendidly."

In the same issue the *Enterprise* copied a comment from Birmingham's *Labor Advocate*: "The burning to death of twenty-eight [later revised to 26] convicts at Lucile mines proves our contentions that Alabama should do away with her convict lease system. Very few states now lease their convicts and the system that Georgia now has should be put in operation in Alabama."

The August 13, 1914, *Blocton Enterprise* reported there were 260 prisoners at work in the Lucile Mine. Convict labor at Lucile ended on January 1, 1918. On that date the prisoners were transferred to Belle Ellen. The *Enterprise* also reported that from that time, free labor would work the Lucile mine. Convicts apparently worked in Belle Ellen until the practice was ended by the 1928 law.

New Area Mine Development

The eight mines that were opened by the Cahaba Coal Company and acquired by TCI were the major operations in Blocton until the end of the nineteenth century. Mines numbered 2, 3, 7, and 8 were operated the longest. Mines 9 and 10 were opened by the TCI Company. The March 12, 1914, *Enterprise* noted "the beginning of work on a new mine under Smith Hill . . . it will be the most modern in the state, better than Mulga and Marvel." Further on April 23, "The new mine under Smith Hill is progressing, the slope at 25 degrees will reach coal at 290 feet. A steam heated wash house is one of the mine's features." Later newspaper reports indicate these mines were not operational until late in 1915.

Other mines were developed as reported by Ellison:

> Around the turn of the century, other mines were established both in and around Blocton. Howell and Jones, fresh from their success in the gold rush of 1897–98 in the Klondyke region of northwestern Canada, sank a shaft on rented T.C.I. property between Smith Hill and McCulley Hill and called it Klondyke. They connected it to "Dago Town" by means of a track laid in the summer of 1903. John Milner opened the Hill Creek mine in West Blocton (McGraw Town) in 1908 and sold it in 1914 to E. D. Reynolds, W. R. Young and others.

The editor of the *Blocton Enterprise* reported the Hill Creek development in his July 16, 1908, issue. He headlined the tongue-in-cheek story:

> GREAT EXCITEMENT
>
> On last Saturday morning at 7 o'clock the citizens of McGrawtown were thrown into great excitement by the blast of what proved to be a steam whistle. Women ran to and fro in search of their children who had taken refuge under logs, in ditches, or anywhere the little fellows could find to hide themselves from apparent danger. Men stood with eyes and mouths wide open while the deafening whistle continued to issue its frightful sound. After a time some of the men with iron nerve ventured forth to find out from whence came the fearful sound. Upon investigation it was found to be the first blast of the new mine in McGrawtown operated by the Hill Creek Coal Company. That whistle is continuing to sound from five to eight times daily, while between blasts the men employed are busily engaged removing dirt from the entrance to one of the best small seams of coal in Alabama. This mine is sure to be one of the best in the state. The seam of coal it will operate on is the No. 1 seam which

Unidentified Miner, Hill Creek mine entrance. (Courtesy of Carolyn Lemley, West Blocton, Alabama.)

underlies McGrawtown, reaching near West Blocton. The seam will average 3 1/2 feet of the best coke coal in Alabama.

Verily, McGrawtown will be the garden spot of Blocton very soon. With its bountiful supply of water, fruit orchards, good gardens, within easy reach for the smallest school child to attend the West Blocton School, and the new coal mine opening in its center, makes it one of the most desirable places in the Blocton District in which to live

Hurrah for McGrawtown!

The Hill Creek mine would operate until 1950. Like other local mining enterprises, it occasionally shut down or employed only a small number in slow market times and during the Great Depression. It supported its own commissary near the mine's tipple and washer, and for many years Marvin M. Martin was the store owner-manager. Cars hoisted from the mine were drawn for more than a quarter mile by dinky steam locomotive to the tipple near the railhead. Once the mine was in operation, the McGrawtown residents frequently heard the loud, shrill whistle of that small locomotive as it plied its way from near the mine entrance to the tipple. On the day in 1950 when operations ceased, Arthur Finnen, Berl Dyer, and Lester McCulley pulled the last trip of cars from the mine. Also working that day were Hulet Reach and the blacksmith-mechanic, "Preacher" Henderson.

CHAPTER 5

Over There and Over Here

While the United States did not enter the Great War until April 1917, West Blocton and Blocton were affected from the opening salvos in August 1914. Local concerns about the work on the state highway, plans for opening two new mines by TCI, and the election of townsman J. B. Davie to the state legislature were local news topics in early 1914. But dominating the news was the war in Europe.

Kaiser Bill went up the hill
To take a look at France
Kaiser Bill came down the hill
With bullets un his pants.

A popular rhyme taught to children during WW I years.

"Little Italy Items," giving local news and personals from the Italian community, appeared in the *Blocton Enterprise's* issues of September 1914 for the first time. Constituting another "first" in the county was the brief sentence announcing that "Mr. E. Castella has received a call to go to his country, Italy, to serve in the army [of that country]. He will leave next week."

The *Enterprise* of October 1, 1914, reported that during the October Yom Kippur services at the synagogue, the Jewish community took up a collection for war sufferers. This was the first reported response in the community to the needs of European war refugees. Many of the members of this congregation had relatives in Poland and Lithuania, countries eventually affected by the war.

The St. Francis Catholic Parish priest at the beginning of 1914 was Reverend S. A. Muller. The local paper reported on May 28 that he would leave for Europe in June to spend the summer months with his parents in Germany. Reverend Muller had been an articulate and dedicated defender of his native Germany in the days leading up to the war. There had been an exchange of letters printed in the *Blocton Enterprise* in which he argued for the position of his country and the Central Powers against England and France, the major Allied countries. His letter-writing adversary, W. L. Quinn, a Protestant, had written in opposition to Muller, but he let a decided anti-Catholic posture, rather than anti-German, become a major portion of his letters to the editor. In early December, the *Enterprise* voiced a guarded defense of Reverend Muller's personal devotion to his country, and later in the month the *Enterprise* also printed a long letter from Muller, then several months in Europe, pleading for sympathy for his nation's cause. It was dated November 14, and was postmarked from Bonn University, Germany. The letter's ending paragraph said:

> I fear this letter is getting too long. But being a member of the Blocton community, I considered it my duty to write to your esteemed paper. Let us be fair minded and just, let us hear both sides before we judge. We must not only read the telegrams from London and Paris, from Germany's enemies, we must hear the other side as well. Hoping all my Blocton friends are well, I send them my best regards and ask them to pray for peace, that peace which God alone can give.
>
> Sincerely,
>
> S. A. Muller, Pastor
> Catholic Church

Reverend Muller did not return from his homeland, and there is no evidence of what happened to him. He likely was caught up in Germany's military conscription or perhaps, he became an army chaplain.

The August 6, 1914, *Enterprise*, hopeful of a quick end to the conflict, offered the following on the war in Europe:

> England and Germany declared war against each other Tuesday, Italy is reconsidering her recent declaration of neutrality, Sweden will side with England, and with the exception of Spain and perhaps Norway, the whole of Europe will be in a state of eruption in a few days. Germany sustained the first serious reverse yesterday when her advance through Belgium was stopped with a loss of several thousand killed. President Wilson has offered the services of this country to the belligerent powers to mediate their differences. The facilities for human slaughter possessed by these combatants are appalling, and will of itself soon suggest a cessation of the useless sacrifice of life and a return to the less gory field of diplomacy.

Hope for a diplomatic success in ending the August 1914 hostilities soon faded. The war raged until November 1918 with the loss of millions of lives. The Blocton District was to supply thousands of tons of coal to the Allied war effort and would send dozens of young men to war. The war years from 1914 to 1918 were times of prosperity, patriotism, the departure of young male citizens for army and navy service, and finally, a triumphal celebration at war's end.

The war caused the same degree of uneasiness in Bibb County as it did in the rest of the country. Within Blocton's Italian community, there was a distinct concern because relatives and close friends were involved in the war. Other local residents of European descent were worried about the dangers their old country relatives and friends faced in the war zones. Their concerns carried over to their friends and neighbors in the whole community. There was a special feeling of patriotism in Blocton and West Blocton because of these family ties to a number of countries on the continent and Great Britain long before the United States' entry into the war.

The *Blocton Enterprise* of August 5, 1915, printed the announcement of the first call from warring nations for bandages to dress the wounds of soldiers. The local United Daughters of the Confederacy accepted the request as a project. Their membership and other ladies from Blocton and West Blocton met at the home of Mrs. J. S. Morgan, a bank official's wife, in Number 1 Town to sew and make bandages. Old sheets, pillowcases, and tablecloths, properly cleaned, were gathered to ship to Europe.

Little Italy Postcard. Residents of Little Italy could send a note on a postcard commemorating their neighborhood in 1909 (Courtesy of Sandra McCool Everse.)

Local issues and developments momentarily took the minds of Blocton residents away from total concentration on the war. For as long as the possibility of reliable electric power had been a part of the Alabama scene, the Blocton area had anticipated the day when lines would extend to all its homes and businesses. Since early in the community's history, generators were run by gasoline engines and provided electric power on a limited basis. Tuggle had his plant at the Theatorium as early as 1907. Harvey's power company supplied several businesses, and the Knights of Pythias Lodge had a generator to operate their lights and overhead fans. The local paper editor mentioned the need for utility-supplied local residential and business electric power and proposed a vote on the matter. However, the attempt to build interest had failed when too few prospective consumers had signed up in April of 1907.

Evidence of electric generators on a limited scale was also evident in the local mining operations. When the TCI Hospital opened in 1912, the operating room was equipped with electric lights. Centreville had a primitive plant that operated from an installation at Lightsey's Pond, and it provided power for several evening hours. However, it frequently shut down because of mechanical problems. Occasional reports in the *Centreville Press* indicated that the plant seemed to create more complaints than illumination. Later, a generating plant on Schultz Creek at Blake's Mill supplied power. In the early 1920s, the electric franchise for Centreville was granted to Alabama Power.

The real opportunity for electrification came to West Blocton and Blocton in May 1914. The *Enterprise* carried stories in its issues that R. A. Mitchell of the Alabama Power Company had "looked over the field with the view of placing electric power in West Blocton." A week after the initial visit, Mitchell met with the town council that granted the right-of-way franchise to his company by unanimous vote. (This is the Mitchell after whom the dam on the Coosa River is named.) If the power company approved the proposal, power was to be ready within twelve months. The *Enterprise* editor welcomed the news enthusiastically, "With sufficient electric energy in our thriving little town the way to additional industries will be opened up and we feel there is a great future ahead for West Blocton and vicinity."

However, it took longer than expected for further development on the electric power project. In May 1915, the *Enterprise* reported that Mitchell was back on the scene, this time with a hydroelectric engineer to make further plans. By June 24, the town council had passed an ordinance governing Alabama Power's right to supply electricity, and the company shortly started work with electric lights due within ninety days. The local paper's story ended with a popular contemporary weekly newspaper endorsement, "Hurrah for West Blocton!"

By August 19, the paper stated, "The work of putting up poles and wires of the Alabama Power Company is nearing completion and the houses are now being wired. It is thought that everything will be complete by September 15 at the latest." Further information was given on September 9, "All persons desiring to use electric lights are requested to have their wiring done at once so that it may be thoroughly tested before the current is turned on. All persons who are not ready when the town makes the test will be required to pay expenses of testing their own wires." An advertisement in the paper read, "Bessemer Electrical Supply Company Contract Work. See us before you have your house or store wired".

The *Blocton Enterprise* reported on October 7 that the mayor and a dozen citizens inspected the Opera House for electric light installation. This 1880s building had been outfitted with a fireproof projection room as a part of readiness for electric power to make it into yet another picture show. The theater's motto was announced as "Safety First." The managers were J. W. Stewart and Pete Saramayou.

The September date for electric power in West Blocton was not met. The reason for the delay given by the *Enterprise* was the war in Europe. As the big day finally neared, an announcement appeared in the paper on November 4, saying:

> Ladies of the West Blocton School Improvement Association have planned to celebrate the turning on of electricity Friday night by running a popular girl contest and serving free refreshments at W. W. Perry's store. Boys, come out and run your girl. She may win. There will be lovely prizes given and the winner will have the pleasure of turning on the electricity. Be sure and come Friday 6–7:30. Votes on sale at M. M. Odess' store also prizes on exhibit in his window—5 votes 5 cents.

When the long-anticipated day arrived, an impressive gathering stood in front of the balcony of the new picture show (Opera House) just after sunset anxiously awaiting the historical moment. An oversized replica of an electric double-switch made of wood and papier-mâché had been put in place on the Opera House second-level landing, sometimes referred to as its balcony, for the evening's events. The contest winner, Bertha Sims, introduced by Dr. G. W. Randall, grasped the projecting switch handle and with a flourish pushed the contraption's home-made, imitation connections together, and on cue by pre-arrangement with the power company officials, the town's new lighting system blazed to life. Also appearing on the balcony was second place contestant, Pearl Dunn, and third place winner, Evelyn Green. According to the *Enterprise*, the Blocton Home Band showed up late, but played several nice selections. In homes wired for electricity, gaslights and oil lamps went out of use. The picturesque kerosene street lamps that had lighted the way for after-dark shoppers and theatergoers on Main Street for just under four decades were suddenly obsolete.

The local paper reported other European war events and developments of 1915. In March the University Glee Club, under the direction of "Uncle" Tom Garner, performed at the Theatorium. Also in March the town ladies organized a civic league to promote a cleaner town. Unfortunately, the group was short-lived, disbanding in late April for lack of citizens' cooperation. The editor admitted "it was a great slam on West Blocton that they have no more pride in their town." A public meeting at Smith Hill School featured a debate on women's suffrage that resulted in "a decision for women."

The Pandora Theater featured high-class vaudeville every night each week, changing companies every Monday and Thursday. Playing in February was the Goldy Comedy Company. In April the TCI Company completed a "splendid tennis court situated just above the office on the hill." Ben L. Edmonds announced in late July that his dry cleaning plant was in operation. His $2,000 installation was claimed to be the only such business in Bibb. In November, W. H. Wright was elected mayor. In early 1915 there were six physicians serving the area: Drs. Trigg, Peacock, Allgood, Davie, Thomas, and Alexander. The Theatorium every Saturday night in July was showing *The Exploits of Elaine*, a spellbinding serial starring Pearl White.

The *Enterprise* reported on two claims to Blocton "firsts" in its June 3, 1915, issue. "A coffee salesman, Joe P. McDonough, was in town, a former Blocton boy, and it is claimed he was the first white child born in the district." Another item announced the death of Bob Hargrove, who had lived in and around Blocton for twenty-five years. "He was said to be one of the first colored settlers in the Blocton District."

Mine expansion in Blocton was accomplished in anticipation of the added energy needs of warring overseas countries. Since May 1914, TCI's Number 9 and Number 10 mines had been sporadic projects, with several premature

Blocton Flood, 1916. (Courtesy of Margaret Parker Fancher, Birmingham, Alabama.)

Old Highway 5. This photograph was taken at what is now the intersection of Truman Aldrich Parkway and Main Street facing south. The bridge spanned Hill Creek. (Courtesy of Wanda Pope Miller, West Blocton, Alabama.)

start-up announcements. Finally, during the week of the electricity arrival celebration, officials from Birmingham made an official visit to the community with the news that work would begin immediately. Wartime demands for coal to power the merchant marine and warships on the high seas had reached the Southern coal fields. Edward Ball, general manager; J. F. Meagher, general superintendent; and Robert Hamilton, consulting engineer made the welcomed announcement that the two mines would be opened soon. The *Blocton Enterprise* editor proclaimed, "Blocton will see a wave of prosperity she has not seen in years."

A special opportunity was available to residents in November 1916. Arrangements were announced for a special train to Birmingham to see D. W. Griffith's great film, *Birth of a Nation*. Round trip fare was $1.50. Tickets for the showing at the Jefferson Theater were available at the A.G.S. Depot. The film made it to the Theatorium at a later date.

In July 1916 there was another newsmaker event, though short-lived. A summer cloudburst brought a deluge to the geographical basin drained by Caffee Creek. Within a short period enough rain fell to put water several feet over the bridge and roadway that crossed the stream near the depot while also flowing into the machinery of a creek-side sawmill and the TCI company mule lot. Reports said the creek was six feet higher than it had ever been. The *Enterprise* indicated that local officials believed a dam upstream had been dynamited causing the water's sudden rise. While there was once a small dam on Caffee Creek at the site of Thrasher's Mill near Big Springs, there is no confirmation that it was still in place in 1916. No serious, lasting property damage was reported, and there was no loss of life. Rufus Lawley of the Hebron Community said in a 1989 interview that the old folks referred to the 1916 Blocton flood as the "Big Flush."

The completion of grading on the new road to Centreville and Woodstock, Alabama Highway 5, prompted A. J. Belcher of West Blocton to inaugurate an "auto-bus" service. In July 1916, he announced that a regular schedule had begun. The conveyance met the L&N train from Birmingham, both morning and afternoon, to take passengers to and from all points from Blocton to Centreville. The bus line was an example of an increase in business activity during war years in the Blocton District. It was also evidence of improved roads. A further example of better roads occurred in December 1916 when the *Enterprise* noted, "Mrs. James Stewart and children motored to Birmingham last Saturday and made the return trip in one hour and fifty-five minutes."

Another July 1916 announcement reported that several young men from town had passed tests at the state capital to join the Army. They were to return to Montgomery in August to enlist. July was also the occasion for the fifteenth annual encampment of the local Confederate veterans' post, Camp Clayton, at Scottsville.

August was election time. W. T. Griffin was elected mayor along with aldermen F. M. Downing, M. F. Pope, Charles Hubbard, W. R. Ray, and Frank C. Wallace. When West Blocton High School opened for the fall session there were 200 students enrolled in nine grades, the largest number in its history. At Number 1 Blocton School there were eleven grades with more than 200 enrolled, and St. Francis Catholic School reported 100 students. These pupil counts were an indication of an increased population. The years 1914-1919 were probably the era of record population in Beats Ten and Eleven (Blocton and West Blocton as well as McGrawtown, Smith Hill, and Little Italy). Other evidence of the community's growth was the report by the Parish Priest Adam Heikel in June 1917 that sixty-five children were confirmed at St. Francis Catholic Church by Bishop Patrick D. Allen of Mobile. Five masses were scheduled to accommodate parishioners attending the Bishop's confirmation services.

The *Official and Statistical Register for 1931* (Alabama) reflects the following population count:

	1930	1920	1910
West Blocton	1,070	1,023	892
Blocton	2,112*	2,811	3,315
Totals	*3,128*	*3,834*	*4,207*

*Questionable figure even though it may include Blocton, Smith Hill, and Little Italy.

The 1916 national election pitted Democrat Woodrow Wilson, the incumbent president, against Republican challenger Charles Evans Hughes. Wilson beat out Hughes 277–254 in the electoral college and polled 49.3% of the popular vote. The difference was California, which Wilson won by fewer than 4,000 votes. Not until Friday following the Tuesday, November 7, voting did the final, official results come over the wires. Wilson won in Bibb by a vote of 1,209 to 217. The following is the *Blocton Enterprise* report on the reaction in Blocton and West Blocton:

> PATRIOTIC DEMONSTRATION
>
> One of the biggest demonstrations that has ever been known in the history of Blocton occurred last Friday night when it was known for certain that Woodrow Wilson was re-elected president. All reports received here had been received with some doubt until Friday when it was flashed over the wires that without a doubt Democracy had won the victory; then all the pent-up enthusiasm burst forth, and men, women and children went wild with patriotic excitement. A torch-light procession headed by the band and about fourteen filled automobiles paraded the streets going to Smith Hill and thence to West End. A wagon containing a goat was the subject of much fun and many jokes.
>
> Speeches were made by Messrs. W. W. Perry, John Golson, Dr. Randall and others. Another significant feature was an automobile filled with Confederate veterans who carried the Confederate flag, signifying the fact that those who fought under that flag would be just as true to the United States.

The year 1916 ended in West Blocton with the automobile speed limit being raised from eight to ten miles per hour; a church play, *Blundering Billy* was produced at the Opera House; and in mid-December, the home of prominent citizen W. A. Harvey was destroyed by fire. In another matter, the town council saw fit in mid-December to pass an ordinance "To prevent the renting or leasing of houses in the Town of West Blocton for prostitution or other illegal or immoral purposes." Violaters could be fined $100 and sentenced to hard labor. Proof that West Blocton was still basically rural, the town council repealed the ordinance against cows being on the streets at night, "as the ordinance was never popular and was the cause of many complaints."

In early January 1917, Mrs. L. E. Peacock and Mariana Green went to Birmingham to see the great Sarah Bernhardt in a performance that was billed as the last opportunity to see this noted actress. The Birmingham appearance was advertised as occurring during her final American tour. Adding to the business scene, S. S. Allen and Ben L. Edmonds boasted of providing auto-truck deliveries. There was a celebration in January of Robert E. Lee's birthday at the Number 1 Blocton School. The United Daughters of the Confederacy, the Yellowhammers, and Confederate Veterans Camp Clayton took part in and sponsored the event. In March, the West Blocton Savings Bank, the largest in the county, reported total assets of $194,686. 60. Meetings were reported in February of the Hilltop Needle Club and the Pollyanna Crochet Club.

Life went on in the Bloctons, but the major concern was what the daily newspapers reported on the war in Europe and the policy in Washington determining the country's involvement in the conflict.

Daniel Harmon, Principal, T.C.I. School, Blocton. (Courtesy of Alabama Historic Ironworks Commission.)

Within a week of the United States' entry into the war, a troop from Alabama's First Cavalry was assigned to Blocton to guard the mines from possible saboteurs. Some members of the unit were assigned to the 60-foot-high, steel, concrete, and wood Bluff Trestle several miles southeast of the depot. Captain Dorsey commanded the troop and among the officers was Lt. E. D. Thames, a cousin of local resident Mrs. C. C. Caffee. The contingent of uniformed soldiers in the community's midst brought a sharp focus on the involvement of the country in the war, and the soldiers were immediately the objects of indulgent attention. During a Confederate Ball at the Elks Lodge, there was a cake and pie shower for the troop, who were invited to attend as special guests, and were also presented a box of cigars. In early May when Camp Clayton held a memorial service at the Methodist Church, the cavalry troop members were there as a unit to salute the old vets. The troop had responded to the community's show of welcome and patriotism by sending white roses to be pinned on each of the Confederate veterans. In mid-June the members of Troop D, First Alabama Cavalry reciprocated by sponsoring a dance at Four Orders Park. The *Enterprise* reported, "the pavilion was prettily decorated with pine boughs, festoons of red, white and blue, small U. S. Flags and Japanese lanterns. There was a band from Birmingham and delicious refreshments were served. It was one of the most enjoyable occasions of the season."

The unit found need to call on community resources within a week of its arrival. Soldier Eli Andress was operated on in the TCI Blocton Hospital for appendicitis. In early June, the troop took part in a town patriotic ceremony. An elaborate "flag raising" was staged near the company store supervised by Captain Dorsey. The company superintendent, James Stewart, ordered a platform built and a sixty-foot flagpole erected on which Old Glory was raised as the band played "The Star Spangled Banner." Professor Daniel Harmon of Number 1 School was master of ceremonies, and Joseph Powell was the featured speaker. To end the special program, an enthusiastic crowd of 300 sang "America" accompanied by the Blocton Home Band.

The town was made dramatically aware of the demands of war when it was announced in the May 24 *Enterprise* that Dr. N. T. Davie had been ordered to report immediately to Ft. McPherson, Georgia to assume his duties with the Army Medical Corps. The community was later to lose another doctor to the war. Dr. H. W. Allgood of the TCI staff left for the Medical Corps in January 1918. In the same issue the following notice appeared:

> All male inhabitants past the age of 45 living in or near our community who feel an interest in organizing a home guard company for the protection of our citizenship during these war times are earnestly requested to meet with Camp Clayton of Confederate Veterans at Wright's Hall 11 am Saturday, May 26. [Signed] T. C. Wallace.

The next week's paper reported that thirty-three men pledged they would volunteer their services to the "home guard."

Camp Clayton leadership drew up resolutions at one of their May meetings offering their services that were telegraphed to the president. The chapter shortly received a reply addressed to Commander T. C. Wallace:

White House
Washington, D. C.
The President deeply appreciates your generous and patriotic proffer of your services, and he wishes in this informal way to express his grateful thanks.

The *Enterprise* listed the Home Guard on June 14: Capt. T. C. Wallace; 1st Lt. Tom Latham; 2d Lt. Pat Jennett; 1st Sgt. W. H. Logan; 2d Sgt. W. B. Dinkins; 3d Sgt. R. C. Lagrone; 1st Cpl. L. J. Hayes; 2d Cpl. J. J. Lucas; 3d Cpl. J. F. Golson; Adjutant W. H. Wright; Quartermaster and Commissary Sgt. W. T. Griffin; and Chaplin H. F. Crim.

In keeping with instructions from the state capitol, conscription registrars were appointed in each of the county's beats. The communities and their registrars were listed for north Bibb, one for each voting box, in the May 31 *Enterprise*:

West Blocton	W. R. Young
Blocton	H. H. Farrar
Belle Ellen	Joe Clifton
Piper	W. T. Crawford
Marvel	George Ellis
Woodstock	John Green
Hargrove	George Archer
River Bend	George Desmond
Red Eagle	G. G. Tate
Primitive Ridge	Mit Cundiff
McGrawtown	W. R. Ray

These committee members had the responsibility of making an official count of young men between the ages of twenty-one and thirty-one. The County Draft Board was made up of Sheriff John S. Ward, Robert Thrasher, and Dr. L. E. Peacock. In the first drawing, seven men were called. Ben F. Jennett of West Blocton was one of those draftees who left for Camp Gordon in early September. Although no farewell ceremony was held for this first group, Ellison's history says there was thereafter public demand for farewell ceremonies and "for the next fourteen months farewell parties honored one group of departing soldiers after another." On an occasion in May 1918, eighty-six Bibb boys were honored at the M&O Station in Centreville on their departure for military duty with patriotic speeches accompanied by the Blocton Home Band.

Dan Gallagher, an early West Blocton enlistee, had been overseas for three months in August when he wrote a letter home saying his location was "somewhere in France." A profound shock came to the town when this young soldier was reported missing-in-action. Then the November 8 *Blocton Enterprise* reported that the Gallagher family had been notified by the War Department that their son was one of the twelve soldiers missing in the first skirmish engaged in by Americans. In a patriotic show of concern and sympathy, a delegation of neighbors and friends went to the Gallagher home for a brief ceremony. The Blocton Home Band played the national anthem, and Dr. Randall made a brief statement to the gathering. In early February, the Gallaghers had a letter from their son saying he had been wounded and captured by Germans during a skirmish on November 4, 1917. He had three wounds affecting an arm and his side. Writing from a prison hospital at Touschell, West Prussia, he sent greetings to his Blocton friends. Gallagher reported in another letter in mid-September that he was in good health and that Red Cross support helped relieve the discomfort of German prison life in various ways. He remained a prisoner until the war's end, when his release was reported in the December 19, 1918, *Blocton Enterprise*.

Grim news was delivered to the home of Mr. and Mrs. James Gillespie. Their son, twenty-one-year-old Matthew Leo Gillespie, had entered service on June 2, 1918, and after basic training in North Carolina, he was sent overseas, where he contracted pneumonia and died in October. His body was returned to West Blocton in October 1920 from Brest, France, and interred in Mt. Carmel Cemetery after last rites at St. Francis Catholic Church. Town businesses displayed flags at halfmast, and the community turned out in large numbers to salute one of its boys who had made the supreme sacrifice in the service of his country.

The list of Blocton area soldiers who died in the war in Owen's *History of Alabama* included Gillespie, Lonnie Morse, Jim Peyton, Jake Polovich, and Doc Smelly.

As a part of the duty of a community making its contribution to the war effort, the Liberty Bond Drive netted sales of $5,500 in June 1917. The bonds were sold through the West Blocton Savings Bank. By October, chapters of the American Red Cross were formed in Blocton with H. H. Harris, chairman; Mrs. Ben Huey, vice chairman; E. D. Lee, treasurer; and Elizabeth Stedham, secretary. The chapter reported 28 members. In the West Blocton chapter, E. M Bishop was chairman; Marianna Green, vice chairman; Bertha Macon, secretary; and W. R. Young, treasurer. There were 32 chapter members. The following May the newspaper reported, "a Red Cross Chapter for colored people was organized on the Heights."

There had been earlier evidence in the *Enterprise* that the Negro citizens were being given more positive recognition. Special send-off parties for soldiers from the Negro community were reported favorably by the paper. In an unprecedented news item in the September 27, 1917, *Blocton Enterprise* a Negro social event in Frog Level was reported, and the Negro participants were identified as Miss, Mrs., and Mr. It was the first such use of titles appearing in the local paper for African-Americans.

In mid-November 1917, an Army recruiting office, headed by Private First Class H. C. Hawken, opened in Wright's Hall. It closed by mid-December after enlisting over thirty recruits.

In June 1917, Blocton Number 3 mine closed and men employed there were transferred to mines 9 and 10. Company officials declared the slope was "worked out." The Number 3 slope had been one of the earliest opened in old Blocton, and many of the miners there were from the close-by Italian community, Little Italy.

While the community's attention was kept focused on the war, other local events of note took place. The annual

Blocton Mine Number 3. (Courtesy of Will Reese.)

gathering of Camp Clayton Confederate Veterans was held July 11–12, in Scottsville. Another sign of prosperous times occurred during August when the *Enterprise* reported that the West Blocton Savings Bank, just west of the Harvey Drug Store, was overhauled and remodeled. Fixtures of fine Italian marble were added, the floor beautifully tiled, and the walls finished in hardwood trimming.

Union Reorganization Movement

In early June 1917, the United Mine Workers held a mass meeting at nearby Seymour attended by 1,500 to 2,000 miners. District 20 president J. R. Kennemer proclaimed "the miners were not organizing to strike but to keep men at work." The following week Blocton miners assembled on Smith Hill to organize. Locals were formed for Blocton, Hargrove, Hill Creek, and Red Eagle where over 500 members were enrolled. There was a subsequent overflow meeting of miners in the Opera House during the next week.

According to family tradition, Harry Adams was elected president of the Blocton local in absentia. He was in Birmingham on a Blocton Home Band contest trip. The date of his election is not clear, however, it was while the war was still in progress. One continuing local complaint that was brought to union local president Adams' door by miners' wives during war months related to the poor quality of the black powder in dynamite used in the mines. It was claimed that fumes from shots made the miners sick. It was generally understood that the chemical mix was not up to standard because of priority at the battle fronts for the highest quality munitions.

These union organizational meetings were the first efforts to reorganize since the failure of the unions to gain benefits in strikes occurring in 1904, 1905, and 1908. The violence in the 1908 strike had contributed to the success of management and government in breaking the union.

The year 1918 did not get under way on a happy note. A windstorm did considerable damage during the night of January 14, destroying the Albert Brown and Frank Vandergraff homes on the Heights. Fortunately no one was injured, though the West Blocton school had windows blown out as did the C. C. White home. Bricks were dislodged from a chimney at the George W. Young home, and there was damage to the front of the Price Building on Main Street. The storm blew trees down in Blocton and West Blocton.

In mid-month the disturbing news came from the mines that water had broken through from old Number 1 mine into the workings of Hill Creek. One miner, A. W. Thompson, was reported missing and presumed drowned. His body was not recovered until late July.

The Blocton miners came out on a brief strike in mid-February 1918. Miners were out at Johns and other places. TCI mines 7, 9, and 10 were affected for less than a week. Reproach by politicians and mine owners declaring that it was unpatriotic and illegal to strike during wartime, and the threat of a government take-over of mine operations ended the walkout. Reasons for the strike included poor working conditions in the mines, as well as a marked rise in living costs without wage increases. In addition, the union argued that an agreement made through the Federal Fuel Board for the Alabama District had been disregarded by the company. The serious management-labor confrontation that surfaced in early 1918 in Blocton would again be faced in 1919 and 1920 on a nationwide scale.

Though the February strike in Blocton was a short one, the *Birmingham Ledger* alleged that it had been a dangerous one. The paper said, "there has been a clash between strikers and guards at Blocton . . . a message to union officers here [Birmingham] declared a guard at the mine fired on a negro [*sic*] miner. There was no damage done." But the *Enterprise* denied this report, responding promptly and emphatically that there had been no "clash" between strikers and guards. It seems a deputy sheriff fired a shot in an arrest attempt and that was the basis for the erroneous story. The editor said everything was quiet in Blocton and deplored that the occasion had given the community the false appearance of being a terrible place. He proclaimed, "There is perhaps no town in the state more law-abiding than Blocton. The small amount of criminal acts committed here is remarkable considering the size of the place and the different classes of people living here."

A number of community support programs relating to the war were announced in the *Enterprise*. In January 1918, the Blocton Red Cross sent its first box of supplies in response to a call from an American Hospital in France. At the same

Blocton Home Band member Will Wyper somewhere in France, 1918. (Author's collection.)

time $100 worth of supplies was sent to the Southern District Red Cross. The local chapter met regularly at the Elks Lodge for work sessions that produced operating caps, convalescent robes, pajamas, sweaters, and socks which were sent to the regional Red Cross headquarters in New Orleans. The Douthit House Hotel was the deposit station for a paper drive, and proceeds went to the Red Cross. Because of fuel conservation regulations, Manager Almon Jones announced that the Coca-Cola Bottling Works would be closed each Monday. In March, the Red Cross chapter met to make hospital garments, and a manless wedding at the Theatorium netted $70 for the chapter. Mrs. E. D. Lee led the third Liberty Loan drive during the first two weeks of April with an overwhelming response. In Beat 10 alone $23,850 in Liberty Loan bonds were sold, with a quota of only $6,800. An additional $12,000 was sold within the TCI Company. A drive in March netted 200 books for the Red Cross to send to servicemen.

In May, the newspaper wrote that a group of the town's Christian ladies had asked that churches ring their bells each evening at six o'clock for three minutes to remind all citizens to pray for victory of our Allied armies. The editor, an Episcopalian, said the Church of Our Savior had been complying for some time. Also in May, the ladies of St. Francis Catholic Church sponsored a dance at the Elks Home with the proceeds going to the Red Cross war fund. At the end of May, a proclamation by President Wilson commemorated the entry of Italy into the war. The Italian community took the occasion to celebrate with a parade led by the Blocton Home Band through all the principal streets and on to Four Order Park where local attorney and orator E. M. Bishop addressed the gathering. Thirty-one dollars were collected at the park and donated to the Red Cross War Fund. On June 14, Blocton received a special, honor flag for its splendid work in the Liberty Loan drive. Bond sales in the company town had many times exceeded the suggested goal. The flag was raised in front of the TCI Company Store, with music by the Blocton Home Band and flags decorating the store and company hospital. The assemblage adjourned to West Blocton where the ceremony was repeated with the raising of an honor flag and the American flag. The well-practiced Dr. Randall spoke to the gathering, and the honor flag was placed in a frame and illuminated by electric lights. The special day closed with the band playing patriotic music.

On the Sunday closest to the Fourth of July in 1918, local churches held special services honoring their members in the military. The Catholic Church reported a blessing of a service flag displaying twenty-one stars, one for each serviceman from the church. The Methodists' flag service honored thirty and listed their names for the *Enterprise*. Other church records have not been found but probably all active congregations held similar programs.

Sheriff Huey Wood. (Courtesy of Bettye Kitchen Hughes, Tuscaloosa, Alabama.)

A county-wide special Fourth of July festival was announced in the June 27 *Enterprise* following a request by President Wilson that all Americans join with citizens of foreign extraction for the event. Alabama's Governor Henderson also issued a proclamation covering the occasion. It was predicted to be the greatest Fourth celebration in Bibb County. Later during the campaign in West Blocton, editor Allen of the *Enterprise* wrote that 5,000 spectators were expected for an event involving five airplanes dropping Liberty Loan brochures. He believed that few people had ever seen five planes in the air at the same time. Plans to have planes land were aborted when adequate space was not available for a safe landing. Also, as a promotion to sell bonds, the Liberty Loan Relic Train came to town with three cars of items from the Belgian and French battlefields accompanied by Captain Edouard Hirsh of the French Army, who had been wounded and awarded the French Army medal for bravery, the *Croix de Guerre*. Hundreds of local citizens visited the exhibit.

With so much going on in relation to the war effort, not much was made of the campaign and election of West Blocton's Huey Wood as sheriff of Bibb County. The local paper announced his qualifying for the office in June and his election in August. His victory was an indication of the shift in voting strength to the communities in north Bibb and the concentration of a large wartime population in the mining towns.

Other matters of concern were printed in the local paper's pages. In mid-May a special notice was issued to the citizens of West Blocton:

> Let us get together and observe two days in which to clean up our town, and around our premises, around our stores, in alleys and on the streets. The days for the observance of "clean up" will be Monday and Tuesday of next week. All failing to observe these days will be subject to fine, as the health of our citizens depends on this. All weeds and grass must be cut, paper and debris of all kinds raked up and put where the trash men can get it and he will come around to every place these two days. If the law is observed we can keep our town clean; otherwise, diseases of all kinds will prevail here.
>
> By order of W. T. Griffin, Mayor

By October, regardless of the May clean-up campaign to stave off disease, the world-wide Spanish influenza epidemic came to the community. Mayor Griffin called a special session of the town council to propose cautionary measures for the town and surrounding area. Resolutions were adopted as follows:

> 1. Schools to be closed for two weeks beginning October 9.
> 2. All places of public assembly: churches, Sunday schools, theaters, pool rooms be refrained from holding public meetings for the same period of time.

The extent of the epidemic was recorded weekly in the paper's personal columns that identified various families reporting cases. Four deaths resulting from influenza complications were reported for the previous week in the October

23 paper. The epidemic was still raging the following January. The paper wrote that many were sick and that "each week brings another death. There is scarcely a house where there is not sickness." By the time the epidemic ended in America 600,000 people had died. A 1998 Public Broadcasting System "American Experience" program reported that some cities ran out of coffins. It was the worst epidemic the nation has ever experienced.

The paper also recorded that the fourth Liberty Loan drive had gone "over the top" in Beat 10 with $75,000 in bonds sold. The chairman of this last wartime drive was local merchant and civic leader Morris M. Odess.

As a result of what proved to be a false report, the town broke into celebration of the war's end on the night of November 6. The band started a parade, bells were rung, firearms discharged, and joyful demonstrations occurred all over the district. It turned out to be merely a warm up for the real thing that was pre-announced for the eleventh hour of the eleventh day of the eleventh month in 1918. The Armistice was finally here and West Blocton and Blocton went wild with joy.

The *Enterprise* reported the event in its November 14 issue:

VICTORY!
Blocton District Celebrates
the Downfall of German Autocracy
Great Enthusiasm Displayed over
Victorious End of Great War

> When the news came over the wires early Monday morning that truly the armistice had been signed, thereby signaling the unconditional surrender of the German autocracy, Blocton and West Blocton went wild with patriotic enthusiasm. Bells were rung, guns were fired, dynamite exploded, women laughed and cried and even the eyes of strong men were wet with tears of happiness that peace was in sight and our boys were no longer targets for German bullets.
>
> A half-holiday was declared, all places of business were closed at 12:00 and at 2:00 PM a mammoth parade was formed on Main Street, composed of people from all over the district. The parade was at least a half-mile long. Numbers of automobiles, trucks, buggies and wagons decorated with the national colors, were filled with enthusiastic men, women and children, laughing and shouting [and] wreathed in joyous smiles at the termination of the great world war with victory on the side representing democracy and justice.
>
> The Blocton Band led the parade playing patriotic airs and was followed by mothers who had boys in the service; next were the fathers, then came sisters and brothers. After these came the patriotic citizens and school children.
>
> After parading through West Blocton and back, then through Blocton to the foot of Smith Hill, the procession returned to the starting point on Main Street where patriotic speeches were made by Dr. G. W. Randall, Rev. J. C. Francis (Catholic priest), Rev. D. Z. Wooley and Dr. L. E. Peacock, all of whom were heartily applauded.
>
> The celebration ended at night with a regular jollification at the pavilion which was brilliantly lighted. Music was featured and dancing was also engaged in by those who were so inclined.
>
> The Italian Band also came later in the evening marching through West Blocton and Blocton playing patriotic music.
>
> The school children both white and colored, paraded in the morning.
>
> Only one accident occurred throughout the entire celebration, and that was when a mule threw a young man whose name we failed to learn, who landed on his head and was unconscious for a few minutes, but was not seriously injured.

At many of the gatherings in the Blocton area the crowd joined in group singing; favorites were "Tipperary" and "Over There," and the lines of a favorite and, incidentally, prophetic, refrain, "How you gonna keep 'em down on the farm after they've seen Paree?"

The war's end brought immediate anticipation of the return of the area's young men. Many had been away from several months to more than a year, either at training camps or in action at the front in France. A number of parades, picnics, and other special celebrations dotted the calendar well into 1919. The *Enterprise* reported the status of the community's service men from week to week. Neil Gallagher was happy to receive notice in mid-December from Adjutant General Harris that his son, Dan, had been released from a German prison and was in France in good health. A letter in January 1919 from the soldier said he was "one happy boy" and that he would soon be home to tell his experience of the last thirteen months. Good news came by way of a cablegram from Lt. N. T. Davie to his family saying he was safe and sound after the last big battle on the Western Front had brought the firing line within three hundred yards of the hospital where he was stationed.

Word had also come in December that Tulon Desmond had arrived safely in New York but was confined to an army hospital in Oteen, North Carolina, being treated for pneumonia. Without naming them, the paper said in early January that "a number of our soldier boys have returned home from different camps, and will take up their positions where they left off when they went away." It did list Charley Wright and Abe Berlin as returning from military training in Jacksonville. They were reported to have "been on their way to the front when armistice was signed." A war relic train appearance was announced, and in May the newspaper reported, "The war tank arrived per schedule and was visited by a large crowd. It was carried to the West Blocton School House. Captain Peters of Centreville and others delivered speeches." This was evidence of continued interest in the war and the desire of the local population to get a feel for what it was like for the boys who fought in the French trenches.

A series of "Letters from France" appeared in the *Enterprise* in the days following the war's end. Private Ben Jennett wrote his mother a letter dated November 19, 1918, as follows:

Main Street Parade, November 11, 1918. The Blocton Home Band was aboard the Pritchett Ice Company delivery truck. (Author's collection.)

> Your letter received a few days ago and certainly enjoyed. We are still in Mansigne attending a school, but I don't know how long I'll be here or whence we'll go. This morning accompanied by a friend I took a short walk. The ground was covered with frost and I was wrapped up well. We passed several Frenchmen all wearing wooden shoes but no sox. Now that the war is over my thoughts are naturally of home. Certainly hope ere long the Jenette family can be reunited. I am well and hoping and trust you are the same. I do want to see the dear ones at home, but until the government is thru with me I remain willingly. I would write you more often and longer letters but there is nothing to write. We are in only a small village and it's impossible to buy anything much. However, I'm sending you a little gift just a remembrance from a boy to his mother to show he thinks of her. A merry Christmas to one and all. The populace in this little village were very much elated over the declaration of peace, and indulged in much hilarity. The whole town is decorated in flags. War is a terrible thing and after four years I couldn't blame them for celebrating. Write to me and tell the others to do so. Love to all. Ben

Shortly after the war ended, Mrs. W. T. Griffin heard from her son Roy who was attached with the American Expeditionary Force Hospital Corps. He wrote:

> We are located in a small town (just about the size of West Blocton) and have a real nice hospital camp, one formerly used by the French, and they have flowers all over the place, and that makes you feel a lot better and more at home. About the hardest job this buck private has had so far is trying to tell a Frenchman something; one that doesn't understand English. Can't say I've progressed very far in the study of the French language. A fellow has to make signs and do all kinds of stunts to make 'em understand. They all seem to think a whole lot of the boys in khaki.

In another letter printed on December 26, 1918, Corporal Truman Aldrich Blake wrote to his father, J. H. Blake, on

Tulon Desmond. (Courtesy of Tulon E. Desmond, Birmingham, Alabama.)

Ben Jennett, U. S. Army, at age 22. (Courtesy of Elizabeth Jennett Brannon, Tuscaloosa, Alabama.)

Route 1, West Blocton, telling him that a special project had all the "boys" writing their fathers. He wrote, "I am in what is known as the Army of Occupation and we are on our way to Germany. I will not attempt to relate the interesting things that can be told. So far I have not had a leave and I am going to wait now until I get back to the States before I ask for one and then I am coming home '*toute a suite*'—that is only a little French that means immediately." A letter from John Forest Shaw printed January 9, 1919, said to readers "at one time in the trenches a man on each side of me was killed. I was on post with them when a shell bursted in the air and I didn't get a scratch. I was nervous for a while from the shock. At another time a shell hit the trench just above me and I had to be dug out." Shaw was later wounded and spent time in a French hospital.

One of the last of the soldiers' letters was printed in August 1920. Jesse Jones was a member of the Army of Occupation stationed at Coblenz, Germany. It was a rambling description of the local German populace and an expression of homesickness for "that little old mining town I call home." He reminisced, "I sit idle and watch myself stroll past the Allen Furniture Company, up by Davie's and Berlin's, on past Shannon's and Dunnabie's on to Keith's where I stop for a shave, then cross the dingy little street that looks like it is forever asking for help from the hands of reckless Ford drivers, and to the pool room where I linger a moment and then go to see what Tuggle has on [at the theater], and then out past Pope's [drug store] and then down to the Depot pass [the pedestrian pathway to Number 1 Blocton]." The writer ended by sending hope that "everything is lovely in dear old Blocton."

In an April 8,1956, *Birmingham News Magazine* story, "Wave No Flags" by Sue Pickett, the walls of the local American Legion Hut were said to display a letter from King George V of Great Britain sent to West Blocton troops for bravery in World War I. Also displayed was a letter from General Pershing to Corporal George R. Blake for bravery during the first World War. The story further said, "it is claimed that West Blocton had more men in service per capita than any other place in America, and the greatest number dead." The article covered both World Wars.

As the community adjusted to peacetime in 1919 there were more mundane matters considered in the local paper. The editor deplored the loss of a local Western Union office noting the inconvenience of sending telegrams by phone to Tuscaloosa.

L & N passenger fares were announced to area destinations: Birmingham, $1.48; Bessemer, $1.10; Garnsey $.53; Marvel $.57; Boothton, $.64; Gurnee Junction $.71; Coleanor $.39; and Piper, $.41. Another housewives group made its displeasure known about the lack of concern by local government for the town's unclean condition. The Thursday Afternoon Study Club was organized to replace the old Three Arts Club, and the Culture Club also reported on its monthly meeting. There was obviously no absence of club opportunities for the community's sophisticated social set. It is reported that "club ladies" of the town had engraved calling cards and strategically placed silver plated trays in their parlors on which callers could deposit them. In May 1919, advertisements for the Paige Automobile appeared. Brown Frederick, as local agent, offered a seven-passenger model of the Paige Essex, "the most beautiful car in America," for $2,010 and a Linwood Six 39 for $1,555. Reflecting continuing prosperity, the West Blocton Savings Bank reported $330,589 in assets. In July, prominent citizen Rufus D. "Rufe" Smith's passing was mourned by area residents. Smith was the founder of Smith Hill and was a brother to Uriah, who is given credit for promoting real estate sales and naming West Blocton. Another community leader, Mayor W. T. Griffin, died in mid-August, and M. C. Davie was selected to continue his term.

While it did not receive wide notice, the *Enterprise* reported a county centennial celebration in June at the Theatorium under the auspices of the Methodist Church. It was directed by Josephine L. Biddleman with Mrs. Kenneth D. (Lillian) Young as chairman. One hundred and fifty local people were to take part in a program that the paper called "unique entertainment." It was said to be a success both as entertainment and financially, the ladies clearing "something over seventy dollars for their share for the church. Everyone taking part seemed to enter into the spirit of the occasion and put forth their best efforts. The children were especially praised."

According to author-researcher Ellison, the proper centennial anniversary should have been held in 1918. Her observation was that county residents had been too absorbed in the action in France to take notice of the county's pioneer history and development. No detailed program appeared in the paper nor has a copy of the script for the production been located. West Blocton could boast that the county centennial was indeed commemorated even if it was a year late.

In 1919, celebrations of one kind or another continued to honor the town's former servicemen. The paper announced a picnic and barbecue in honor of returning veterans for Flag Day, June 14. Shortly after that came a gala Elks Picnic for July 4. The advertisement read: "A BRASS BAND WILL LEAD THE LODGE PROCESSION FROM WEST BLOCTON AT 9 AM AND THE BIGGEST PICNIC EVER HELD IN BLOCTON WILL BE ON." However, the largest of the several gatherings came in August at Four Orders Park to honor returned Bibb County soldiers. It was announced that an official Navy Band had been arranged by Congressman W. B. Oliver. It did not arrive in time for the parade so the old reliable Blocton Home Band stood in, performing its familiar music. It was to be the last major appearance of this prominent, sixteen-member ensemble, which would be a victim of the company's strike-breaking moves within a few months. The Navy Band did perform an open air concert in the afternoon. The paper said there was an eighty-five-foot-long table loaded with barbecue and other "toothsome victuals." For the grand parade, Maude Wallace led the procession from the depot to the Knights of Pythias Hall and back to Four Orders Park. The participants were the old Confederate veterans, members of the United Daughters of the Confederacy, and the Yellowhammers, an UDC girls' club, all in automobiles and dressed in red, white, and blue. There were 150 soldiers and sailors in uniform at the gathering. The featured speaker was Lt. R. H. Wright.

On a Saturday in late August, the "Colored Soldiers Picnic and Barbecue" was held at Number 2 picnic grounds. A parade featured a band from Birmingham and 100 soldiers and women and girls carrying red, white, and blue streamers with autos following. In a 1983 interview, seventy-six year old, Mem Wells told Willie Jo Parker that he remembered the special day with the soldiers all in uniforms. It was his recollection that the Waddley Band from Birmingham took part in the celebration. The *Enterprise* reported that Lt. Melton H. Gray and Professor W. B. Driver were the guest speakers. Dr. Randall, a frequent visitor to the community's African-American institutions, also had a speaking assignment.

A defining explanation is lacking but on August 23, 1919, the *Enterprise* printed an article titled "Colored Citizens of Blocton Clear Up False Rumor." It is only speculation, but this was the period in regional history that witnessed the resurgence of the Ku Klux Klan. There may have been local meetings relating to that possible anti-Negro movement. The

View of Number 1, Blocton. The rear of the company school (white) is visible in the background. (Courtesy of Patricia Clinner Deerman.)

Dunnabie's Store, *circa* 1918. (Courtesy of Annabelle Dunnabie Martin, Birmingham, Alabama.)

story opened with the statement that "someone holding meetings with Negroes at their halls and churches untrue." Further, "Feeling between races shown on our celebration day since whites donated funds to that event." The item ended with the following statement signed by prominent citizens in the African-American community:

> We guarantee that should any strange wolf happen into our circles with such stuff as has been rumored, we will take charge of him.
>
> C. M. Williams J. R. Banks
> A. S. Jackson Ed Boom
> O. W. Branch I. S. Dooley
> Frank Vandegraff P. S. Jackson
> M. Lanier A. Mixon
> Ed Ward Scott Coleman
> J. H. Wells Frank McCannon

Patriotism remained at a high level long after the 1914-1918 war ended. In October, there was a meeting at the Elks Lodge to organize an American Legion Post. Leaders were B. H. Gatlin, T. A. Blake, and Tulon Desmond. In years to follow, though somewhat low key during the Depression, annual celebrations were held to commemorate the nation's role in the war.

In 1980, Charles Clarence Kindig, a resident of Russellville, listed in order the business and professional offices on Main Street as he remembered them from his teen years. His father, Charles Kindig, had been an engineer on the M&O Railroad. The following closely resembles the Main Street of West Blocton that the soldiers of World War I returned to in 1918:

North Main Street (east to west)

Shannon Brothers Hardware (wire storage)
Ben Edmonds (tailor, dry cleaners)
T. C. Tuggle (photographer)
Dr. W. W. Griffin (dentist)

Florida Street
Harvey-Wright Drug Store
Uncle Mel Hayes Hamburgers
City Restaurant
Brown Frederick Pool Hall
West Blocton Savings Bank

Alley
Douthit House (hotel)
Old Post Office
Econami Brothers Bakery
Fire Station

South Main Street (east to west)

Woco Pep Service Station
Alabama Power Company
Masonic Lodge/Post Office
M. F. Pope (drug store)
W. W. Perry (women's wear) & Dora Lewis (millinery)
Israel's I & N Quality Store
M. M. Odess Department Store
Florida Street
Tovich Fruit Stand
Klotzman Dry Goods
Walters & Gamble Ladies' Wear
Ogletree Grocery
Bob Keith Barber Shop
Joe Israel Department Store

Zavelo's Shoe Fixery
Mose Zavelo Jewelry
Dunnabie Grocery

Ammons Barber Shop. (left to right) James Thomas Ammons, his older brother, Bill Ammons, and Jennings. (Courtesy of Wanda Pope Miller.)

North Main Street (east to west) continued

Lagrone Market
Will Ammons Barber Shop
Stewart's Opera House/Y. S. Smith
General Store (caskets)
John Kren Grocery
W. H. Logan, Attorney, Justice of the Peace
Manring Hotel
Jeanette Fruits & Notions
Pope Brothers Cleaners
Gentry Hardware
Stanley Grocery
Stanley Residence
Grocery
Residence

Alley
Oscar Light Grocery
Joe Israel Residence
Grocery
Finnen Meat Market
Hayes Roycroft General Store
School Street
Knights of Pythias Hall/Dr. Bloomer's Theatre
J. W. Jones Drug Store
Jones Residence
Glover Spring Branch Bridge
W. L. Ambrose Store

South Main Street (east to west) continued

Dunnabie Residence
Shannon Brothers Hardware
Quinn Barber Shop
Pilgrim Shoe Repair
I. M. Faulk (novelties)
T. A. McCool & Sons Grocery
Berlin Brothers Sundries (downstairs)
Dr. Eugene Edmonds (dentist) (upstairs)
J. B. Davie (buggies & wagons)
J. B. Davie Department Store
Alley
Allen Furniture/Undertaker
Riley's Café
Ogletree Residence

Residences (3)
W. H. Wright, Attorney

Wright Residence
Hitching lot/Blocton Ice Company (1927)
Dyer's Blacksmith Shop
African-American Restaurant
African-American Barber Shop

Just off Main Street on Florida Street going north (uphill) was the Theatorium and Dr. Randall's dentist office. South on Florida Street was found the offices of Dr. Peacock and Dr. Jones, the Coca-Cola bottling plant and the Elks Lodge. The Main Street business district was constantly changing. By 1927, the year of the fire, many of the above listed businesses still existed but many had also changed hands or disappeared. Pittman P. Pritchett's Blocton Ice Company opened in 1927, providing locally made ice. The Town Hall was next to the Davie enterprises on a 1927 parking scheme map, and Mel Hayes' Hamburgers was probably next to West Blocton Savings Bank. At the time of Kindig's youth, other offices and stores were in Number One Blocton. The TCI Company Store was a large concern and at one time it had a branch in West Blocton on Main Street. There were other businesses near the depot and several stores in Little Italy and on Smith Hill. W. L. Ambrose had a store on Highway 5 and there were other businesses nearby. There was also Marvin Martin's Hill Creek Commissary. It is estimated there were at least eighty business enterprises and professional offices operating in the Blocton-West Blocton-Smith Hill-Little Italy area just after World War I.

J. W. Stewart announced in May 1919 that he was moving to Louisiana and was selling the Opera House to divide the estate among heirs. The property was advertised as "three store rooms on front, one large stock room in the rear. The upstairs is fitted complete for the show business and is now running pictures." The record does not say who purchased the building, if indeed it was sold. The following November, the Opera House still advertised its billings. At that time the Billy Senior Stock Company was offering a line of all new plays with high class vaudeville between acts. Adults paid fifty cents, children twenty-five, and ladies were admitted free on Monday evenings if accompanied by one fifty-cent paid ticket.

Shortly after this snapshot was taken, little Harry Adams' father was blackballed by the company for union activities, and the family was forced to leave Blocton. The company town, Number 1, is visible in the background. (Author's collection.)

A colorful anecdote about the Opera House was told by Elmo Ellis of Atlanta. At a point in the building's history during a full-house event, a local gentleman who believed himself to be the Messiah and who had the habit of informally appointing angels and apostles in random street meetings, had seated himself in the front row center of the balcony. At a moment in the evening's entertainment when a dramatic quiet had settled over a rapt audience, the demented fellow removed the stopper from a large flask of cologne and sprinkled it onto the audience below shouting, "I baptize you all in the name of the Father, Son and Holy Ghost!" Just how this played out is lost to history, but the surprised audience certainly got more than the admission price had promised. The episode has to be recognized as an efficient, creative, mass baptism.

In early August 1919, a brief, interesting statement appeared in the *Enterprise* indicating that the fall session of the West Blocton School would add a tenth grade. Enrollment was free to any child living in the north end of the county. This may have been a preliminary move by the local school board toward plans that were announced the following April concerning a fully accredited high school.

The spring of 1919 brought the threat of union labor and management trouble. The brief strike of February 1918 gave notice to mine operators that the work force was becoming restless because of the lack of a pay increase to correspond to inflated living costs that had occurred during the war years.

Though details are scarce, the 1919 Blocton strike scene was recorded in the pages of the *Blocton Enterprise*. The April 3 edition reported that a miners' meeting under the auspices of the Blocton local was to be held on the following Sunday. J. R. Kennemer, UMWA president of District 20, was the speaker. The gathering was at the pavilion and the speakers promoted the issue of "Six hours a day, five days a week." On October 23, the paper announced that the miners came out on Friday night and that "Everything is hot at Blocton." On November 13, it said the strike had been called off, and miners were congratulated on law-abiding loyalty.

The Harry Adams family was directly affected. In a typical union breaking tactic, the company presented Adams, president of the union local, with an affidavit signed by several "company men" and a non-English-reading, foreign-born worker charging that Adams had called an illegal strike during a time of national emergency. It was after the war had ended, but the peace treaty had not been signed. The foreign-born miner later confessed he was told the official-appearing paper was a formal commendation for Adams' good work and leadership. Following company policy, the TCI superintendent at Blocton, James F. Stewart, ordered Adams fired. This was an example of the "victimized-and-fired" union officer's fate so often used by company officials as a weapon against organized labor. Adams was given 48 hours to remove his family from company housing and his children were to be withdrawn from the company school on the same schedule. Non-compliance would lead to prosecution by the courts. Blackballed by the company as a UMW leader and member, he could never expect to be re-hired. The local paper gave no details on the matter. The personals column of January 22, 1920, said, "Mr. and Mrs. Harry Adams moved from Blocton to West Blocton the first of last week." Other union officers also experienced varying degrees of company punishment. Adams contemplated court action but was persuaded by family friends that his winning a case against United States Steel would be impossible.

At some point after the family move, likely during the 1920 strike, Adams, a member of the Methodist Church and a cornet player in its small orchestra, was walking to a Sunday service with his small son astride his shoulders when a shot was fired over their heads. The church was located in the Number 1 Blocton company town area where National guardsmen, aware of Adams' union affiliation, had been encamped during the walkout under the directive "to protect company property." The unsettled times were not only stressful for union men, they took their toll on local management as well. Superintendent Stewart died of a stroke in April 1921.

In early September 1920, the District 20 miners, including those in Blocton, came out on strike. By the end of the month a contingent of state militia arrived under order of Governor Kilby. The editor of the local paper said the move was a surprise to the law-abiding citizens of West Blocton and Blocton and further, "if the state wants to spend $3,000 a day to keep them—some of it may as well be spent here as anywhere." Shortly after their encampment, a squad of soldiers was reported drunk to Mayor Davie, who took the matter of punishment to their captain. The mayor wanted permission to try the soldiers in municipal court but a telephone call to General Steiner, state commander, confirmed that action would have to be under military courts martial. Later, a miners' meeting was threatened by the troops. However, the mayor and the town council intervened to grant permission for the assembly. Minor conflicts continued until the strike ended in February 1921.

Brown and Davis, editors of *It Is Union and Liberty* (1999), summed up the strike:

> In the end, the Union miners of Alabama went down in defeat, unable to prevail against the awesome power ranged against them. But they did not go quietly; and even in their defeat, miners of

> the Birmingham District left behind a stirring example of solidarity. Many vowed they would live to fight another day. Angry letters poured in to Governor Thomas Kilby's office in the aftermath of the strike. Wylam miners informed Kilby that they would "not be responsible for any conditions that that may arise from this unhappy situation. You have spent thousands of the people's money to protect the coal operators." But the sharpest threat came from a miner's wife at Blocton. (Mrs. Nora Bates) "This is our state," she warned Kilby. We will fight for freedom. There will be a revelution [*sic*] start right in Alabama [and] it won't be weeks about it either."

While the strikes of 1919–1921 were a major community concern, life went on as usual. Those interested in furthering their education had the opportunity on July 31 through August 2, 1920, by enrolling in the Radcliff Chautauqua, an entertainment and educational festival under a big tent. The traveling program was held to raise money for the School Improvement Committee, but it failed to show a profit, perhaps because of the strike and summertime heat. To support the West Blocton School, M. M. Odess, a local merchant, donated a set of historical pictures after displaying them in his store window. The paper editor commended Odess' public spirit. In April the local leadership decided that the business community could be enhanced if the state highway were routed through Blocton and West Blocton and a local committee pursued the project without success. Also in April, Thomas J. "Cotton Tom" Heflin spoke at the Number 1 Blocton School. Silver's Economy Store was offering army shoes, pants, raincoats, and ponchos, and an advertisement appeared for the Brennan Hotel (briefly used name of one of the local facilities) which promised reasonable rates.

In July 1920, the Jewish War Relief Committee, composed of M. C. Davie, M. M. Odess, Joe Israel, T. C. Tuggle, and S. B. Israel, collected $300 for the starving millions in central Europe. The Old Soldiers Reunion was held July 14–15 in Scottsville, and the paper editorialized for a local "ice factory" since the Birmingham supplier could not always deliver. According to a rare minority business advertisement in the *Blocton Enterprise*, housewives could call Phone # 34 for the delivery of fresh fish from C. M. Williams of West Blocton Heights on Tuesday and Friday each week. Another prominent contemporary African-American business operation was N. L. Wilson, Undertaker. His advertisement in the local paper read, "Have motor and horse drawn hearse. Can render service in almost any part of county, am agent for tombstones, fencing cemetery lots. Thanking you in advance for your patronage. Am located at (Smith Hill) Blocton, Ala. Phone 90 or 35." The prominent advertisement carried an illustration of a Ford Model-T hearse with up-to-date headlights, decorative side lanterns and whitewall tires.

In addition to the usual July old soldiers' gathering at Scottsville, the Confederate Veterans and their auxiliary organizations also sponsored an annual Fiddlers' Convention. In late March 1920, the event was proclaimed the largest ever with music in charge of Mrs. Jesse Walters and overall management by L. J. Hayes who was said to be "a master of the art." The featured speaker was Judge W. W. "Plain Bill" Brandon, and his "great speech won for him hearty applause." Brandon was the consummate politician traveling around the state proclaiming his perpetual candidacy. He was finally elected governor in 1923. A motion picture shown as part of the Fiddlers' Convention program was said to be one of the best ever at the Theatorium, and "the band playing during the show made it very effective."

In May, the West Blocton U. D. C. Chapter won first prize at the state convention in Decatur for having the best report on the year's work. The local bank, in flourishing condition, reported $444,252.42 in assets, still the largest bank in the county. In the fall municipal election, local attorney E. M. Bishop was elected mayor but resigned in August of 1921. W. H. Wright was appointed to replace him temporarily. The Episcopal and Presbyterian churches announced the installation of electric lights.

The nation had adopted two new national constitutional amendments by the fall of 1920, one on the prohibition of the sale of alcoholic beverages, a local issue that had been before the voters of Bibb County a decade earlier, and the other to allow women to vote. In the November election, the *Enterprise* listed Mrs. C. C. White as the first lady to cast her vote in the Blocton box and Miss Garvich as the second. In West Blocton, Ovelle and Olera Griffin were the first and second to cast their votes.

Business activity, in the doldrums during the strike, showed only modest growth afterward. J. H. Eubanks opened the new West Blocton Bakery in February 1921, and the Opera House was still in business. *Freckles*, the Broadway Amusement Company's big song play, was offered to the theater crowd on March 18 for 50c, 75c and $1 plus war tax. The Campbell-Baily-Hutchinson Circus train arrived at daybreak on March 31 for a glittering, mile-long street parade at noon and performances under a giant tent at Four Orders Park, at two and eight o'clock.

In mid-February 1921, the last Jewish immigrants to make their home in West Blocton arrived from Poland to be with Ben Silver and Mrs. Ed Pont, their relatives. The *Enterprise* editor said they had experienced many hardships and that the united family was very happy. The newly-arrived European relatives were Zavel Kaler, Mrs. Kaler, Claire (Clara), and Melvin (Morris) Kaler and Calmon Silver. In later years, Claire was asked about her European experiences but professed inability to remember. She and the family had immigrated through the port of Boston and did not experience the hardships of those entering the country through Ellis Island and other entry points. Her son, Sidney Mazer of Birmingham and Atlanta, believed she chose to blot out those unfortunate and difficult times from her memory. In 1998, she was in her mid-nineties, living in Florida.

The 1921 Easter program at the African-American Peace Baptist Church was reported by H. B. Banks. I. S. Dooley, superintendent of the Sunday School, and Mrs. Mamie E. Daniel, teacher, were identified. The titles Miss, Mrs., and Messers were used before participants' names, an unusual practice in reporting minority citizens' news at the time.

An Accredited High School for West Blocton

The news causing the most talk in the community came in early April with the visit of Sheriff Huey Wood and county school Superintendent Edward H. Jones of Centreville in an attempt to arouse enthusiasm for a high school in West Blocton. It was an easy task to drum up support, a case of preaching to the choir. The paper editor commented, "The people in this end of the county need a high school and it is hoped it will become a certainty in the near future." It had been eleven years since West Blocton had lost the intra-county fight for the location of the state supported county high school, but the desire for a high school was just as strong as it had been in 1908.

Professor A. J. Brown was appointed principal of West Blocton High School and first visited the community in July. He made an enthusiastic report on the school's plans to offer the same course of study as Bibb County High School. Graduates would be admitted to all state colleges without limitations or examinations. At the time there were only eleven grades in both schools in the county, as in most state schools. To offset the cost of the new program, the tenth and eleventh grade students paid $45 tuition; junior high students paid $18; and elementary enrollees paid $9.

The accredited West Blocton High School opened on Tuesday morning, September 7, 1920, with its first senior class enrolled and with a state-approved curriculum. Rev. Roy Hewlett, the Methodist preacher, opened the first day with an invocation and appropriate statements followed by other local leaders, businessman T. C. Tuggle, Mayor M. C. Davie, and county Superintendent Jones.

Principal Brown had written glowing reports in the local paper covering the plans for the school and had invited the community to the opening day program. The school's teachers included Alberta Chambers and Winnie Lou Laird, high school subjects; Lucille Harkins, fifth and sixth grades; Olene Ricketts, third and fourth grades; and Eloise H. Wright, first and second grades. Brown also had teaching duties.

The school year ended on a high note when the State High School Inspector found West Blocton High School measured up to state standards. Accreditation was reconfirmed allowing graduates to enter any state college without examinations. Principal Brown predicted a new handsome brick building to replace the present one, a prophesy that remained unfulfilled for four decades. The first West Blocton High School graduating class members, the class of 1921, were Miss Willard Keith, T. C. Tuggle, and Nat Frazier. A larger class of fourteen graduated in 1922. It is interesting to note that Latin was a feature of the school's curriculum. Historian of the class of 1922, Ester Light, mentioned the Latin teacher in her notes about her senior year.

In December 1920, the first local varsity athletic contests, basketball games, were held between West Blocton High

Blocton's Girls' Basketball Team. (Courtesy of Bessie Robinson Collection.)

School and Bibb County High School. Games were played on the dirt court located behind the schoolhouse. The girls' basketball game ended in a tie. However, the boys' game was not concluded because of a difference in opinion between the umpire and referee concerning a foul. The local editor reported that the first half ended with Blocton leading 4 to 3. Eleven fouls had been called on the local boys and five against Bibb County. The story added, "After the game goodwill prevailed and the West Blocton School served refreshments to the teams and other visitors, displaying their usual hospitality to the visiting team. All had a jolly good time."

The following school year was announced by a bold face advertisement in the *Enterprise*. "ATTEND THE WEST BLOCTON HIGH SCHOOL. BROAD ATHLETIC SPORTS. ALL REQUIREMENTS OF AN A-ONE SCHOOL." It was also noted that the school building had been renovated inside and out. The major improvement was a metal roof that remained on the structure until it was razed in the 1940s. The 1921–22 session opened with an enrollment of 290. In October, the first reported club was announced with the organization of the Tennyson Literary Society. In obvious approval of the way school development was going, the community voted overwhelmingly for a school tax. Sensing possible enrollment competition, the Bibb County High School advertisement for the 1922–23 school year in the *Blocton Enterprise* proclaimed it was "the only high school in the county accredited by the Southern Association and the state." West Blocton High School was state-accredited but not a member of the Southern Association.

In June 1921, an editorial by *Enterprise* editor C. E. Allen decried the departure of many long-time citizens because of unsettled work conditions. "It is much to be regretted that so many miners, most of whom are old-timers and good citizens of the district—in fact, some of them were born and raised here, have married and raised families of their own in Blocton or Bibb County—are leaving the state to hunt employment elsewhere." Worker migration was a statewide

West Blocton's first honorary club, the Tennyson Society, 1922. (Courtesy of Bessie Robinson Collection.)

phenomenon in the coal mining districts and continued to be a trend as miners moved from Alabama after each major strike to better opportunities in the nation's other coal fields.

In February 1922, a plan to build a $10,000 Masonic Hall was announced. The cornerstone was laid in June during a "Big Mason Jubilee." The celebration began with a parade by a brass band that started at the Knights of Pythias Hall, proceeded to the new building's site next to Pope's Drug Store and ended at Four Orders Park for a picnic. When completed, the hall housed the post office and the Guarantee Cash Grocery Company on its ground floor. The grocery opened in November 1922 and the post office at the end of December. W. W. Perry, the postmaster, announced new equipment that was modern in every way. Some citizens complained that the location was not central enough to meet all needs.

In June 1922, the *Enterprise* recorded the mine operation situation:

> We are authoritatively informed that operations at the Red Eagle mine will begin as soon as necessary repairs can be made on the tipple. We also learned the Red Feather Company will resume operations at full capacity next Monday. The mines of Braehead Coal Company, Jones Brothers and Moffatt Brothers, all of them on the M & O branch line, are running on full time. Hill Creek mine has had on a full force of men for some time, while the Klondike, Tyco, Evans and No. 4 mines are doing good business. Nos. 9 and 10 are the only mines of the T. C. I. Co. running and they are on short time.

In early June, the opening of the old TCI Number 7 shaft was announced. William Evans, the promoter, called it the Alabama-Blocton Coal Mining Company. This operation was not mentioned in later newspaper stories, so it must have been a short lived operation.

Swindal House. The house where Col. J. B. Swindal was raised is now the home of Mamie Foster Kuhns. (Courtesy of Cahaba Trace Commission.)

West Blocton's Colonel James B. Swindal

The *Enterprise* frequently recorded birthday parties in its local news page columns. In February 1922 it wrote that Master J. B. Swindal had entertained twenty-five little guests at a birthday party hosted by his mother, Mrs. S. Y. Swindal and Mrs. J. M. Shaw. J. B.'s mother died later that year in a return of the flu virus that had paralyzed the community in 1918–1919. His father, a local building contractor, moved the family to Lowreytown to live with grandparents. Shortly thereafter, the family moved to Birmingham. Early in World War II, J. B. joined the army and became an Air Force pilot. He gained a place in the country's history on November 22, 1963, as the commander of Air Force One when he flew President John F. Kennedy into Dallas where the President was assassinated by gunman Lee Harvey Oswald. On the tragic return trip, with Colonel Swindal at the plane's controls, Vice President Lyndon B. Johnson took the presidential oath of office as the plane taxied out of Love Field to take the widowed Jacquelyn Kennedy, her husband's flag draped coffin, and the new president and first lady back to Washington. He remained on the presidential pilot assignment during the Johnson administration. Col. J. B. Swindal and his wife reside in Cocoa Beach, Florida. Mamie Foster Kuhns, his first cousin, lives in the house on Main Street, West End, West Blocton that was Swindal's birthplace and childhood home.

In 1998 Colonel Swindal wrote his recollection of the events in Dallas on November 22, 1963:

> Before we left Washington November 21st several of President Kennedy's staff advised him to skip Dallas because it was a hot bed of far-right radicals and Adlai Stevenson our U.N. ambassador had been roughed up there the month before. But Pres. Kennedy felt that he was president of all the U. S. and should not be afraid to visit any U. S. City. He also overruled suggestions that the bubble top be installed on his convertible automobile. As the trip moved through San Antonio, Houston, and Fort Worth, the President received such a warm welcome from the huge crowds everyone began to relax and enjoy the trip to Texas.
>
> When we arrived at Love Field in Dallas I remained with Air Force One to help get ready for our short flight to Austin later that afternoon. I was listening to the Secret Service radio following the

progress of the motorcade when I heard the president had been hurt, at first I hoped this was a mistake. Just minutes later I heard that part of his skull had been shot away by what sounded like a high powered rifle as they passed the Book Depository on Elm Street. I felt as though I had been hit on top of the head by a ten pound sledge hammer. I just did not believe that something like this could happen in our country—at the same time I realized that we would be leaving Dallas right away and we only had fuel aboard on A. F. One for the short flight to Austin. So we had to get busy refueling for the flight to Washington and also to arrange for more security around AF I.

When we were told that Pres. Kennedy's casket would be returned to Washington aboard, A. F. One, we had to remove some passenger seats to make room for it, certainly did not want to put it in the baggage compartment.

All of this kept the crew very busy so we did not have time to think about what had happened until we arrived back in Washington.

On May 19, 1998, the news media announced the retirement of the Air Force One plane Number 26000 that was involved in this history-making event. It was taken out of service and flown to Wright-Patterson Air Force Base, Ohio to be put on display in an Air Force museum. Swindal also was the pilot assigned to fly President Kennedy to Germany where he made the famous "*Ich bin ine Berliner*" speech during the Cold War era. He reported that the president asked him to give Chancellor Konrad Adenauer a walking tour of Air Force One. The NBC *Today Show* featured the plane in a story on May 20, 1998, and included a brief interview with Colonel Swindal.

In early July 1922, the *Enterprise* reported that Governor Thomas E. Kilby had registered at the Douthit House while visiting the Belle Ellen Convict Prison. He was accompanied by state mine inspectors. The Old Soldiers annual reunion was held in Scottsville on July 11–12 with "Plain Bill" Brandon, a candidate for governor, as the featured speaker, and the Elks held their annual picnic to celebrate the Fourth of July.

Col. J. B. Swindal. (Courtesy of Col. J. B. Swindal, Cocoa Beach, Florida.)

Perhaps encouraged by the success of fellow townsman Huey Wood who had won a county-wide election for sheriff, Dr. George W. Randall announced his candidacy in March 1922 for state senator representing Bibb and Perry Counties. He laid out his platform in a two-column paid advertisement in the Blocton *Enterprise* of March 30. In essence, he was against special privileges and for the uplifting of the common man. He promoted new ways to fund education in the form of royalties paid on natural resources, and he believed that the state administration must get down to "brass tacks." Randall proclaimed "duty" and "enthusiasm" to be the most beautiful words in the English language. He said he had long ago, when a student at Marion Military Institute, adopted this motto: "That we recognize no title to Nobility but Merit, and no claim to greatness but the greatness of accomplishment." The voters of the early 1920s appeared to understand what the doctor was saying. The north end of the county gave him an overwhelming vote and with family connections in Perry County, he won in a close race with planter W. H. Cooper of Centreville. His district plurality was 320 votes. Bibb County gave him the nod by less than seventy-five votes, the south Bibb beats heavily favoring Cooper.

Another interesting West Blocton candidate in the election was Mary Curd Allen, a local businesswoman (Allen Furniture and Undertakers Company), who was a candidate for the state House of Representatives. According to the local paper, she was the first woman from Bibb to run for state office and the second in the state. Mrs. Allen received the least number of votes in her bid for office. The following September, she ran for the town council, again receiving the least number of votes. West Blocton and Bibb County evidently were not ready for a woman in elective public office.

Striking Miners. Throughout the history of Blocton strikes, miners would often be forced to leave the company housing and set up tent villages nearby on private lands. (Courtesy of Martha Fancher Kelley, West Blocton, Alabama.)

Editor Allen wrote in August:

> The coal miners' strike has been settled and the men have gone back to work. When they went out the main complaint was that the operators would not "sign up," that is, make a contract with the union. This point was gained and arrangements made whereby further contracts will be made, as well as some arbitration machinery agreed looking to permanent peace. On the whole it looks like the strikers came out very well, but they paid dearly for what they gained.

For all practical purposes the union had lost. The 1921 strike had ended in failure. TCI and other companies published the names of UMW miners so they could be blacklisted and denied employment. Miners deserted the UMW en masse, and membership dropped from 25 percent in 1917 to 1 percent in 1927. The 1920–21 strike in Blocton was to no avail. It gained nothing for the miners.

For years, Wright's Hall in West Blocton had been the meeting place of various groups. It had the historical distinction of being the "Court House" for the Blocton Division of the County Court. The upstairs accommodated the meeting room and downstairs housed a grocery store and, briefly, an African-American managed dry cleaners. In October 1923, because of age and structural problems, it was torn down. Once, when it creaked and groaned with a full house, court had to be dismissed and moved to the Opera House. A small metal-covered law office was erected on a corner of the lot and occupied by attorneys W. H. and Charles Wright. It was one of the last "metal sided" buildings to be removed from Main Street in the 1990s. Another similar structure was removed in the 1980s at the eastern end of town to make

Ludlow's place. Located where McKinney's is today, the Ludlow place was a landmark on old Highway 5. (Courtesy of Julia McKinney, West Blocton, Alabama.)

room for expansion of Bell Telephone equipment housing. This small building had been the Alabama Power Company office prior to 1927 and later was occupied by a grist mill. For a short time it was the African-American facility for R. F. D. Funeral Home, and in the 1940s, it housed the Black Diamond Coal Company cafe and later served as the office and work room for the local chapter of the Red Cross.

The fall of 1922 was the high point in recovery from the months of the strike. All mines were back on regular working schedule, and the West Blocton Savings Bank reflected the better times. Its October statement showed assets of $533,365.59, the largest in its history. It was still able to boast it was the largest bank in Bibb. The paper carried an announcement of the operation of the West Blocton-Bessemer Bus Line. Passengers could leave Blocton at 7:30 a.m., arrive in Bessemer at 9:00; leave Bessemer at 3:00, and arrive back in Blocton at 5:00 p.m. Fare to Woodstock was 50c and to Bessemer $1.25. The bus ran seven days a week and was managed by Chester Black.

The area long had supported its entertainment houses and local musical ensembles. In September, McCall's

Main Street, West Blocton before the 1927 fire. Jean Israel Farrar in foreground. (Courtesy of Jean Israel Farrar, West Blocton, Alabama.)

Comedians played for a week at the Opera House. In November, a group of young men practiced for a local band led by Andy Wyper. The Blocton Home Band, the old, reliable musical group begun in 1913, had lost its company support during the long strike period, and this new band attempted to fill the void. The West Blocton Band was active as late as April 1923 when a concert on Main Street was reported. However, the band appears to have been short-lived because many of its members moved away. In November, the Theatorium was the site of a demonstration of a "loud sounding Radiophone" and in December, the first Ford automobiles were unloaded at the depot for sale by the Universal Motor Company in West Blocton, managed by A. V. Thompson.

Early in 1923, school principal A. J. Brown introduced the idea of a school to be identified as North Bibb County High School. His ambition was to obtain a respectable, modern brick high school building. It appeared to Brown that the word "county" needed to be part of the school title to get full county support. The proposal fizzled.

The year brought a variety of activities and events. On February 3, a "disastrous blizzard and ice storm" which brought down telephone, telegraph, and electric lines and closed the mines hit the community. The February 22 paper reported from St. Francis Catholic Church that "at the Lenten devotions appropriate hymns will be rendered in English and Italian as well as Latin . . . there will be orchestra rehearsal every Friday evening after the devotions." Continuing with

I & N Quality Store advertisement. (Courtesy of Alabama Historic Ironworks Commission.)

more business activity, Mel Hayes rented the old post office building to set up an up-to-date refreshment parlor. The town put plans in the works for hitching posts and watering troughs on Main Street to cater to the farmers' trade. The community hosted the state convention of the Children of the Confederacy at the new Masonic Hall in April, the first state convention of this sort to be held in West Blocton. On June 28, the paper carried an advertisement for Brown Frederick's Chevrolet Sales and Service. The 1923 Old Soldiers Picnic and Reunion was announced for July 25–26. Because the facilities used by the reunion at Scottsville had fallen into disrepair including the pavilion and a bridge across the creek, the group met at Red Eagle. Unfortunately, during the celebration the dance floor fell, injuring several; one seriously enough to be taken to the TCI Hospital at Fairfield. Drs. Tucker and Peacock attended at least five of the injured. Senator "Cotton Tom" Heflin had been a featured speaker at a daytime session but was not present to witness the dance floor collapse. In September 1923, the I&N Quality Store opened a branch in Centreville. Tuggle and Harvey replaced their old Theatorium building with a "fireproof structure." At the Opera House in November, Jimmie DeForest presented a troupe for a week's run, the Cosmopolitan Players featuring Miss Jyles Larue (one of the last vaudeville entertainments at this facility). Too, Will Fuller and A. McKinney opened a new filling station on Main Street across from the Knights of Pythias Hall.

Earlier in the year, the *Enterprise* did not have much to say about the re-born Ku Klux Klan that flourished in the 1920s. The editor wrote in March 1923, "A representative of the KKK spoke at the Theatorium Sunday afternoon. We didn't learn what good he did as we failed to attend." That was as close as he could get to a statement of condemnation. The political atmosphere of the 1920s found many of those who aspired to elective office bound to the Klan. In Birmingham all the Klans together enrolled an estimated 18,000 of 32,000 registered voters by 1924. It is likely the same proportion of Bibb County voters was involved with the white robed knights. The Blocton-West Blocton KKK has left no record. In the Masonic Hall there are still a series of niches in a hall closet that are numbered. The "number one" compartment was the storage place of the white robe of the highest-ranking member. Some recall that the main focus of the local Klan was to harass the Italian community over alleged illegal liquor and gambling. It is also said they would threaten men throughout the whole community not properly supporting their families. Other stories relate that the organization was known to interrupt Protestant church services, especially during night revival meetings, by marching in single file down one aisle and out the other leaving an offering at the altar. It is said that many males in the membership who were regular church-goers would be absent and thus probably associated with the Klan. Others were easily identified by a special shoe style or size. One oft-told story allegedly involved prominent citizen W. R. Ray, who at the time was mine superintendent at Hill Creek Mining Company, and his pre-school son, Billy. Ray often wore leather leggings. During a Main Street demonstration by the local Klan, an observant Billy is said to have yelled, "that's my Daddy, see his leggings!" Harry Adams who was six feet, six inches tall was told by an Italian friend after one of the Klan "marches" to Smith Hill and Little Italy, "I believe you were one of the big devils in front, nobody in town is as tall as you are." In July 1924, the local paper reported that the KKK presented a floral cross at the close of funeral services for 32–year-old Ben McCool, who had accidentally drowned in the Cahaba River. The only further remark about the KKK in the local paper appeared in July 1925: "A large delegation from this place went to Centreville Tuesday night to attend a big parade of Ku Klux and a barbecue. Several thousand were present." According to tradition, the First Baptist Church, built in 1925, had among its stained glass windows sponsored by local lodges, one contributed by the KKK, carrying its seal. That window is no longer in the building. Perhaps it was one of several windows removed as the result of a number of renovations.

The window's existence was supported by a Texas resident who grew up in West Blocton, a son of a prominent family. Telling of his boyhood memories of the 1920s, he said, "I recall a plump, frog-faced man who was Kleagle of the local Klan. One of my earliest memories of him is when he came to the dedication of the new Baptist Church [1925], dressed in a white sheet and tall conical hat, with his frog face and glasses in between. He came in an open Ford Model T, leading his followers to the dedication of the new church to which the Klan had contributed enough so one of the dedicatory stained glass windows noted their generosity."

The national Klan movement of the 1920s faded toward the end of the decade because of public indignation over some of its practices, bold newspaper opposition to the organization, and corruption within its ranks.

The West Blocton High School opened its session for 1924-25 under the state's new 6–3–3 plan that added the 12th grade. The previous year had seen added enrollment and over-crowding so two new rooms were built on the north side of the 1903 structure. This annex, which did not match the other part of the school, appeared to be a stopgap arrangement, and there was still talk of a brick building. The local editor put the best face possible on the construction. "The school was too crowded last year, and with these two new rooms added the congested condition will be relieved." It would have been an opportune time to add indoor rest rooms but the old outdoor privies across bridged Glover Branch on a rise fifty yards to the rear of the school remained in use until the building was abandoned in the mid-1930s.

Playing at Tuggle's Theatorium in September 1924 was the film, *Bella Donna.* Its cast included the star Poli Negri supported by Conway Tearle, Conrad Nagel, and "sweet" Lois Wilson. According to editor Allen, "at one time Lois Wilson was a little girl of our town living with her parents here, a brown-eyed, curly-haired girl who attended the West Blocton School." Lois Wilson has the distinction of being the first "Miss Alabama" chosen at a Birmingham pageant in 1918. A graduate of Livingston, she did not like her first teaching assignment and opted for Hollywood after her pageant success.

Early Boy Scout Troop, West Blocton, *circa* 1925 (Left to right) J. D. Clements, Charles Eubanks, Otto Miller, Harry J. Adams and Herman Eubanks, scoutmaster. (Author's collection.)

She and her parents were residents of Birmingham at the time of the prize-winning event, but at some point in his career, her father was associated with the Blocton District coal mining industry. It has not been determined when Lois Wilson and her parents were town residents, but it was after 1906 when C. E. Allen started the *Blocton Enterprise.* Lois Wilson acted in many moving pictures, both silent and sound, and she also had a career on the stage. In the mid-1930s she had the role of Shirley Temple's mother in *Bright Eyes* the film that featured the popular song, "On the Good Ship Lollipop." She also appeared in a Ronald Reagan movie in the 1940s that a latter day critic labeled, "forgettable."

The pages of the local paper recorded interesting developments in the community during this period. The Methodists added a new room to their Number 1 Blocton Church in October 1924. The new temple-style Baptist Church, a magnificent brick building with imposing white columns, had its opening service and dedication in April 1925. The Allen Furniture and Undertaking Company built new quarters in August and advertised a contest for a wedding planned to be held in its big front show window. A lucky couple was to be chosen from applicants for the occasion for a 9 p.m. event on grand opening day. The couple was to be awarded a handsome piece of new furniture, as well as their marriage license and a preacher to perform the nuptials. There is no evidence that the wedding was held. Perhaps the town's engaged couples were not ready for such a daring, public display. The building, erected in 1925, is the last of the old Main Street structures. In subsequent years it housed the Yellow Front, a chain general store, and later housed grocery stores, was used as a warehouse and is now a thrift store. The Allen furniture concern also opened

West Blocton First Baptist Church. (Courtesy of Cahaba Trace Commission.)

a store in Centreville. The new Gillespie Furniture Store was completed in July 1925. Not to be outdone by fellow merchants, S. B. Israel remodeled his I&N Quality Store, adding modern, lighted show windows across the entire front. This store for many years had the distinction of having the only elevator in the county, moving from basement to main floor to second floor, hand operated with ropes and pulleys. The biggest business news of the year came in the fall. Pitman P. Pritchet announced that he would build an ice factory for West Blocton. No longer would his customers be dependent on ice shipped from Birmingham. However, the "ice factory" was not in operation during the summer of 1926, and the *Enterprise* editor commented he had heard nothing of it recently, after learning that an ice-making facility would soon be in operation in Centreville. At last, the Blocton Ice Company was built in 1927. It operated into the 1940s. In 1997, the 70–year-old, dilapidated structure was demolished to rid Main Street of an eyesore. The town acquired the lot for a proposed public library building.

Other business changes of 1925 found Oscar Light moving his store and family to Birmingham. Jim Shannon, hardware store operator, also moved his family to Birmingham, and J. H. Eubanks, the bakery operator, and his son Charles, moved to Andalusia. Joe Israel moved his family and store to Florida in July. His business there was ruined in the 1926 hurricane, and he returned to open at his old location in West Blocton only to be burned out in 1927. World War I veteran Pat Murphree, pharmacist for Pope's Drug Store, bought out Brown's Drug Store at Harvey's Corner. The Standard Bottling Works was renamed the West Blocton Coca-Cola Bottling Company in May. The West Blocton based Black Brothers Bus Line acquired a 21–passenger Reo Sedan Bus. In August, *The Flapper Grandmother*, advertised as "the peppiest, cutest play ever given here" was scheduled for the Theatorium. The traveling company had 50 people, all in colorful costumes. In November the Theatorium booked Arthur Callahan's Comedians for a three day stand. The company had fifteen musicians and actors with special scenery and costumes. Their production was *The Little Mother*. Admission was 25c and 50c. The silent film spectacular, *The Ten Commandments*, Cecil B. DeMille's "cinemasterpiece," was shown at the Theatorium, August 24-25, 1925. The Theatorium's competition, the Opera House, had closed by this time but was occasionally used by civic group productions and school events.

In December 1925, Dr. L. E. Peacock and Dr. Clyde White Jones announced the opening of their joint office practice. They were to remain the town's doctors until the 1940s. Dr. Peacock had been in the community since 1897.

Blocton Methodist Men's Sunday School Class, 1926. (Author's collection.)

Jones first studied pharmacy at Auburn having entered there at age sixteen and graduating in 1907. He was awarded a medical school scholarship and subsequently studied at the University of Alabama Medical College in Mobile. He graduated in 1912, opened practice in West Blocton and later was appointed the state prison doctor at Belle Ellen. When Jones began his partnership with Dr. Peacock, they were under contract with TCI and had private practices as well. Dr. Jones' untimely death at age 54 in 1944 was a great loss to the community. He served all his patients regardless of their station, often during Depression years without pay.

Bibb County Sheriff Huey Wood opted not to run for a second term in 1925. He had been offered the position of chief detective in Miami, Florida, and moved his family there in August. Wood later became the City of Miami Beach's police chief.

Local businesses had to be encouraged by the September 3, 1925, item in the *Enterprise* stating, "Mines of Hill Creek and TCI are now working six days a week."

The high school basketball program met with considerable success in the 1925 season. Their record put them in the regional finals at Montgomery where they won games against Ramer High and Bibb County High only to lose to Sidney Lanier in the finals. The newspaper said of the team, "they were little but loud."

Merchant M. C. Davie won the mayor's race in the fall of 1924 over Attorney W. H. Wright. By the spring of 1925, he and the council had offered bonds to cover the cost of extending water lines. In an effort to add further civic improvement, the mayor and council passed an ordinance covering the paving of Main Street and part of School Street. A town meeting was called for the Elks Hall in late March to put the plans before the public for a vote. The paper editorialized for the project saying, "Everything is in shape so the streets of West Blocton can be paved, and the property owners will have ten years to pay for it. There is nothing that speaks better for a town than to see nice paved streets. They also add to the value of property." This argument was not strong enough to sway property owners, however. The proposal was voted down "because of poor industrial outlook." The conservative merchant leadership used an old argument, continuing the perception of the town's impermanence. Periods of labor unrest had forged this position of caution about long-term town or personal indebtedness. There also may have been rumors about projected plans being made in the boardrooms of Pittsburgh and Birmingham concerning the future of TCI in Blocton. The reluctance to act on substantial improvements in the public school naturally carried over into the proposed street development sector.

Ten Commandments handbill, 1925. (Author's collection.)

Though the community might send its children to school in a sub-standard, poorly equipped facility, it took great interest in the development of the institution's athletic program. The success at the state level of a poorly equipped basketball team brought a feeling of pride to the town. In the summer of 1925 plans were made to inaugurate football. (Centreville's Bibb County High School began the sport as early as 1915.) Organization and practice after school started in September. The practice field was the old Four Orders Park, renamed the TCI Park. Incidentally, Bill Carroll, then a small boy watching the practices, reported that players from Belle Ellen and Piper-Coleanor rode the train to school each day. Those commuting players at practice would scramble for the depot a quarter mile away when the daily passenger train blew for Ashworth's Crossing two miles up the railroad. The first game was a practice session with local "Old Timers," won by the high school neophyte gridsters 19–0. The first players were recognized as green at the sport (some had never seen a football game) but under the leadership of Coach Emmett Durette, the three-game season was launched against McAdory in early November. The result was a 0–0 tie. West Blocton fans, numbering approximately 250, traveled by automobile caravan to the out-of-town game. Those were the days long before "team busses" so special

Dr. Clyde White Jones. (Courtesy of Jewel Jones Dailey, Aiken, South Carolina.)

registration of volunteer's cars was requested to transport the team. The second game was scheduled at Verbena and even though the team lost 32 to 7, the first report of a WBHS football contest in the local paper read as though the young eleven had won. The headline said, "Boys Play Real Football." The item reported that the men of the town were backing the boys financially. Money was solicited to pay for the new Hill Creek Ball Park. The story ended with, "Our boys are learning fast, look out Bibb County High, we have your number." The first game with Bibb was played at the new West Blocton park in November, the first high school football contest in the town, and only the third engaged in by the WBHS team. The "Little Tigers" did well, holding a 6–0 lead until the last few minutes, when Bibb County scored a touchdown and extra point to win.

School colors in the early sports activities were gold and white. The girls' basketball team was called the Golden Tigerettes. In February 1926, the boys' basketball team again made the trip to the district tournament in Montgomery where they won over Cloverdale 36–28, and Verbena 39–21, and lost again in the finals to well-coached and much larger Sidney Lanier, 43–30.

Even with business activity improving and growing excitement about the high school's new athletic successes in football and basketball, there was still reluctance on the part of the town's leadership to support its school properly. In April 1926, the School Board issued a plea in the local paper for funds that were urgently needed to pay the teachers their last month's salaries. It read, "Payment of tuition has been very slow for the last half of this term, and it has become necessary that the School Board take drastic steps in order to finance the remainder of the term." Those who had not paid their fees were to be prevented from taking exams. Twelfth graders would not be allowed to graduate nor would there be promotions for those not having paid. Those quitting school without paying up would not be allowed to re-enter the following fall until all payments due had been paid in full. At this time, some of the leading families in the community still opted to send their children to the Number 1 Blocton TCI Company School where a better environment was perceived. The company school had a teacher's home, high standards for their teaching staff, adequate supplies, indoor plumbing, and special extra-curricular programs for its students. As late as December 1927 in an *Enterprise* article promoting a school bond issue, principal Balch described the local public school.

> Here in West Blocton we have approximately 400 children who come to us seeking admission into our local school. The building from a standpoint of size is totally inadequate to take care of these children. The building is unsanitary, poorly ventilated, and poorly lighted. In fact it is a mere firetrap, which endangers all who occupy it. The mildest thing we can say about the West Blocton school building is that it is totally inadequate and unfit for our purposes as an educational institution.

J. T. Balch had been principal since the fall of 1926. He had hired an assistant principal and coach, Pelham J. Merrill, a bright young man with aspirations to be an attorney. Merrill's "Little Tigers" played a more ambitious schedule of six games. The season looked especially good to the team's fans after an opening win over McAdory, 30–0. The rest of the season included a 25–0 win over Dallas County; a 6–0 loss to Marion Institute; and an 18–0 win over Tuscaloosa County. A second game with Dallas County was won 13–0. At the season-ending big game with Bibb, played in November in Centreville, the "Little Tigers" won 13–6. Two pass plays, one by Fred Wood to David Pruett and another by Pruett to

Blocton Basketball Team, 1920s. (Courtesy of Howard Prather.)

Harris Warren, accounted for the Tigers' touchdowns. Coach Merrill told the story of that game many times after leaving West Blocton. He said he dreamed of a pass play the night before and used it for the winning touchdown.

The win over Centreville caused considerable celebration. The West Blocton Hunting Club gave the team an oyster supper, and Dr. and Mrs. Randall honored the team with a "course dinner" at their home. Coach Merrill was given a gold watch chain by the team at a special assembly at the school. Merrill later wrote a letter in the local paper addressed to the community stating his appreciation for their support. In a manner giving evidence of his eventual long tenure as a successful politician, he gave credit to the community for the team's success because of their generous backing. He wrote, "Never have I come in contact with a more loyal and likable group as these boys in our school. They fought hard and never gave up, and deserved the honors bestowed upon them. May they fight in life as hard as they did in WBHS. Everyone on or concerned with the team owes the people of this town a debt that can only be repaid by giving back to the town the best we have in service." The basketball season in early 1927 was another successful one with Merrill having the coaching responsibility. The team and its coaches were bold and confident enough to schedule Tuscaloosa and Bessemer High Schools. The West Blocton Tigers lost both by 2 and 3 point margins.

Coach Merrill was to remain in the community for five years. Then he returned to law school at the University of Alabama, eventually practicing law and finishing his legal career as an Alabama State Supreme Court Judge. However, he continued an interest in the community for the rest of his life because in 1936 he married a local girl, Gladys Morrison. His last visit was in 1986 for the 50th reunion of the WBHS class of 1936, some of whom he had taught in junior high.

From time to time the editor of the paper felt the need to boost the morale of the community. In the February 10, 1927, issue he wrote:

TCI Instructors. In 1925, company teachers included Evelyn Scott, Frances Ruth Johnson, Marion Brown, and Birdie Whiteside. (Courtesy of Alabama Historic Ironworks Commission.)

TCI Spring Festival. In the 1923 company pageant, Blocton girls traveled to Muscoda dressed as Norwegian dancers. (Courtesy of Alabama Historic Ironworks Commission.)

> Despite the fact that a lot of calamity howlers claim that Blocton and West Blocton have seen their best days, we are glad to note the fact that new houses are being built, and old ones being remodeled and repainted, a new brick ice factory is being erected, and there are no empty houses. The mines are working every day and new business houses are being opened up. We are surely growing and the best is yet to come.

On February 21 and 22, local movie-goers had the opportunity to see Rudolph Valentino in what the paper called his last performance, the film *Cobra*. In April, the Howard College Glee Club performed at the Theatorium, and later in the month the annual Decoration Day honored Confederate and World War I soldiers. During the early part of May, the TCI school children went to Fairfield to compete in the annual Spring Festival and Pageant sponsored by the company for their schools. At the same time, the community was excited over a local athlete, David Pruett, being invited to Birmingham to join the Baron's pitching staff. He was to be coached by the renowned Dixie Walker. Pruett eventually had a contract with the St. Louis Browns. His last residence of record was Tuscumbia.

In mid-June the Theatorium advertised one of the outstanding pictures of the year 1927, *Flesh and the Devil* with John Gilbert and Greta Garbo. The feature was scheduled for Saturday and Monday since it was considered unseemly to show films on Sunday, and the "blue laws" of the era did not permit the Sabbath to be corrupted with moving pictures.

The city council took action in early June to alleviate traffic problems on narrow Main Street by passing an ordinance governing parking. There was to be parking only on one side of the street in alternating blocks in the congested major business locations. A diagram was drawn for the newspaper showing the specific restricted spots. On some blocks parking was allowed only on the street's north side and others only on the south.

In mid-May, the town acquired a new undertaking business. The RFD Funeral Home boasted of having a new big Studebaker combination ambulance and funeral coach. Their new facility site was being readied for construction just back of Tulon's Place service station. The proprietors were Roland, Fuller, and Desmond. The I&N Quality Store advertised a summer clearance sale to begin in early July.

Announced for a week's run on Monday evening July 11, 1927, was Donegan's Show, a big tent entertainment that was to open opposite Tulon's Place, but what Donegan and his troupe were to witness accidentally on Tuesday morning probably surpassed anything they had experienced on any of their previous stops. At Tuggle's Theatorium on Monday night, the last film to be shown in that house was *The Show* starring John Gilbert, one of the giants of the silent screen. *Sandy* scheduled for Tuesday would not be seen. The weekend of July 9–10 passed with the usual church services and family gatherings but nothing special to talk about except the anticipated tent show downtown beginning Monday night. The Monday Birmingham paper predicted temperatures in the low nineties with a southwesterly breeze for Tuesday. But on that Tuesday morning, things were destined to get much hotter for West Blocton, as the town faced one of the county's worst disasters.

CHAPTER 6

The 1927 Town Fire

The West Blocton fire alarm sounded shortly before 10:30 on Tuesday morning, July 12, 1927. Flames had suddenly erupted at the Pope Brothers Cleaning and Pressing Shop when an open vat of cleaning fluid had been ignited accidentally by a spark from a malfunctioning electric motor. Three hours later, most of Main Street and houses immediately north of the street lay in ashes. The toll was forty stores, several professional offices, and twenty residences.

In a 1996 interview at his nursing home just a month before his one-hundredth birthday, Macon Pope, one of the owners of the shop where the fire originated, said: "I have thought many times that had I grabbed a sack of flour from Ben Stanley's grocery store next door, I could have smothered the burning tub of cleaning fluid. I was just too upset to think clearly."

There had been several major fires and a number of near-disasters during the town's half-century history. One of the earliest recorded fires in West Blocton was the warehouse fire at the rear of the Cottingham and Suttle store on Main Street.

At the time of the fire, Blocton and West Blocton were less than a half-century old. Much of the Main Street business section had been built in boom-town haste in the mid-1880s with little attention to fire safety. For years local leaders had voiced their concern about the danger created by the congested area. A February 1894 issue of the *Blocton Courier* carried an ominous warning from Thomas McCumber, its prescient editor:

> The fact that Blocton is entirely without fire protection is the cause of frequent anxious comment among the people. If fire should break out in any of the more densely populated or business portions of the town during a dry, windy day or night, there is absolutely no telling where the destruction would end. Some plan, if possible, should be devised to guard against this constant danger, but the *Courier,* like everyone else, is unable to suggest anything practical, at least until the dawn of the long deferred "better times." One thing can and should be done, however, by everyone, that is, to keep barrels of water and buckets standing conveniently about the premises. A timely bucket of water has often saved thousands of dollars worth of property; and this much protection, at least, is within the reach of all at a very trifling cost, either in money or labor.

There had been several major fires and a number of near-disasters during the town's half-century history. One of the earliest recorded fires in West Blocton was the warehouse fire at the rear of the Cottingham and Suttle store on Main Street. The uninsured loss of $1,200 worth of feed was reported in the *Centreville Enterprise* in February 1890. This blaze may have been remembered by the town council in later years when ordinances governing feed storage space were conceived.

A major fire in May 1900, discovered at a pre-dawn hour, destroyed the Louisville and Nashville Depot and four loaded freight cars. Said to have caused $20,000 worth of damage, it was a compounded disaster since fourteen cases of dynamite were stored in the Tennessee Company's portion of the building. When the explosives detonated, wakening the whole terrified village, it left a hole several feet deep below the room where the dynamite was stored.

Pope Brothers Dry Cleaners. (left to right) Macon, Harvey, and Frank Pope. (Courtesy of Wanda Pope Miller.)

Two decades before the West Blocton fire, on March 17, 1907, the nearby community of Smith Hill was all but destroyed by a fire that started in a restaurant. Ten businesses including the Smith Hill Dispensary that carried an $8,000 liquor stock, and thirty residences burned to the ground. The total property loss was estimated at $125,000. This fire was one of Bibb County's major disasters, however, it has gone unnoticed, perhaps because there was no loss of life. Smith Hill had no water system nor fire fighting equipment so the TCI Company's mine superintendent Deloney and nearby mine employees tried in vain to fight the fire by forming a bucket brigade taking water from several wells. In desperation, dynamite was also employed in an effort to halt the flames. The destructive fire and the futile attempt to stop it prompted the *Blocton Enterprise* editor to editorialize in favor of improved fire fighting equipment. He recorded his thoughts in the paper's March 27, 1907, edition:

> The lesson of the destructive Smith Hill conflagration is that disasters often become great only through lack of simple means to cope with them. When first discovered this fire could have been put out with a garden hose, unfortunately not to be had. Those having in charge the apparatus of the West Blocton fire department should see to it that everything is in readiness in case of fire breaking out here. Inspection should be made at stated intervals of fire plugs, hose and tools, as the slightest delay on account of damaged or mislaid fire fighting material may sometimes entail disastrous consequences.

Over the next decade, a series of town ordinances was passed having to do with merchants' storage of combustible feed for cattle and horses. Separate buildings a safe distance to the rear of the stores was a requirement. Livery stables were not allowed on Main Street but even in their removed locations were to have special fireproof storage areas for hay and fodder. Fireproof metal roofs and iron siding were mandated for business concerns on Main Street. To emphasize the reason for the laws, the street was identified as "the fire district."

Main Street, West Blocton, *circa* 1903. (Courtesy of Carey Scurlock, West Blocton, Alabama.)

The town had experienced a number of near-disasters along its central business street. In September 1906, the West Blocton jail located but a short distance from Main Street mysteriously burned. A Negro prisoner, his life endangered, was rescued by the forced removal of iron window bars. The fire could not be put out because of a faulty fire plug. The *Blocton Enterprise* editor wrote that if the fire had begun elsewhere than the jail the whole town would have gone up in flames. He added a sarcastic comment: "If the water plugs are to be used simply as ornaments, they should be nicely painted and have flowers planted around them; if they are to be used in case of fire, they should be kept in order and working shape."

Borum's Jewelry Store burned in March 1908 threatening an entire block. The local paper's report suggests that in this instance there was an adequate fire fighting capability. "A steady stream of water was laid on from all sides which gradually had its effect. The effort went on for a half-hour until the flames were controlled." Readers of the weekly were reminded of the Smith Hill fire of the previous year. (Incidentally, this had been a case of arson. Borum was arrested for setting the fire.)

In the spring of 1910, there was another close call at the mid-town Douthit House. On March 16, while the family and guests were at breakfast, Maude Douthit heard the crackling sound of burning shingles from a blaze making headway on the roof. Several hotel residents checked the fire with buckets of water until volunteer firemen could arrive. Though the dining room and an adjoining room were badly damaged, the paper editor said that the timely appearance of the West Blocton Fire Department may have saved the town from a serious loss because there was a strong breeze blowing.

That same year there was another fire scare. The town had experienced an unusually late spring cold wave at the start of the last week of April. A three-inch wet snow covered blooming roses and early gardens on Sunday night and into Monday morning. The cold weather had descended suddenly on the Saturday before and many had found it necessary to fire up their stoves to counter the chill. At Wright's Hall on Main Street someone in the African-American cleaning plant downstairs had started a heater fire with paper sending sparks to the roof and the sharp wind caused leaves there to catch fire. The *Enterprise* editor commented, "If it had not been for the prompt response of the local fire department, the building and a good part of the town would have been burned, as the wind was blowing a strong gale and nothing could have saved it, had the fire gotten a good start. If our property owners would put fire-proof roofing on their houses it would prove a great safeguard to their own property and their neighbor's property."

After a fire destroyed a home and damaged another in the northern section of town in November 1913, the *Enterprise* editor wrote: "Because a fireplug was out of order, the hose line had to be extended an extra distance to fight the blaze. Those having the care of the affairs of the city in hand should have fair warning. The city officials must examine all water plugs once a week and have defective ones repaired at once." In what proved to be a prophetic observation, he added: "West Blocton is so closely built up, and there are so many frame houses, that should a fire get headway in the main part of town and water fail to work, as it did the other day, there would be a repetition of the Smith Hill disaster."

In March 1914, Mr. Krenshaw, the town night watchman, discovered the Tuggle and Harvey's Theatorium on fire. The handsome new stage curtain was burned and water damaged the piano, chairs, and other articles. It was fortunate, the local paper editor said, that other buildings were not burned and lives threatened. This was not the first fire scare at the Theatorium, the county's first moving picture theater that had opened in 1907. The projector was run by a gasoline powered generator. Once, in the early days of the entertainment house, Tuggle had doused a small, yet dangerous, fire in the motor with only minor burns to his hands. The newspaper report said it could have been much worse.

Ordinance #135, passed by the city council in November 1914, required erection of fireproof buildings in the fire district. It stated: "Any new building hereafter within the fireproof building district in said town of West Blocton shall be built of bricks, stone or cement and the roof covered with tin, iron roofing, rubber roofing or some other recognized fireproof covering." In July 1915, Ordinance #140 permitted the town marshal, acting as inspector, to ask owners of unsafe buildings to remedy the problems or remove them (the buildings). In December 1916, the city council passed Ordinance #154 prohibiting storage of hay, fodder or other combustible feed in business houses on Main Street. But though the council continued to put laws on the books, it did not take action to update fire-fighting equipment.

In an unwise move in March 1917, local merchants voted thirty to seven in favor of repealing the building code and allow erection of wooden buildings. However, the council at city hall did not act on the vote so the protective ordinance

Buildings from south of Main Street, 1906. (Courtesy of West Blocton Library.)

stood. In June 1918, the coming of automobiles to town in significant numbers prompted the council to enact ordinance #166 preventing garages or repair shops in the fire district. Obviously, the town council believed such businesses to be fire hazards. This action may have resulted from a fire at Herron's Garage the previous May that the newspaper reported was out of control until the "fire laddies" arrived just in time to put it out. (The *Enterprise* editor frequently called the volunteer firemen "the fire laddies.") In December 1918, Ordinance #171 provided for a fire tax, presumably to support the updating of fire fighting equipment. The following February the tax was protested as "discriminatory and illegal." The result of a vote on the matter is not recorded, and the tax was probably voided. A potentially dangerous fire in mid-town occurred in August 1920 when the C. C. White residence burned. The Ben L. Edmonds house had been threatened and sparks caught fire to the roofs of the Manring and Douthit House hotels. Again and again, the old Main Street structures were at risk but concrete action to remedy the danger was not taken.

A harsh reminder that fire was an ever-present threat to the community occurred in August 1922. The Baptist Church in Number 1 Town was completely lost to flames. A fire that started at the residence of W. W. Logan spread to the C. C. Huckabee residence and then to the church and pastorium. The lack of adequate fire fighting equipment again proved that a major fire could not be contained. Five years later the inevitable and long feared horror came to West Blocton.

Tuesday had dawned hot and dry as is typical of July days, but that morning a steady breeze from the southwest gave some hope for relief from the sun's heat. The alarm on that awful day heralded the conflagration predicted by the local newspaper editors. It was the disaster that the city fathers had tried to prevent by legislation. It is unlikely the *Blocton Courier* editor's suggested barrels of water and buckets nor Macon Pope's afterthought sack of flour would have been an effective "first aid" deterrent in stopping the resulting destruction.

Main Street's buildings, many in place since 1884, closely resembled those seen in old western movies. Half-story high wide board facades carried the stores' names in bold, broad brush lettering. Most had wooden or tin awning-like shelters over entranceways. To add to the combustible material along the street, some of the business area's sidewalks were made of planks. Most of the older one-and two-story buildings were constructed of heart pine.

Not all the structures on Main Street were wooden. The West Blocton Savings Bank and the two-story Murphree Drug Store were brick structures. The Odess Department Store, the I&N Quality Store, and the Masonic Lodge were also made of brick as was the Davie store on the south side of the street to the west. Two of the larger wooden buildings were hotels, the Douthit House, dating from the mid-1880s, and the twenty-room Manring Hotel, built in the same era. The old wooden frame constructions, though complying to a degree with the various fire prevention ordinances, were but giant tinderboxes. The frame buildings' timbers had four decades of drying and curing time. One of the largest buildings was the unoccupied, three-story Opera House. Thin metal siding over pine boards to prevent the spread of fire had little effect when flames from neighboring structures threatened. Another hazard was the number of storehouses with common walls. When one business caught fire, the next door structure was doomed. Many roofs had been water-proofed with tar, completely disregarding the ordinance requiring metal or other fireproof construction.

Despite all the warnings and passing of fire proofed building ordinances, as well as a short-lived collection of a fire tax, the town did not own a pump truck. Firefighters still depended on pressure from two wooden water tanks situated on the highest elevation in the community, with gravity providing the force for the water flow. Thus when several fireplugs were tapped, the result was equally diminished streams of water.

The *Blocton Enterprise* had reported that a fire in December at a local church was put out promptly by volunteer fire fighters and added that soon the town would be better prepared to fight fires by the arrival of a new fire truck. In early February the paper proudly reported the following:

> The new fire truck for the Town of West Blocton arrived last Thursday afternoon, and is fully equipped with hooks and ladders, hose and everything necessary. It has a loud sounding gong and a siren whistle. It is painted red and has "West Blocton Fire Department" printed on the side and is quite an innovation. The Mayor and Council are to be congratulated on their selection of so complete an

Wooden Water Tanks. (Courtesy of West Blocton City Hall.)

> equipment. We understanding they are installing a fire siren fire alarm in the central office [telephone exchange], so when an alarm is turned in it will be sounded from there.

Editor Allen failed to say the truck was not pump equipped. It could only transport hoses, nozzles, chemicals, and other equipment to fires; the volunteers still depended on gravity for water pressure. Usually, only one stream of water could be used.

Lewis Deerman, William "Billy" Smith, and Jerry Clements all remember the "fire truck" owned by the town. They recall it being a customized T-Model Ford vehicle with ladders, fire hose, and other modest equipment on its bed. Clements confirms that the truck was lost to the town fire flames. Dave Israel, a teen-age member of the volunteer fire department in 1927 and a resident of West Blocton until 1991, frequently told that the T-Model Ford truck could not be "cranked" on the morning of the fire and was lost to the flames. The destruction of the town's "fire truck" in the fire did not make the newspaper accounts of the disaster.

The explosion at the Pope cleaning plant scattered flammable cleaning fluid that immediately ignited turning the modest wooden structure into an inferno. Next door, Stanley's Grocery was quickly engulfed and then the Gillespie Furniture Store which had undergone renovation and enlargement only two years earlier. Furniture stores make hot fires and this one lived up to expectations. Within an hour of its start, the out-of-control fire was approaching the empty Opera House, an old structure made of rosin-laden timbers. In desperation, the volunteer firemen called on the management of Hill Creek Mining Company to send explosives and workmen to blow up the doomed building hoping to lay it into the approaching fire. Dynamite was hastily set and detonated, but the ploy backfired. The opera house, its west wall ablaze, fell across Main Street. Burning timbers were blown onto adjacent storehouses spreading the flames farther to the east. Once across the street, the fire was like a demon with a mind of its own. The blaze backtracked to the west and destroyed

The Roycroft building survived the fire. (Courtesy of James M. Roycroft, Jr., Birmingham, Alabama.)

the J. B. Davie Department Store and damaged the adjacent funeral establishment. The Davie building was brick but had a vulnerable roof. Next to it, the Allen Furniture and Undertakers with a tin roof, corrugated iron siding and a brick facade, somehow survived, along with its contents, though its windows were broken in the intense heat.

Less than an hour after the cleaning plant explosion, the situation appeared dire enough for Mayor M. C. Davie and fire chief Sam Israel to telephone a plea for help to Bessemer and Birmingham Fire Departments. The obliging city neighbors dispatched fire fighting units immediately. Unfortunately, the help from Jefferson County was more than two hours away over crooked and unpaved roads.

Main Street, a narrow road today, was narrower before the fire. The cross streets and alleyways were even closer. A steady breeze from the southwest helped create a firestorm in the closely built, canyon-like street. By early afternoon the fire had run its destructive course through businesses, professional offices, and dwellings. The north side of the street had burned completely from the point of the fire's start some two and a half blocks to the west. On the south side of the street the west wall of the brick Masonic Lodge, where the professional fire fighters with modern equipment from Birmingham and Bessemer laid on streams of water, served to stop the blaze. Its downstairs, which housed the post office and a grocery store, was heavily damaged by water and heat.

There are conflicting stories about the effectiveness of the fire equipment from Bessemer and Birmingham. In a *Birmingham News* article in July 1992, Dave Israel declared that the West Blocton fire hydrants were not standardized. They had "wildcat" threads that would not fit the city firefighters' hoses. Other reports said the equipment was useful but had to make numerous trips to nearby Caffee Creek for refills of their on-board reservoirs. A newspaper photograph of a blaze at the rear of the Masonic Temple shows no fire-fighting equipment in use. Help may have arrived after the photo was taken. Other recollections by those who experienced the day's events say the Birmingham and Bessemer units were very helpful in preventing more residence fires north of Main Street.

With the failure of the dynamite charge, it was obvious the whole business section to the east was doomed. Several residences closely built on small lots on Catlin Street, a parallel narrow roadway just north of Main Street, had caught fire quickly. The first flames from the explosion swept through a six-room, frame house, vacant only since early June. One after the other, the entire row of houses went up in smoke. Twenty houses were destroyed as the wind-swept blaze moved eastward. At the end of the row of small houses was the two-story Randall home and when it burned, the fire jumped Florida Avenue to consume the Crowe, Turner, and Brown residences. Virginia Crowe had help from family and friends who removed her household furnishings to safety on a nearby hillside. For years following, her upholstered dining room chairs carried telltale scorches made by flying sparks from her home, one of the last of the dwelling fires.

The West Blocton fire was one of those traumatic life experiences that many remembered the rest of their days. What they were doing and what they did that dreadful day became etched in their memories. Some of those personal stories have survived.

Dr. George W. Randall was the town dentist and in 1927 was serving a term as state senator from Bibb and Perry Counties. His teenage daughter, Martha White, and four-year-old son, Henry Clay, were sprawled on the floor of their living room thumbing through the pages of the newly arrived fall edition of the Sears, Roebuck and Company's catalogue. The catalogue and the two-story Randall home were consumed by flames less than two hours later.

Youngsters Steve Bodnar and B. B. Desmond, fast neighborhood friends, were playing in the Bodnar's yard when the fire started. Steve remembered that the first house to be enveloped by flames from the cleaning plant explosion was empty, having been vacated by the Harry Adams family only six weeks earlier. Bodner's house was barely a hundred yards up a steep hill from the row of burning businesses and residences. One of the largest and most menacing blazes had occurred

Martha White and Henry C. Randall. (Courtesy of George H. Young.)

when the three-story Opera House burned. It was the structure blown up by dynamite in an attempt to stop the fire's progress. The blast of heat from this eruption was overwhelming. On the side of the Bodnar house that faced the fire, made of heart pine clapboards, hot rosin cascaded down the surface. The same effect was experienced by neighbors whose homes were facing the fire. A half-dozen endangered residences close to the blaze were saved by alert and fast-working homeowners and volunteers dousing sparks and airborne burning debris.

Dorothy Adams, part-time sales clerk at the I&N Quality Store, had turned eighteen on the day before the fire. At mid-morning that Tuesday, she and Frank Israel were cleaning the store's light fixtures. There had been only a few early customers, so store owner Sam Israel had decided it was put-up-stock and clean-the-fixtures time. Israel was the town's volunteer fire chief. When the alarm sounded he hastily made his way to the firehouse to assume fire-fighting duties. At the point the flames appeared unstoppable, he left the fire scene to return to the store. Dazed and exhausted, he found his other children, nine-year old Elmo and teenage daughter Libby Jean, had arrived to help transport goods in the store's delivery truck to a safe location. But Israel determined that removal could not be accomplished without considerable damage to the merchandise. The Hart, Schafner and Marx suits, Florsheim shoes, and Stetson hats that his store carried would be rendered worthless by even the slightest damage. In addition, possibly he hoped the brick building could withstand the fire. He removed the cash register's contents, placed the store records in the safe, tucked a box of his favorite cigars under his arm, sent his children and clerks home, and locked the store. As she left the doomed building, Dorothy remembered looking longingly at a rack of silk hosiery near the front exit, thinking of how badly she needed new stockings.

Samuel B. Israel. (Courtesy of Elmo Israel Ellis.)

The heavily tar-coated roof of Israel's building could not withstand the fire, and the I &N Quality Store met the same fate as its neighboring business places. Like many of his fellow merchants, Israel was only partially insured and the road back to solvency was not easy. Despite this, the I&N was one of the first businesses in the town to be rebuilt, reopening in time for the 1927 Christmas season.

High school coach Pelham J. Merrill had living quarters in the Douthit House, one of the town's two hotel-boarding houses. When the alarm sounded, he and several of his football players were at the high school checking on equipment for the coming school year's athletic program. They rushed to Main Street and were quickly joined by other team members to help fight the fire. When it became apparent that flames would eventually reach the Douthit House, the crew retreated to Merrill's room to remove his belongings. The coach was custodian of the team's donated funds. He had placed recently contributed currency for temporary safe-keeping in his Bible. He handed the volume to one of the boys as he instructed others what items to save. In the excitement of the moment, the Bible was placed aside while its guardian gave assistance in removing a heavy parcel. The Bible, with the team's fan-contributed funds folded between its pages, was consumed in the flames.

In his reminiscences of his boyhood in West Blocton, Jerry Clements of Georgia related that the fire alarm that day rang very loudly and was unusually long lasting. He remembered the sound as "unmistakably catastrophic." Clements recalled that the town's fire alarm system was the steel rim of a locomotive drive wheel about five feet in diameter, suspended on a platform atop a building in the center of town. A large metal clapper affixed adjacent to the makeshift gong was rung by pulling an attached rope. The *Blocton Enterprise* reported that this alarm device had been replaced by a siren a few months before the fire. Plans announced in February in the local paper said the alarm was to be operated from the telephone exchange console. Perhaps Clements heard the new alarm but his description of the earlier device is interesting. A mid-town converted storehouse was the town's fire station. It housed reels of hoses on caissons, ladders, and several chemical fire extinguishers. These items made up the basic inventory of the town's fire-fighting equipment. Clements remembered the volunteer department's fire truck as identified earlier, and that it was a victim of the fire

Clements's mother had a small grocery store with living quarters attached adjacent to the Smith-Kuffner Hardware Company. As the fire approached, she and her sons worked as fast as possible to remove its stock a safe distance to the store's rear. It soon was clear the spot was not secure enough but before further removal could be accomplished, the hardware store's supply of ammunition began to explode in the flames. Bullets whizzed in the Clements' direction, and the family fled to safety abandoning their goods to the advancing flames. They had, he remembers, the clothes on their backs, one quilt and about $150 cash at the tragic day's end.

In 1927, Jack Lawrence was a nine-year old lad, living with his parents about two blocks west of downtown. Upon hearing of the seriousness of the spreading fire, he corralled his pet goose, Howard, into a safe spot on the Lawrences' back porch. The bird had been a gift from a family friend, Howard Cooper, a Centreville planter. Jack remembered waving to his cousin George Young as he and other workers from the Hill Creek Mining Company sped by the Lawrence home in a pickup truck. He later learned they were taking dynamite to town to use in an attempt to halt the fire's progress.

Carl Oglesby, a clerk at the J. B. Davie Department Store, had stayed at the business until flames crossed to the south side of the street, working as long as he could to save the store's merchandise and to assist in fighting the fire. When it became hopeless, he raced to his sister's residence two blocks north of downtown, pleading with her to take her infirm husband and all they could carry to the home of black family friends, Mary and Henry Caldwell, a safer distance away from the fire. Carl's hair was singed and his sooty and sweat-streaked face was blistered. He was still wearing his clerk's attire, a black bow tie and a white, though now scorched and tattered, shirt. He told his sister that he believed that "the whole damn town was going to burn." Carl said he had seen huge rats scampering up the hill to escape the flames. He pled with his family and neighbors to put more distance between themselves and the threatening fire, however, they chose to remain. Fortunately, the fire did not reach over the hill.

Lewis Deerman was twelve years old. His father and grandfather managed a mule lot south of Main Street near the Coca-Cola Bottling Works, with hay stored in an adjacent barn. Even though he was not an experienced driver, Lewis was told to go to their home in West End to get the family's T-Model Ford truck to transport the hay to a safe spot. Driving the truck, he took an open route north of the fire and down Florida Street. About halfway down the hill to Main Street, with billowing yellow and black smoke to his right and overhead, he encountered a fire hose across the road. He made the decision to drive over it. Fortunately, it did not burst. After loading and transporting the hay to a spot near the Baptist Church, young Lewis remained with the truck and its cargo to snuff out threatening sparks.

Mrs. Frank Morrison, a Southern Bell Telephone operator working in place of her vacationing sister, Annie Keyes, stayed at her exchange console on the second floor of Murphree's Drug Store as long as she could. An account in the *Enterprise* of July 21 indicated the burning roof of the building was falling in when she finally left her post. Special emergency calls to Bessemer and Birmingham fire departments had been placed through her office by Mayor Davie and Fire Chief Israel just before fire reached the long distance telephone lines along Main Street. At the end of the day another operator, Goldie Thrasher, managed one hastily installed long-distance line in temporary quarters set up at Tulon's Place service station at the eastern end of the devastated Main Street. The fire had finally stopped at the Masonic Hall, a two-story brick structure some one-hundred yards west of Thrasher's temporary quarters. She remembered reporters from major state daily newspapers taking turns phoning their stories in for late afternoon and next morning's editions. After sunset that day, Goldie recalled seeing the eerie glow of red-hot bricks and smoking embers as she looked westward down the street of rubble and ruin. Because of her brave act, a reward fund for Mrs. Morrison was initiated by Benjamin Sherrod of the Blocton Cahaba Coal Company of Coleanor.

W. H. "Pinkie" Wright, Jr. remembered that he had been visiting cousins in Atlanta for a few days. With Birmingham

Ruins of the Odess Department Store and the I&N Quality Store. The two-story Masonic Temple that survived the fire is visible to the rear. (Courtesy of Jane Greene Collins.)

relatives, he had begun the drive home early that Tuesday over old unpaved U. S. Highway 78. Ironically, in mid-morning about half way to Birmingham, they had to stop their automobile because of a summer cloudburst. The rain came down in torrents and the road was so obscured it was impossible to drive. After a lengthy wait for the storm to abate, they resumed the trip and arrived in Birmingham too late for Wright to take the afternoon passenger train home. He learned of the fire through the *Birmingham News* late edition. He tried again and again to phone his parents at their Main Street residence from his cousin's home but because of downed telephone lines in West Blocton, he was unsuccessful. He caught the early train on Wednesday not knowing what to expect at the end of his trip. At each train stop en route to Blocton, he remembered, more and more passengers came aboard—sightseers going to West Blocton to view the ruins. By the time the train pulled into the depot, there was aisle-standing room only. As the train arrived, he was elated to see his father waiting on the platform. The Wrights and their home, just west of where the fire had started, had been spared. Unfortunately, there had been no downpour in West Blocton on that Tuesday.

A Birmingham newspaper wrote lengthy stories about the fire in its Wednesday morning edition. The following are two paragraphs that reflect the essence of the reports:

> The residents of West Blocton face the task of rebuilding. The smoking ruins greeted the bewildered populace, who seemed unable to realize the magnitude of the fire, other than it was disastrous, the loss being estimated at more than $500,000.
>
> As the fire moved swiftly up the street, frantic shopkeepers carried their stocks into the yards back of the stores only to see them burned to cinders there. Some voiced no plans to rebuild. Without insurance they had suddenly found themselves in financial ruin. More than a score of families were left

homeless and the affected business section was left a conglomerate mass of charred debris as the aftermath of the disastrous fire.

The *Birmingham News* staff correspondent, W. F. Fry, with a 1:30 deadline and the inferno nearing the Masonic Temple and Post Office, phoned in a report stating the temple had also been consumed, adding there was no indication at that hour when the fire would be brought under control. As a matter of record, the *Enterprise* reported on August 4 that Ira Tidwell and Huey McCool had stayed and worked to save the Masonic Temple from being burned after all had left, their efforts perhaps being the difference in its escaping destruction.

The *Blocton Enterprise* plant had been a victim of the fire but managed to publish the next Thursday by using the printing presses of Birmingham's Western Newspaper Union. The *Enterprise* editor, C. E. Allen, voiced doubts about the paper's future, indicating he had no insurance. The paper did manage to continue for a few more years but ceased local publication in the early days of the Depression. The paper recorded a list of the other burned businesses: West Blocton Savings Bank, Pope's Cleaners, (where the blaze began) the Manring and Douthit Hotels, Gillespie Furniture Company, Murphree's Drug Store, Pope Drug Store, J. B. Davie Department Store, Royal Supply Company (owned by the Silver family), Worthington's Grocery, I. M. Faulk General Store, City Shoe Shop, Smith and Kuffner Hardware, Dunnabie Grocery (the family residence next door was also lost), Smith Undertakers, Keith Barber Shop, Joe Israel Dry Goods, I&N Quality Store, Walters and Gamble Millinery, Joe Klotzman's Dry Goods, Tovich Fruit Stand, Triangle Dry Goods (the former Odess store), Dora Lewis Millinery, the office of Drs. Peacock and Jones, Dr. G. W. Randall's dental office, Stanley Grocery, Green Shoe Shop, Kren Grocery, Charles Rice Market, West Blocton Bakery, Theatorium Theater, E. E. Forbes Piano Company, Posey's Jewelry, Elite Cafe, Ben L. Edmonds Dry Cleaning Plant, Albert Woods Billiard Parlor, Tom Quinn Barber, W. W. Perry Sewing Machine Office, Floyd Ambrose Ford car shop and filling station. Not listed in the fire story was the Clements' store, and there may have been other small operations that were not included in the newspaper listing. Among several unoccupied buildings that went up in flames was the old Opera House, dating from the late 1880s, formerly the town's vaudeville theater, briefly showing silent films and serving as the local concert, recital,

Blocton Masonic Temple. Scribbled on the back of the original snapshot was the terse reminder, "All that was left - Masonic Temple." (Courtesy of Elizabeth Jennett Brannon.)

Ruins of the Randall home. (Courtesy of George H. Young.)

and town assembly hall and place for amateur play productions for almost four decades. It has been said that everything except "opera" was performed there.

The *Enterprise* editor wrote: "Let us hope that out of this misfortune all will feel more closely bound by ties of brotherly love and sympathy. Let cooperation be our motto, and let us rebuild our town on a firm foundation that will reflect credit on all."

The paper's July 14 edition further stated: "The devastated section of West Blocton will be replaced by buildings more substantial than those burned. Already, day and night, forces are working at clearing away the debris and making ready to begin the actual work of rebuilding." The paper also apologized for reprinting a Birmingham paper report saying lack of sufficient water, worn out hoses and out-of-date chemicals had contributed to the fire disaster. Mayor Davie and the council reminded the public that new hoses and chemicals had just been purchased the previous February. The town leadership believed that after the fire gained headway nothing could have saved the business section in its path.

The Birmingham *News* editorialized about the fire:

> Blocton will come through
>
> Tomorrow she will build again
>
> Blocton has been terribly hard hit. Her people have been stunned by the damaging fire that swept through the business section, exacting its costly toll.
>
> Thirty-five storehouses and a score of residences have been twisted, felled, erased from what was just now one of the most thriving towns in Alabama.
>
> Succor from Bessemer and Birmingham prevented worse havoc than occurred, but it arrived too late to halt the first high wave of leaping flames.
>
> A goodly portion of the bustling mining town lies in ashes.
>
> The only hope by Blocton sufferers and by neighbor towns that are ready to give material aid in this dark hour is that the unexpected blow may be largely neutralized by insurance coverage—that one point of light appearing when material disaster like this occurs. Birmingham stands ready-and this newspaper feels that it speaks the whole thought of this people-to do what may be done to temper this almost mortal blow.
>
> Birmingham knows the temper of this Alabama town-Blocton-as nearly as one neighbor knows another.
>
> We have grown up together.
>
> Blocton is a vital part of what we trippingly call the Birmingham District. It is a vital part of the throbbing unit we call the industrial district of the South.
>
> This newspaper knows that upon the ruins of these wrecked storehouses and residences will appear a larger and fairer city, more fit to carry on toward greater achievements.
>
> In this grave emergency, Blocton, though seemingly frustrated in growth, "will laugh at a fall, and baffled get up and begin again."

This fanciful editorial, sprinkled with archaic language, may have given the West Blocton populace a morale lift by predicting a comeback. The quoted line of poetry, incidentally, is by the English poet, Robert Browning. The poem is "Life in a Love" written in 1855.

Up From the Ashes

The Blocton *Enterprise* reported in its first issue after the fire, "The devastated section of West Blocton will be replaced by buildings more substantial than those burned." In the July 25, 1927, issue editor Allen wrote a noteworthy editorial:

> The fire was a knockout, but not a deathblow. Our people suffered great loss and a terrible shock, but it is characteristic of them to rally from any misfortune and they will rally from this, our greatest one. Somewhere the money will be found, each unfortunate loser will get it in his own way, and soon we will be carrying on again. We have the same resources that have only partially been developed. There is demand for our coal, and most of our people know how to get it from the earth. When West Blocton was named there was a poor piney woods wilderness here, and it was supposed by the scattered population to be too rough for even railroads to come here. By hard labor and much determination much money was wrested from the rough hills, great wealth in comparison to that [which] was destroyed by the recent heart rendering disaster. We now find ourselves much better equipped to force fortune from the [now] familiar hills than were the pioneers. Better buildings will be erected, and the building [of them] will furnish employment and a rallying start for those who have lost their all, and, best of all, there will be an even more kindly and friendly and neighborly feeling, more genuine human sympathy. Let us all clothe ourselves with determination and [get the job done].

The remainder of the summer of 1927 was rebuilding time. The town merchants quickly removed most fire rubble and ashes and began building a new Main Street business section. John Kren's store, and other buildings financed by Kren in the block just to the east of his store and Sam Israel's I&N, were begun immediately.

Starting in the late 1920s, a special Christmas season attraction at the Israel store was a top-of-the-line Lionel electric train displayed annually in the front, center show window. Continually circling the arrangement of toys and gifts in the window, it made its endless journey over an oval track along the inside margin of the window. The engine, trailing passenger cars, and Pullmans passed by a miniature village, ignored tiny but authentic signals, and ran through a papier-mache' tunnel with a painted-on landscape. Every child in town yearned for the not-for-sale marvelous toy train. (Since 1970, the train has been packed away in its original storage boxes in Elmo Ellis's attic in Atlanta.) Also during the holiday season, inside the store's entrance, stood a windup, clockwork Santa. The four-foot figure in his traditional red suit trimmed with cotton piping, wearing a long snowy-white beard, had an all-knowing expression accentuated by its constantly clicking, nodding head. The figure held a decorated tree in one hand and a bundle of switches in the other. No child had to ask the icon's meaning but was told by parents regardless.

Immediately after the fire as proof of a determination to continue, interim business locations were announced in the *Enterprise*. John Kren sold groceries and took phone orders for immediate truck delivery from his home near the water tanks on the hill. The I&N Quality Store had space in the Allen Furniture Store. Murphree's Drugs was located in the Elks Lodge, as was Ben Edmond's Dry Cleaners that set up operations in the basement. Gillespie Furniture was also on the main floor, and several other shops were in temporary quarters in the lodge resulting in its being dubbed the "West Blocton Arcade." The bank resumed business in the back room of the post office at the Masonic Hall, the lone surviving edifice in the fire's path. Maude Douthit relocated at the home of the A. W. Wynnes to offer meals to her former customers. The paper reported the Douthit lot was cleared of debris in August preparatory to rebuilding, however, today that lot remains empty. Smith and Kuffner Hardware opened in the building opposite Wright's Law Office. J. B. Davie took temporary quarters in the J. F. Jones Building "near the bridge." The Royal Supply Company opened temporary quarters in the Price Building. Dora Lewis had space in the Guarantee Grocery Store where she offered "an advanced line

Ruins of the drug store. (Courtesy of George H. Young.)

of new fall felts, hosiery and notions." Walters and Gamble, ladies' wear, located temporarily in the Allen Furniture Store. T. T. Dison advertised his new lumberyard behind Frederick Motor Company to supply builders with a full line. Keith's Barber Shop did tonsorial duties at the two-story Roycroft Building.

Some long-time business owners elected not to rebuild. Among those who soon departed was Mrs. Manring whose hotel was lost. Jerry Clements of Georgia, a pre-teen errand boy for Mrs. Manring at the time, said that after a short tenure in a private residence offering room and board to some of her clients, she moved her few remaining household effects to Birmingham.

The *Centreville Press* commented in its July 28 issue that "the citizens of West Blocton held an enthusiastic meeting last week and determined that West Blocton would be rebuilt. The fire in many ways will prove a blessing in the end. Brick buildings will take the place of the wooden structures that have always been a menace to the town. We glory in West Blocton's spunk and hope her come-back will be at an early date."

It was a remarkably upbeat time in the business community. Each week another rebuilding plan was announced. As the new buildings were completed, interesting descriptions made the newspaper pages. An early October story reported

Opening the ruined West Blocton Savings Bank vault. (Courtesy of George H. Young.)

that twenty new buildings had been started. "From one end of the street to the other, excavators, carpenters and masons are at work on the score or more of buildings [now] going up."

One of the projects was Tuggle's theater with 675 seats including a balcony for "colored" clientle that could be entered only by way of an external stairway. The theater was to have a fireproof projection room and modern heating and ventilating systems (that translated to a big electric paddle fan at an opening on the east alley and a huge coal stove near the left front entrance). Sound films, "talkies," had not been perfected in mid-1927 so the building was outfitted for "silents." Across the front of the building's interior, just under the screen and stage area, a state-of-the-art player piano was installed to play piano rolls produced to accompany the silent films. The player piano rolls were a part of the rented film package. A special mechanism within the piano console could reproduce sounds of all kinds such as train and steam boat whistles, sleigh bells, Chinese gongs, drums, galloping horses, and others to enhance the piano accompaniment. Most of the sound effects were activated by a series of hand-operated straps attached to tear-drop-shaped wooden pulls. Watching the player piano operator at work keeping the music and sounds in synch with the image on the screen was sometime more interesting than the film being shown. Because Tuggle had been unable to keep patrons from standing in the aisles

Theatre Reconstruction. (Courtesy of Harmon W. Frederick.)

in the old building, he had the new theater constructed so the public had to walk up the aisles. Upon entering the swinging doors from the small lobby, individuals found the screen at their backs, an unusual movie house configuration.

Other facilities nearing completion during the early fall were the I&N Quality Store, Gillespie Furniture Company, Davie's store, and Kren's building. Building had begun on Rice's Market, Quinn Barber Shop, the bakery, and Edmonds' Cleaning and Pressing Shop.

Drs. Peacock and Jones financed the building known today as the old drug store building. It was a handsome edifice with the Murphree Drug Store in the corner space and another shop to the left. The drug store was furnished by the Georgia Showcase Company and included a marble soda fountain counter and a mirrored console behind the counter. The furnishings stayed in place for more than fifty years but were eventually sold to several buyers by an absentee owner when the business closed in the 1970s. Upstairs were the offices of the doctors, the Southern Bell telephone exchange, and several rooms for hotel use—eleven rooms in all. The 54' x 50' building was equipped with central steam heat. Pat Murphree was in his new drug store by Christmas. On the northeast Florida Street corner, also being built and financed by the doctors, was a one-story building to house Keith's Barber Shop. Two business spaces to the rear would be the location of Wood's Pool Hall and a small eatery, once called the "Blue Goose Cafe."

Across the street on the east corner was the beginning of the new West Blocton Savings Bank. The colonnaded facade and trim of the 70' x 25' brick structure was constructed of Indiana buff limestone, and the floor was marble tiles. The Mosler walk-in safe was of the latest design, the door weighing four tons and equipped with clock locks. The public was served from four windows. The decorative, plastered ceiling was painted bright greenish-gold, and small decorative brass chandeliers provided illumination in the public area. The bank, like the drug store building, had steam heat. The new bank took several months to complete, opening in early April 1928.

Today the bank building, owned by an absentee landlord, is used for storage. Most recently it was used as a pharmacy. It also has served as the Hill Creek Mining Company office and, during World War II, as a government rationing and price control program office.

The local newspaper commented on other developments along Main Street. The widening of the thoroughfare from the point where the fire started to the area of Tulon's Place and the installation of sidewalks was more evidence, the editor wrote, of "our come-back."

The fire loss had been too much for the C. E. Allens, the *Blocton Enterprise* owners since 1906. They were among the town's best and most prominent boosters and citizens. Their farewell editorial said, "it wrenches our heart strings to tear loose from the many warm personal and business friends and attachments of all these years. But the late fire caught us without insurance, and to continue the business would involve us in financial responsibilities that we do not consider it prudent to assume under the circumstances." It was said that Mrs. Allen's personal touch and her persistence in finding news items were the real strength of the paper. The new owner was Cooper Blanton, once associated with the *Birmingham Age-Herald* and a commercial rabbit breeder. He wrote in his first issue, "it was a pleasure for him to announce we are among Blocton's citizenship."

The Blocton Methodists had a homecoming on the fourth Sunday in September 1927. Among the visitors were Mr. and Mrs. Rock Contri of Birmingham, former governor W. W. Brandon of Tuscaloosa, and Senator-elect Hugo LaFayette Black. While serving as senator, Black later was to sponsor various New Deal legislative measures including the Fair Labor Standards Act that endeared him to the Blocton District labor movement leaders. Appointed by President Franklin D. Roosevelt in 1937 to the United States Supreme Court, he became known for his learned interpretation of the nation's constitution. He served until September 1971, dying just a week after his retirement.

As new buildings were completed, businesses moved in and by mid-November many were reopened in permanent quarters. Joe Israel moved into his shop next to the Masonic Hall. Walters and Gamble left their temporary quarters to move into their new store, and I. M. Faulk moved into part of the rebuilt Davie Building. M. F. Pope's Drug Store and Keith's Barber Shop also were in new quarters, and Ben L. Edmonds Dry Cleaners, claiming to be "the most modern and

New Drug Store. (Courtesy of J. Turner Crowe, Birmingham, Alabama.)

Rebuilt West Blocton Savings Bank today. (Courtesy of Cahaba Trace Commission.)

Rebuilt Davie Store, 1928. (Author's collection.)

up-to date plant in the state," located just to the right of the new theater was almost complete. Attention to the matter of safety at the cleaners was apparent. A separate building housed the dry cleaning vats and the coal-fired furnace that powered the steam presses.

The town had been without moving pictures since the fire. The new theater was opened on Wednesday, November 16, 1927, with five showings of *Her Second Chance* starring Anna Q. Nilsson. The advertisement said the admission would be the same as before, 10c and 25c. It might be said the movie title was a metaphorical statement about West Blocton, which was being given "her second chance."

Soon after taking the helm of the *Enterprise*, editor-publisher Cooper Blanton announced the paper's program for West Blocton would be a new high school building, diversified agriculture, hard surfaced roads, extension of trade territory, an organized business club, and industrial development with special attention to mining enterprises. His program met with modest success even if only briefly. In January 1928, considerable column space was devoted to the announcement of the organization of the West Blocton Chamber of Commerce. Some forty-eight citizens, including

Frederick Motor Company. J. M. Duncan and Huey Frederick in front of car lot, February 18, 1930.
(Courtesy of Huell Lane Frederick Martin, West Blocton, Alabama.)

professionals, merchants, and clergymen were listed in the paper asking for support. The advertisement carried a number of promotional phrases: "West Blocton Shoves to the Fore," "Become a loyal citizen voluntarily!" "Neighbors are Neighborly in West Blocton," "Make and spend it in West Blocton, this is a great community and working together we can make it better." At the Chamber's organizational meeting on January 11, J. R. Young was elected president; R. M. Bearden vice president; and Cooper Blanton, secretary-treasurer. Blanton eventually gave up this position and S. B. Israel took his place. Over one hundred had assembled at the Masonic Lodge for the opening meeting and twenty-five were enrolled, a constitution was adopted, and committees chosen. Monthly meetings were scheduled for Tuesday nights. Initiation fee was $1, with dues of 25 cents a month. Blanton wrote enthusiastically about the organization:

> One of the first things the present editor of the *Enterprise* began agitating in West Blocton was an organized club to promote civic pride among its citizens and to furnish information to outsiders who might consider establishing firms or manufacturing plants in West Blocton. It is with a great deal of pride that I refer to the newly organized Chamber of Commerce. Now that we are organized, what do we do? The answer is spelled in four letters: W-O-R-K.

Another of his suggestions met with less local support. The relative newcomer commented in February, "Cowpens and Pavement: When we have Main Street all nicely paved and our final touches on almost a half-million dollars worth of new buildings, are we to continue the old custom of making a cow pen of our downtown section? I am not advocating a stock law. That is not the trend of this article. We want those people who own cows, pigs and horses to try to keep them off Main Street." Editor Blanton had some adjusting to do. The semi-rural, independent citizens of the town had not chosen to change all of their old habits. Most families were used to having a small barn or simple shed in their back yards for a milk cow, and Bossie was free to roam in the nearby woods or open spaces to graze. If Main Street was on the animal's foraging route, so be it.

Paving was scheduled to begin in March 1928 by the Phillips Company of Birmingham. By June the street surfacing was completed and reached from a half-block east of Tulon's Place to the bridge over Glover's Branch to the west.

A stock law may have been unpopular but fire prevention by legislation was still in style. In February 1928, the mayor approved and adopted action by the council to put Ordinance # 223 on the books. The ordinance read: "It shall be unlawful for any person, firm or corporation to build or repair the roof of any building within the Town of West Blocton, Ala. with wood shingles." A more concrete answer to fire prevention and control was soon to be in the works.

Reopenings continued in 1928. On March 8, Frederick Motor Company opened opposite the post office. Including the owner there were sixteen employees. The bank in new quarters reported assets of $382,092, still the largest in the county.

Causing some stir in county politics in 1928 was the run for the probate judge office by M. C. Davie. He lost to long-time holder of the office, W. L. Pratt. The Pratt judgeship era would continue for two more decades.

An end to another era in Blocton and West Blocton was at hand. Early in 1928 the nation's iron and steel industry had begun to react to a faltering demand in the national and world metals market. The decisions made in the board rooms in Birmingham and Pitsburgh reached dramatically into the Blocton District. Dire consequences followed.

CHAPTER 7

The Great Depression and Rebound

In spite of the rumors of several months that TCI would close its mine operations in Blocton, the positive business atmosphere after the fire held together long enough for merchants to replace three-fourths of the destroyed commercial district. Then, in mid-June a news release from TCI hit the streets like a gigantic sledgehammer. The company's last day in Blocton would be Saturday, June 30, 1928. The edge was taken off the blow with the announcement that a deal had been struck to continue the Number 10 mine under lease and management of the Blocton Red Ash Coal Company.

The *Enterprise* of June 21 printed the content of the statement by George C. Crawford, President of the Tennessee Coal, Iron and Railroad Company, Division of U. S. Steel:

> Birmingham, Alabama, June 11, 1928
>
> Due to the Blocton mine being detached from the other plants of the Tennessee Coal, Iron and Railroad Company and the excessive amount of freight that had to be paid for coal used from this mine, rather than from its mines adjacent to its plants in Ensley and Fairfield and on the tracks owned by the Tennessee Company, it was necessary that the Blocton properties be either sold, and the mines leased, or the properties abandoned and salvaged and the mines be allowed to fill up with water.
>
> Realizing that if these properties were abandoned and the mines filled up with water it would be a great loss to both the Tennessee Company and the citizens of Blocton and would throw a great many of the Tennessee Company employees out of employment and eliminate practically all the freight now being shipped over the railroads, which would curtail train service and inconvenience smaller operations, it was decided to sell this property to the Blocton Red Ash Coal Company and [TCI] has accepted a lower price than the value of the property rather than to abandon these properties and work a hardship on the entire community.
>
> Therefore, having made this sale which will take effect on July 1, 1928, the Tennessee Coal, Iron and Railroad Company commends to its old employees and the city of Blocton the new company known as the Blocton Red Ash Coal Company and its officers, W. R. Young, president and William Evans, general manager and wishes for the new company and its officials every success in their new undertaking. [Also associated with the new company was V. P. Terry, secretary and treasurer.]

Recently, Dave Evans, Jr. of Ashville, who was born in Number 1 Blocton, related that his grandfather, William Evans, lived in the old company superintendent's home during the life of the Red Ash Coal Company and that his father, Dave, Sr., managed the company store, once TCI's commissary.

The paper editor wrote that this deal saved Blocton and West Blocton, saying that it had been known for three years that TCI did not contemplate using the Blocton mines as a fuel source. It was also reported that enough contracts were in force to assure good work in the mines.

Welcome to West Blocton–Harvey Pope. (Courtesy of Wanda Pope Miller.)

The impact of the loss of the company that had been the economic core of the community since 1892 was not fully articulated in the local paper. The excellent company school and store were continued by the new enterprise but on a reduced scale. Company office workers, teachers, and many of the town's prominent citizens who were company personnel and provided leadership in churches, fraternal lodges, and other parts of the community, would depart. It was the beginning of an economic and civic decline that was to be further accentuated with the advent of the Great Depression heralded by the stock market crash of October 1929. These three blows in three years—a disastrous fire, the loss of TCI, and the Depression had an undeniably serious impact on the town.

The first anniversary of the fire was noted in the July 12 *Enterprise* which carried front page photographs labeling the views as "Here's Rebuilt West Blocton." Editor Blanton called on his connection with Birmingham papers to produce the photo plates, an unusual feature for local weeklies. The edition carried a brief town history and identified several of the prominent rebuilt businesses. The article claimed West Blocton was the county's most important town with one of the most modern business sections in the state. A special item reported on coal output at the new Red Ash Coal Company. On the first day of operation, 972 tons were pulled with some 4,000 tons produced in the first week.

The special issue carried a half-page advertisement promoting the West Blocton Chamber of Commerce. It proclaimed: "ALMOST A MILLION DOLLARS SPENT IN WEST BLOCTON IN THE LAST TWELVE MONTHS." The promotional further said, "We Have: Cheap, good farm lands, good through highways, natural drainage, access to markets, good schools and churches, six strong lodges, four railroads, and paved streets." The chamber's board of directors was: W. R. Young, Sr., T. T. Dison, W. A. Jones, John Kren, E. G. Gillespie, Dr. L. E. Peacock, T. C. Tuggle, W. T. Alexander, Rev. C. B. Martin, Dr. George W. Randall, M. T. Smith, W. R. Ray, J. T. Balch, P. P. Pritchett, and J. H. Blake.

The merchants and town leadership did not push the panic button. After all, Hill Creek and Klondyke were operating and the Red Ash Company held forth the promise that Number 10 Blocton would continue, and West Blocton was still the shopping center for Piper, Coleanor, Marvel, Belle Ellen, Lucile, Moffat, and other communities, all still actively involved in coal mining. Too, there were a number of farm communities to the north and west whose residents shopped in West Blocton.

The September 1928 town election featured a strangely labeled slate of candidates. Here is the way the event was reported in the *Enterprise*:

> An even dozen candidates are in our midst. Due to a lack of interest a good many men have had their names placed upon the ticket by friends. And the more evident showing is that there are two opposing tickets, one for the Democrats and one for opposition. The *Enterprise* ventures this assertion that no matter who wins we will have a good mayor and council. Here is the line up and you may split your ticket.

Democratic

T. A. McCool (For Mayor)

W. H. Wright

B. L. Edmonds

L. H. Deerman

R. G. Moore

S. B. Israel (For Councilmen)

Anti Democratic

Marshall Davie (For Mayor)

Dr. C. W. Jones

W. R. Ray

W. T. Morrison

E. G. Gillespie

T. T. Dison (For Councilmen)

> West Blocton needs to get busy with the new city officers. Our streets are in very, very unkempt condition, we need a new high school and our sanitation needs attention. The rubbish from the fire over a year ago still lies around in unsightly heaps and there is plenty of room to shine up the town. Now we know there is no money—but with the right leadership from each department, there is a chance for some cooperation. Let's have a clean-up and paint-up campaign and let's have the town officials back of it. If it takes it, let's have a public street working. The idea is—let's have results. Weeds are tall and mosquitoes are numerous and that is not a necessity among well-bred and civilized folks.

The editor asked for leadership that would work and said it was not meant to be unkind to the present mayor and board to whom he extended good will. The 1928 mayor's race was won by M. C. Davie.

The odd opposition party identification may have been related to the national election in which the Democratic Party had nominated Al Smith, Governor of New York (a Roman Catholic) for president, not a popular decision in the South. Candidates might run, then, on an "Anti-Democratic" ticket who were not willing to adopt a "Republican" label.

Positive news in November 1928 showed the bank still in good shape with $395,210 in assets. The unpleasant news was the WBHS football game loss to Bibb County, 13–6. The merchants and others continued to make adjustments to their new quarters, and the Randall and Turner residences were being rebuilt on Florida Street as well as a new home for the Dunnabies. None of the dozen or so small frame houses just north of Main Street was rebuilt. Randall was in his new home by March 7. An interesting feature of the dentist's huge living room was the fireplace mantel that proclaimed the family motto carved in limestone, "HOME THE PLACE OF GOOD CHEER AND BEST THOUGHT," still in place in the residence owned today by John Crowder.

The mayor and city council initiated plans to purchase a fire truck. The building across a narrow street just west of the newly built Kren store was leased to house the mayor's office, and next door to it space was rented to house city hall, fire apparatus, and jail. The "jail" was a steel cage that had been moved from place to place as a holding space for prisoners until they could be transported to the county jail.

Early in 1929, the WBHS Tiger basketball team met its first defeat ever by BCHS. Without an indoor court, competition with larger and gymnasium-equipped schools was especially difficult. Basketball was discontinued as a varsity sport within a few years and not re-established until 1936. By 1929, the school colors were identified as crimson and white. An old timer reported that "gold" had been used but was dropped because the locals wanted a different color from their county rivals. An interesting item appeared in the February 9, 1929, *Enterprise* under the heading, COLUMN FOR COLORED, which carried announcements for church events and a plea for help in a fund drive for Peace Baptist Church. The column was signed by Rev. W. M. Wagstaff.

The new Theatorium offered a piece of "beautiful china ware" to each lady attending, starting February 19. Lady moviegoers were encouraged to attend regularly to obtain a full set. The advertisement pictured the dishes in the "Parrot Pattern."

The March inauguration of President Herbert Hoover was heard by a special radio set-up at the high school. Local Majestic radio dealer Tulon Desmond made the arrangements for this first time special radio use in the school. The Chester Black Bus Company sold its franchise in March to the Alabama Bus Company of Birmingham, an enterprise offering service between Selma and Birmingham. This arrangement soon led to the establishment of Greyhound bus service to West Blocton. Considerable local excitement was reported on March 14 as a result of a small, single engine airplane landing in Caldwell's field on Smith Hill. Hundreds visited the site to observe a perfect takeoff on the following Sunday. It has been reported that the pilot was a card playing hustler and that he flew away with a good roll of cash won from the unlucky local poker playing set. The community took pride in the thrice-weekly appearance of the West Blocton String Band led by George Deerman on Station WAPI in Birmingham. Other members were Henry Fayette, the Fitts brothers, Fred Henderson, Curtis Martin, and Will Finnen. Their introduction to WAPI was in January 1929 by T. T. Dison, Majestic radio dealer in West Blocton.

On May 23, the paper reported that the new American-LaFrance fire truck, manufactured in Elmira, New York, was in daily tests by the volunteer fire department members headed by fire chief S. B. Israel. The article was topped by a photograph of the truck, a type 99 pump and hose car, with a sixty-one horsepower motor, four wheel mechanical brakes, and a 500 gallon per minute pump. It carried a Foamite extinguisher; two lengths of six inch suction hose; one twenty-foot extension ladder; one twelve foot ladder with folding hooks; an electrically operated siren; and a hose compartment carrying 1,200 feet of two and one-half inch hose. At last, a real fire truck was available to protect the town's businesses and homes. Arriving two months before the second anniversary of the town fire, the grand vehicle was painted fire engine red with nickel-plated trimmings and West Blocton Fire Dept. imprinted in gold giant block letters on its hood. The doors of the station were left open for visits and a proud and enthusiastic public came by the dozens to see the magnificent new fire engine. The newspaper said, "Congratulations to the city fathers."

The 1929 Fire Truck and Volunteers. (Courtesy of Susan Riley Gafnea.)

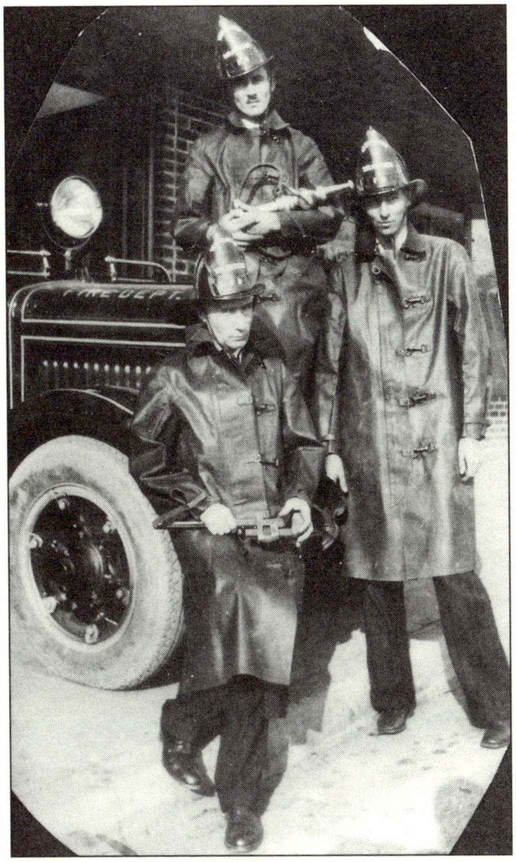

West Blocton volunteer firemen Morgan Smith (left) and Dave Israel (right) were outfitted with modern gear in 1929. (Courtesy of Jean Israel Farrar.)

The July 11 newspaper headlined the second anniversary of the fire. Brief articles on prominent businesses that had been rebuilt were featured, and a benefit dance for the fire department was announced for the Number 1 Blocton School. The item stated, "had we had the equipment two years ago we have now, the story would have been quite different."

An adult education program identified as the "Opportunity School" was offered starting in July at the Number 1 Blocton School sponsored by the county and the Red Ash Coal Company. Coach Pelham J. Merrill and an assistant were expected to have an enrollment of thirty pupils to offer "opportunities perhaps denied in one's youth."

When school opened on September 3, 1929, students found a newly painted school, interior and exterior. The "Old Green Schoolhouse" was a landmark for years. The unoccupied West Blocton Baptist Church (officially, Second Baptist) next door to the school had been acquired as classroom space to relieve the crowded old building. For a few years, it housed the chemistry department and later, fourth and fifth grades. At some point in the early 1930s, the fourth grade was housed in the upstairs of the Knights of Pythias Hall. The paper reported that a department of home economics would be added in the second semester, 1930. The annual big intra-county football game was lost to Bibb 6 to 0, a blocked punt figuring in the defeat.

The October 1929 stock market crash was not mentioned in the local paper. Though some of the West Blocton and Blocton investors were hard hit, the grim and devastating details of this historical event were left to the Birmingham dailies. This worldwide financial crisis eventually affected the community almost as a killer blow. The district bankruptcy court in Tuscaloosa handled a heavy case load from north Bibb during these bleak years.

T. C. Tuggle gave up the Theatorium management to J. P. Upchurch. The change was the result of diminished profits at the theater, evidence of worsening economic times. Stockholders in the new arrangement were Marvin Martin and J. F. Golson. By March 30, 1930, talking pictures were operating at what was renamed the West Blocton Theater. The first "talkie" shown was *Why Bring That Up?* The paper commented, "The tone is good and the Vitaphone has been adjusted to the building so well that the acoustics are good and the talking very distinct. A good picture *Dance Hall* is on today. Friday and Saturday, *The Lady Lies*, both all talking." Those who attended West Blocton's first talking pictures reported that the sound was "tinny and squeaky." If the most famous first talking picture, *The Jazz Singer* starring Al Jolson, ever played locally it is not on the record. In the early 1930s, the theater was acquired by the Neely family of Marion and renamed the Strand Theater. By 1933, the manager was Earl Philpot. During the 1930s, "Mr. Earl" offered Saturday egg matinees to children. For a candled (electric light examined) egg to assure its freshness, enthusiastic children could see a newsreel, comedy, serial, and full-length western. The theater sold the eggs to local grocery stores.

Just a year after its acquisition, the fire truck was in danger of being lost to debtors. The paper headed a May 29, 1930, item:

WE MUST KEEP THE FIRE TRUCK

A large group of citizens met at the picture theater Monday night to see what steps could be taken to retain the fire truck and apparatus. $1,728 is the [payment] amount due on the equipment and $900

Strand Theatre. (Courtesy of Carey Scurlock.)

must be raised immediately or the outfit will have to be returned back to the American-LaFrance Company. The methods of raising the funds were discussed pro and con and a subscription was briefly circulated, resulting in $340 being subscribed. The town council gave the following figures:

Annual revenue received	$ 7,987
Running expenses	$ 6,850
Present indebtedness:	
Water bonds	$ 2,300
Fire hose	$ 1,800
Alabama Power Company	$ 890
Current bills	$ 525
Fire apparatus	$13,328

[Within the latter figure, $6,764.87 was owed on the fire truck]

Do your share. Keep the fire truck. The city treasury is exhausted. Help now.

One of the worst years of the Great Depression for the Blocton District was 1930. Further evidence of economic problems came in July with the announcement of the end of the Mobile and Ohio Railroad passenger run on the Eoline and Blocton Branch. However, coal trains were continued until 1945. Uncharacteristically for the community, there was no gala Fourth of July celebration. A line in the personals column read, "The Glorious Fourth passed very quietly in Blocton this year."

A note of desperation could be read into the paper editor's brief statement on the occasion of the fire's third anniversary:

> We are calling attention to the fact that Saturday will mark three years since the fire. We do not feel that an elaborate discussion of this sad event and West Blocton's period of depression following would be in place. Markets are down, mines practically idle, unemployment is general and the best we can do is to hope for better times.

Paralyzing news came in the week prior to the September 4, 1930, issue of the *Enterprise* which reported that Number 10 Mine, the Blocton Red Ash Coal Company, had closed. The company that was formed in mid-1928 to continue the operation of the town's largest mine succumbed to the lack of a coal market. Also noted as a problem in keeping the mine open was the abnormal overhead expense for around-the-clock pumping operations to keep the mine from filling with water. The property was returned to the ownership of the TCI Company. That was seen as the final death knell for old Blocton, the company town. By October 9, wrecking crews were at work salvaging equipment and taking down properties such as the company store, hospital, the fifty-year old company office, and other buildings. The paper said, "This will not help [West] Blocton but it does not spell doom. Hill Creek and Klondike are still operating."

It is not clear from reading the local paper what the issues were in the mayoral race in 1930. Perhaps frustration with fiscal matters, which no one could control, was the paramount factor. T. A. McCool was again on the ballot opposing M. C. Davie and won in a close race, 109 to 98. McCool was an elderly former merchant. The *Enterprise* editor simply wrote the obvious, "the attitude of the voters changed during the present term of the local officials." In November the paper published a statement showing the financial condition of the Town of West Blocton. Cash on hand was $64.49 and debts were $18,843.41. Fire hose, the fire truck, water works extension bonds, and a loan from the West Blocton Savings Bank made up the largest part of the debt.

As economic times grew harder, other evidence of a rising poverty-stricken class appeared in the paper's columns. The PTA pled for funds to buy books for needy children. The West Blocton Theater that had closed for lack of patronage announced that Dr. Clyde Jones's brother Earl and a Mr. Vick of the Empire Theater of Birmingham would reopen the local facility to show silent movies on Saturday nights.

A statistic revealing the impact of hard times was the number of telephones in the town. In 1925 there were 144; in 1930, 122; and in 1935 only 68. *The Literary Digest*, a widely read magazine in the 1930s, following previous practice, made a telephone sampling on how the populace would vote in the 1932 election. Their results predicted Hoover winning easily. However, the telephone survey reached only the affluent; the less fortunate, likely Democratic voters, had disconnected their phones. The magazine shortly ceased publication as a result of lost prestige and reliability.

The January 7, 1931, paper reported: "The wrecking and salvaging gang of the TCI finished closing Number Ten mine and now Blocton appears as the deserted village. All the property has been removed excepting the residences and these are under the land department and no announcement has been made as yet by the land agent." The salvaging was under the direction of Henry Henderson of Ensley who had been the local superintendent when the company departed in mid-1928. The notice appeared in the local paper that Number 1 Blocton houses were being sold to Tulon Desmond. Rearrangement of the neighborhood's water supply was made so West Blocton waterworks would provide service to the system in the old Blocton area. Desmond later advertised fifty houses for rent. At the height of mining operations in the last days of the Cahaba Coal Mining Company and during the beginning of TCI's tenure (1892), there were nearly 400 company homes in Blocton.

As a part of the dismantling of the TCI properties, the small water works filter plant that had served the company town since the mid-1880s was destroyed. Even the concrete foundation walls were dynamited. After the hook-in with the West Blocton water system, the wooden water tanks near the Baptist Church that had served the Blocton area for years were pulled down and the wood salvaged. Ironically, the Blocton Post Office operated until 1936. Perhaps the government was reluctant to close down any job opportunity during these depressed years. The last postmaster there was Henry H. Farrar.

Later in January the editor-publisher of the *Enterprise* announced: "This paper is for lease to [a] reliable local party. If interested see Cooper Blanton personally." Blanton later announced a leave-of-absence. The paper was briefly under the management of J. I. Nunn and then by March, the editor-publisher was Clarence Martin, the Baptist preacher's young son.

Meanwhile economic conditions worsened. The February 26 edition of the *Enterprise* indicated that Moffat Mine, only a few miles south of town had closed and that the rails had been pulled. By late June the train service to Blocton had been cut to one run a day. Mail was then brought in by bus on regular runs through the town allowing for two deliveries daily. Mrs. Douthit, a long time hotel operator in West Blocton, moved in July to Marion to operate a hotel. The new drug store's upstairs had served as the town's hotel since Main Street's reconstruction in 1927–28.

The community took pride in April 1931 upon learning that local scholar Libby Jean Israel had been elected president of the Women's Student Council at Alabama Polytechnic Institute (now Auburn University). She was also vice president of a literary society and had been admitted to a national honorary speech fraternity. In May, the town bade good-bye to Pelham Jones Merrill who returned to the University of Alabama to pursue a degree in law. Merrill had spent five years in the town's schools as athletic coach, teacher, and administrator. He spent one year as principal of the Number 1 Blocton School when it was operated by the Red Ash Coal Company and had charge of the Opportunity School of Bibb County in the summer of 1930. The biographical sketch of Merrill compiled for his induction into the Alabama Academy of Honor in 1978 notes his tenure in the West Blocton-Blocton schools. Merrill had endeared himself to all his students and to the town. The editor said, "We give Prof. Merrill a hearty handshake and bid him God's speed in his chosen line of law, where we hope for him a brilliant career." Bloctonites followed Merrill's career seeing him eventually elected an associate justice on Alabama's State Supreme Court.

The local paper in May also reported that recordings by a West Blocton musical group, Mike Shaw's Alabama Entertainers, were released in April. Associated with Okeh Phonograph Corporation of New York City, their first record carried the titles: "The Birmingham Special" and "Tennessee River Bottom Blues." Those interested could order from Williams Music House, Birmingham.

Bank failures were sweeping the country in 1931. The West Blocton Savings Bank, founded in November 1892, failed on Thursday, October 1, 1931, when its doors were closed for liquidation. It was another bitter blow for local businesses and individual depositors. The State Banking Department took over for the collection of accounts. George Howard Young of Centreville, grandson of the former West Blocton bank owner, related an interesting story about the bank's failure told him by a member of his family. W. R. Young, Sr., the bank president, unfortunately delayed making a trip to New York for the complicated procedure of selling the West Blocton Savings Bank's investments in the Chase Bank. He stated his intention to do so because of stock market developments during a September 1929 family gathering. Had the transaction been accomplished the West Blocton institution might have survived. The bank's losses due to large investments in the Chase Bank of New York during the 1929 crash eventually led to the 1931 closing. West Blocton did not have another bank until 1965.

Justice Pelham Jones Merrill. (Courtesy of Alabama Supreme Court.)

Other evidence of a community in economic retreat was the demise of the Chamber of Commerce that had been enthusiastically launched in 1928. Its mention simply disappeared from the local newspaper's pages. Also victims of the times were several of the long-time fraternal lodges. After the early 1930s, there is no evidence of the viability of the Knights of Pythias and by the mid-1930s, the Red Men had

W. R. Young, Sr., mine owner and banker, with granddaughter Peggy, and wife Lennie. (Courtesy of George H. Young.)

gone inactive. In the late 1920s, the Cumberland Presbyterian Church, a local congregation since 1892, closed its doors because many of its lay leadership and members had moved away. Its preacher, church records, and membership were moved to its parent institution at Greenpond.

The only cheerful incident in the fall of 1931 was a rare victory by the WBHS Tigers over the Bibb County Choctaws, 14–0. In the series, begun in 1925, the local team had won only one previous game, in 1926, and tied one in 1927. The scores were always close in these early contests. To celebrate the November victory a giant pep rally was held in the Theatorium (the theater had been renamed but the old label held on), with 350 attending.

At the start of 1932, a local organization was created to help those suffering the most for lack of employment. The Relief Committee of West Blocton was said by the January 28 *Enterprise* to be "making all efforts to relieve unemployment. The present plan is to exchange food and clothing for work in the West Blocton Cemetery [Mt. Carmel]. Any contribution of money or other means will be greatly appreciated. Make all contributions to Dr. Peacock, Treasurer." Evidence of a community pulling together was displayed in a program at the Theatorium that featured the area's African-American church choirs; Hope Hill, New Hope, Peace, Bethel, Liberty, and Banks church groups sang to a mixed audience on Sunday afternoon, April 3. A $42 offering was taken for local Red Cross relief.

Hard times added to the problem of thievery. On March 24, Sam Israel discovered his store had been entered through the east front door on the previous night by robbers using a pinch bar. Israel offered a $60 reward for the arrest of the burglars. Only the best suits, shoes, and dresses had been taken resulting in a major loss. It appeared to be a professional job with re-sale of the goods in mind. Because of its reputation for having top of the line goods, the I&N was not a stranger to robberies. In his youth, when his father ran the store, Sam Israel had slept in the store to ward off possible break-ins.

Somehow, either by small payments or by pleading to the American-La France Company for more time, the fire truck remained with the town for several years into the 1930s. A report on the Fuller residence fire in March 1932 indicated the truck was still in use. While the Fuller home was lost, the Smith residence next door had caught fire but was saved by

1931 Football Team. Members of the 1931 team were Bob Ivey, Rucker Goodwin, Joe Wheeler McSpadden, Otto Miller, Charles Mize, Roy Frederick, Burt Waddell, Jim Wood, R. G. "Pug" Moore, Frank McElroy, and James McElroy. (Courtesy of Jake Wood, Northport, Alabama.)

the well-equipped fire engine. The paper said a brisk breeze was blowing, making the fire a danger to the center of town. In September 1932, the American-LaFrance Foamite Company had taken the fire truck matter to the Bibb County civil court asking for payment of the "promissory note." The truck was gone by mid-October when the Davie store was lost to flames. The exact date of the equipment's removal is not recorded.

The publication of the *Blocton Enterprise* locally ended in 1932. Editor-publisher Blanton, after a leave of absence, returned to offer the paper for sale. On July 23, a few days over five years since the town fire that led to the departure of the Allens as paper owners, Blanton sold the paper to Jim Oakley, owner-publisher of the *Centreville Press*. Oakley had been at the head of the Centreville paper since the summer of 1927. The special local flavor of the home town paper was subsequently reduced. Oakley declared his intention to continue the paper to serve West Blocton and north Bibb County. Rather than combining the papers, the plan was to run them as separate units. The new edition was a smaller version of the *Press* containing much of the syndicated weekly material and north Bibb County personal columns. Occasionally pertinent area articles appeared within its pages. What was lacking was a civic promotional spirit that the old Blocton area papers had so effectively and dramatically reflected.

The new *Enterprise* owner-editor reported on his visit to Bibb's only bottling plant in September 1932. The West Blocton Coca-Cola plant had an interesting history. W. A. Jones was the owner-operator and had been for a number of years. The West Blocton plant was described in detail in an early issue of the Centreville based newspaper by the visiting writer.

The community needed something to boost its spirit in the fall of 1932 and found the means in Coach J. G. Burrows's WBHS football team. The local heros beat all opponents by significant scores except down-county rival Bibb which it tied 6–6 in a downpour in Centreville. The rain neutralized the WBHS Tigers' vaunted passing attack. One of the players said the referee found it necessary to put his foot on the ball between plays to keep it from floating off the field. Bibb school reporters proclaimed the tie a victory for the Purple and Gold, who were judged 14 point underdogs. On Thanksgiving, upstart rival Piper gave the Tigers a tough last game of the season that the local team won 7–0. For the year, the Tigers scored 272 points and gave up only 13, one of the school's best records.

The Jewish community had decreased by 1932. The Sam Israels, Joe Israels, and the Klotzman families were the remaining members of what was once a significant population that had arrived in the early days of Blocton and West Blocton. Most had become enterprising business owners on Main Street. Hebrew services had been held as early as 1892, and in 1905 West Blocton's Jewish citizens organized the Agudas Achim (League of Brothers) congregation and built a modest frame synagogue. As late as the mid-1930s, the last few families continued to observe the High Holy Days of autumn. To do so, Melvin Israel brought several fraternity brothers from the University of Alabama to supply enough adult males (ten) to have a minyan, the proper number for services. In October 1932, the *Enterprise* identified those visitors as Jack Winkle, Dave Wallick, Sam Morris, and Nat Novgrad on one weekend and Nathan Katz, Jordan Mitchell, Dan Emos, Louie Fisher, and Morton Riter on the following weekend. After his entry into the radio broadcast profession, Melvin Israel assumed the name, Mel Allen. He achieved national notoriety as the "Voice of the Yankees" and a place in baseball's Hall of Fame. Until Sam Israel's death in the 1960s, Jewish family services were observed during the special fall holy days in his West Blocton household with immediate family, relatives, and visitors taking part. The synagogue has not survived. As a result of mid-1930s storm damage and the poor economic times, it was demolished.

November 6, 1932, marked the departure of the last two passenger trains from the Blocton depot: Number 40 left at 11:55 a.m. and Number 43 at 1:33 p.m. The Depression continued to take its toll on the community.

The December 1, 1932, West Blocton paper carried the following story:

> RED CROSS TO SPONSOR PLAY
>
> Proceeds To Be Used To Help The Needy
>
> The local unit of the Red Cross Organization will sponsor two plays in the coming weeks. The plays are coached and acted by the colored people. On Friday night, December 2, a play will be given at the Bucktown Methodist Church. Admission 10c and 15c. Reserved seats for the White People.
>
> The second play will be given December 9th at the West Blocton Theatorium. This play is entitled, "Object Matrimony." A comedy full of laughter and music. Your time and money will certainly not be wasted if you see both plays and you will be helping a good cause. Admission to this play is 10c and 15c. Balcony for colored people.

Riley's Café. (Courtesy of Susan Riley Gafnea.)

Growing problems with lack of governmental revenue was dramatically illustrated with the announcement by County School Superintendent W. Y. Hayes that Bibb County Schools would close after December 16. The *Enterprise* story said BCHS would continue only through mid-term exams. The Brent School was to run on solicited funds. No mention was made of the plans for WBHS. The superintendent ended a special letter to teachers and administrators, "Thanking you for the spirit shown in continuing thus far in spite of inconvenience and hardship, a spirit over and beyond any reasonable expectation, and trusting this is the "darkest hour" before the dawn of a brighter day in education."

Subsequent plans were initiated to allow a second semester with a system of "scrip" payments to educators and students required to pay tuition. This way, both BCHS and WBHS seniors were able to finish the year. However, at Bibb, as reported in the *Press*, several seniors had to drop out because of inability to pay. This likely occurred at WBHS, but it is not recorded. Mamie Foster Kuhns, a 1933 WBHS graduate, remembered that fees were required for the second semester and that while she could not remember who, several classmates fell by the wayside. Her younger sister had to stay out of school the remainder of the year so family resources could be applied to Mamie's senior year. Harry J. Adams, a member of the Class of 1933, worked at Murphree's Drug Store afternoons and weekends for $5 a week. Without his own income, it would have been a difficult challenge for his family to pay the fees. There were other classmates who worked part-time to stay in school. Otto Miller, the class valedictorian, worked at the town movie as a projectionist and assistant manager to help pay his costs. Before sound films were run, Otto also operated the player piano for silent film music and sound effects accompaniment.

Minimal effort was made by a failing Republican administration to alleviate the distress of the unemployed. Editor Oakley reported in January 1933 that Bibb County officials were asking for $2,500 for relief of the unemployed. Other amounts of something under $20,000 were due to arrive in the county within six months. Citizens who needed employment were required to make application to Miss Jack Mitchell, Bibb's Child Welfare Worker. The request then went through several more hands. One could starve waiting on government relief. A committee went to Montgomery to make application for the funds available under the Hoover administration's Reconstruction Finance Corporation. The members were Judge W. L. Pratt, Miss Jack Mitchell, George E. Ellis, and P. P. Pritchett.

Editor Oakley wrote about the prospects for 1933 in an editorial titled, "Bibb's Job in 1933." Some of his thoughts follow:

> Bibb has had its share of the hardships of the present national disaster. It has seen its people on part time and without work, as they struggled to provide for their essential needs. It has seen hunger and poverty stalk through its midst as no one expected for many years. Let us hope for the best, but, at the same time, make up our minds that we will meet the new year with hope, and that the people of this county will face it together.

It is not surprising that local Democratic Party leaders were calling for an overwhelming vote for their candidate Franklin Delano Roosevelt in the 1932 November election. Their memory of the town's plight during the Hoover years was easily turned to make the Republican candidate a scapegoat for local troubles as well as those of the whole country. When the votes were counted in Beats 10 and 11, there were 237 votes for Roosevelt and 17 for Hoover. The countywide vote was Roosevelt 1,690; Hoover 133. On March 4, 1933, Bloctonites gathered around their Philcos, Atwater-Kents, and Majestic radios to hear the new president's famous first inaugural lines. The most quoted and memorable were: "This great nation will endure as it has endured, will revive and prosper. So, first of all, let me assert my firm belief that the only thing we have to fear is fear itself." Authur S. Link's *American Epoch* speaks for the people of the time. "They did not have to know or understand these things in order to sense that this day, March 4, 1933, was a fateful moment in the nation's history, that one order was dead beyond recall and that another, ill defined and uncertain as it was, had been born."

It was the era of the "New Deal." West Blocton coal miners were especially interested in the National Industry Recovery Act of June 1933 which made it legal for them to organize unions and bargain with management for better wages and working conditions. The area's unions had been dormant and implicitly illegal since the unsuccessful strikes of

1920–1921. One day in the summer of 1933, miners, still apprehensive about the sincerity and workability of the Act, gathered at dawn in small groups in the heavily wooded areas north and south of town and cautiously moved eastward to an outdoor meeting on a shaded hillside near the Catholic Church to organize and elect officers. Other union organizational meetings were held at Hill Creek ballpark at sunrise. In 1998, Arthur Finnen remembered that he and his brother William and a third person met with a Birmingham UMWA organizer to whom a petition was presented declaring the intent to organize. Will Finnen was later to serve as local president of the Hill Creek union. Union activities and officers' names were not announced in the local paper. Wayne Flynt's *Poor But Proud* gave evidence of the date and the magnitude of the movement: "When William Mitch and William Dalrymple came to Birmingham in June 1933 to reorganize the Alabama District, only 225 miners belonged to two UMW locals. A month later, they had enrolled 18,000 in eighty-five locals in one of the most amazing union campaigns in state history." Among those eighty-five state locals were at least two in West Blocton.

After the passage of the National Labor Relations Act (the Wagner Act of July 1935) that prohibited unfair practices against unions, potential union men were less apprehensive about organization.

In the more difficult days of the Depression, Arthur Finnen recalled that the mine would work on Mondays, Wednesdays, and Fridays. On other days at Hill Creek, the miners would go into the mine to dig coal so they would have a larger number of carloads to claim on the official workdays. According to Nelle Thrasher Battle, her father, Charles Thrasher, and other miners who worked at Belle Ellen and Klondyke during these lean years called this labor "idle-work." The small cars at Hill Creek would carry about a ton of coal each. If a miner could get credit for four or five tons a day it would be considered a pretty good shift. Arthur Finnen remembered that pay was 40c a ton. For a week's work, the most one could expect to make was $10. This likely represents recollection of the Depression's later years at Hill Creek. Evidence from other local miners indicates the pay was only 20c per ton and sometimes they had to work ten or more hours to eke out 75c a day. Miners supplied their own tools and blasting powder and carbide for their mining lamps. Costs for these supplies had to come out of their meager earnings.

Other evidence of hard times in the community is found in Methodist Church records. Reverend J. H. Pass was assigned to West Blocton-Blocton in the fall of 1930. The previous preachers had been paid at least $2,000 for the year, a fair Methodist preacher's salary for the 1920s. By 1932, Pass' salary had been reduced to $1,200 but only $945 had been paid him that year. By 1933, his salary was reduced to $600, and only $525 had been paid. He wrote in the quarterly report, "It is hard to know just how we are going to get along. The success of the church is so dependent upon the civic conditions of the community. The condition of the community does not improve. We are going ahead and mean to do the very best we can."

W. R. Ray. (Courtesy of Nancy Ray Sansing, Bibbville, Alabama.)

After T. A. McCool served a two-year term, William R. Ray, former TCI mine boss and superintendent of the Hill Creek Mine, was elected mayor in 1932, serving until 1944. Through Ray's efforts, government programs were brought to bear on West Blocton's problems, but first, things had to sink lower. With the town unable to keep up payments, court orders demanded the return of the fire truck to its creditors. West Blocton was back to a reel of hose on a caisson pulled to fires by whatever means possible and dependent on gravity-driven water pressure from the water tanks on West Blocton Heights off the highest point on Florida Street and one stream of water was its capacity.

Another retrenchment was the removal of all residential streetlights. Only the Main Street business section remained lighted.

Mayor Ray and the council did insist on keeping the town's credit rating even if at a minimal point. Some thought the town had defaulted, but it never did. It was a time of hard decisions and the decisions made were the best the leadership could render under the circumstances.

Also indicating the measure of desperate times were relief programs. Reported in the *Press* was the amount of food sent by the government to aid Bibb County's needy: 7,200 pounds of dried salt pork, 20,825 pounds of flour, 3,400 pounds of butter, 2,554 pounds of ground beef, 850 pounds of cheese, and 122 pounds of eggs. Miss Jack Mitchell, superintendent of the county child welfare department, managed distribution.

The financial report of the Superintendent of Education for 1932–33 printed in the *Centreville Press* of January 4, 1934, reflected the grim financial status of the county's educators. The amounts show only what teachers were paid for the period September through December. Senior classes at both WBHS and BCHS graduated at the regular time, so those schools continued under some local arrangements, mainly by student tuition payments. James P. Floyd, WBHS principal, was paid $916.65 for the school term. The coach, J. G. Burrows made $618.75. Regular classroom teachers' salaries were far less. Gertrude Head was paid $361.95; Jean Israel and Buleah Cundiff, teachers in rural schools, each made $211.25. The amounts paid to "colored" teachers were dramatically less, several being paid less than $100. Further salaries listed were Blanche Wells, county colored supervisor, $400 while the county white supervisor made double that amount. Louise Blake Burrows, now a resident of Tuscaloosa, remembered that the school at Primitive Ridge, where she taught, closed in December 1932 and did not reopen for the remainder of the school year. William J. Murphy went to a rural school near his home that closed at midyear, causing him to lose a year. Generally speaking, the teachers at Bibb County High School were paid more than the teachers in other schools. For example, the principal, G. H. Stacey, was paid $1,533.32, and Mrs. W. R. Young, Jr. earned $412.50. The total allotted to county white teachers was $37,000.00; to African-Amercian teachers, $4,714.88.

The *Centreville Press* of January 11, 1934, reported the Civil Works Administration (CWA) in Bibb County employed 2,000 creating a payroll of $10,000 a week. One of the projects was a new school building for grades seven through twelve in West Blocton. The high school student reporter, Robert Davie, wrote in his March 15, 1934, *Press* column that "construction is well under way on the new school building." The government agreement for the high school building's construction had called for local contribution of materials. Lumber from the Number 1 Blocton TCI school building and the church just north of the old school (West Blocton Second Baptist) that had once served as the chemistry department and later as classrooms for elementary grades was used in the construction. After the dismantling of the old church, the fourth and fifth grades met in the Free Will Baptist Church and the Wesleyan Methodist Church. Brick from the abandoned coke ovens was used for the foundation. The use of local materials at the new school continued for several years. Harmon Frederick reported that during his junior high days in the early 1940s the sidewalks that circled the abandoned Number One TCI School site were broken into pieces and laid as mosaic-style sidewalks at the new WBHS. Manual arts teacher Lester Hosmer directed the project. The record does not show that county school funds were used in this school construction project. Since the county did not have funds to pay teachers, it follows there were no funds for new school construction.

While construction continued on the new school, the 1903 school house had been condemned as a firetrap, so student fire drills were more frequently practiced. When the building was vacated, several rooms were converted to one-room apartments. Ben L. Edmonds, Jr. reported that he and his bride lived in the old building for a short time.

The new school was a wooden frame building because the desperate economic times did not allow the construction of a first-class brick building. The new building had neither gymnasium nor indoor rest rooms. Jokingly, former students said it was equipped with the best out-door facilities available. In the 1940s, the boys called their outhouse "the capitol" and the girls' facility was dubbed the "county seat." Classrooms were heated with old style pot-bellied coal stoves. Some said the reason for a dozen stoves was that they used more coal than one central furnace, thus aiding the local economy. The seniors of 1935 planned to dedicate the new school building during their graduation ceremony. Details on that event are not on the record.

West Blocton High School, 1935. (Courtesy of Nancy Ray Sansing.)

A temporary outdoor basketball court was improvised. Later, however, the new auditorium doubled as a combination gymnasium and assembly hall. However, the floor space did not allow for a regulation-size playing surface for basketball, the free throw circles overlapped the center circle. The auditorium's ceiling was not high enough to allow a long, arching shot, causing visiting teams some disadvantage. There was no seating for the auditorium until the fall of 1936. Elmo Israel reported his class graduation in May 1936 was held at the Strand Theater because there were not enough chairs available for the exercises to be held in the new auditorium.

Possibly spurred on by a new school building and new principal, L. S. Vanderford, the composition of a school *Alma Mater* came about in the mid-1930s. Music teacher Mary Davie said the tune was from an old Boy Scout camp music book. The words are reported written by Mrs. Davie. Also an innovation, the Class of 1937 published the school's first yearbook, *The Westonian*, a mimeographed publication with pasted-in photographs. Special order pages and covers were provided in kit form for the publication.

Among local developments reported in the county paper in 1934 was the organization of a tennis team at the high school. New history and social studies teacher L. S. Vanderford volunteered to be the team's coach. He found willing and talented players. First, an old court was refurbished on the synagogue property, and later a second court was built on the lot where the Davie family home had burned. Eventually courts were setup on the old TCI ballpark. When the new Civil Works Administration financed high school building was completed, courts were built on the north and east sides of the structure with appropriate backstops and regulation nets. Vanderford's tennis Tigers did not shy away from major opponents, often playing Birmingham city school teams. In 1935 the team of Elmo Israel and John Parks stayed in a tournament until ousted by state champs Ramsey High School, one of the larger city schools. The West Blocton netmen remained in the contest longer than any other team outside Birmingham.

Other gleanings from mid-decade in the 1930s reflected a community trying to survive hard times. The Gillespie Funeral Home, successor to Clifton Undertakers, moved into a new building. Today that structure, with bricks from the coke ovens in its exterior wall construction, houses the West Blocton Pharmacy. The local Elks Lodge Ritual Team was declared state champions and made a number of trips over the state inducting new members. Notable trips were made to Demopolis, Mobile, Montgomery, and Birmingham. The Exalted Ruler, Frank S. Israel, represented the local group at the national convention of Elks in Columbus, Ohio. In June 1935, Mayor W. R. Ray and Councilman John Finnen went to Centreville taking data needed to apply for a paving project. This was likely the beginning of the job that finished the blacktopping of Main Street from downtown through West End to the state highway. The *Centreville Press* identified a number of work relief programs for the county on July 11, 1935. Among the projects was the bituminous surfacing (black topping) of the highway from Woodstock to Centreville. At last there was to be a paved road to Birmingham, old State Highway 5 near West End that ran through McGrawtown.

King and Queen Rule at West Blocton 1938 May Day Celebration. The king, Frank Ferrire, and queen, Jewell Jones, are shown seated on the throne. Attendents are (left to right) in the front row Margaret Kren and Felix Bradford. Second row–Margaret Parker, C. W. Lucas, Louise Duncan, Fred Lyles, Gladys Thrasher, and Bill Carroll. Back row –Mary Logan Desmond and Howard Moon. (Courtesy of Jewell Jones Dailey.)

The special industrial and business edition of the *Centreville Press* of June 11, 1936, identified the county's mining operations as essential to its economy. Two major mines were operating in West Blocton, the Blocton Mining Company (Klondyke on Smith Hill) and Hill Creek Mining Company. Other large mines were at Belle Ellen, Marvel, Piper, and Coleanor. In 1934, the paper reported 1,178 mine workers in Bibb, and estimated that a million dollars was paid in a year to coal mine employees.

In October 1936, it was announced that Public Works Administration funds had been approved for the construction of a new elementary school in West Blocton. Cass and Daugette Construction Company of Jacksonville started work in November 1936, and the brick school opened in the fall of 1937. A central heating system and indoor plumbing were again disregarded, as had been the case in 1935 for the high school structure.

In September 1935, the county newspaper reported the funeral of Alf Star, one of West Blocton's oldest Negro citizens. The rites were held in the Methodist Church in Bucktown, evidence the building was still in use. In January 1936, the paper carried notice that Episcopal services were scheduled in the Church of Our Savior every second Sunday at 3 p.m. officiated by Reverend Buchannon of Birmingham. In October 1935, the high school reporter announced that a six-piece school band, directed by Mr. Winkenhoffer of Birmingham, had played for chapel. As far as can be learned, this was the first high school band in the county. The WBHS Class of 1936, with 18 members, was remarkable in many ways. Nine of the graduates were college bound, an unmatched statistic by any class before or since. Achievements by those Depression-toughened boys and girls in their adulthood are stories much like the fabled Horatio Alger success-adventure tales. They had experienced the most difficult of economic times and were to be tested by the country's entrance into total war. All males in that class were to serve the country during World War II.

Prospects of better business times were indicated in April 1936 with the opening of J. F. Wood Motors in the Frederick Motor building. Wood sold Dodge and Plymouth vehicles, and mechanic Johnny Leonard was in charge of parts and repair. Frederick Motors, Chevrolet dealers, moved their business to Knights of Pythias Hall. The Methodist congregation's move into a new brick church in November 1936 was further evidence of recovery. Confidence in what was going on in Washington was voiced at the polls in the national election of 1936. In Beats 10 and 11, 292 votes went for Democrat Franklin Roosevelt; 30 for Republican Alf Landon.

July 4th celebrations returned in 1936 with a big labor union program at Hill Creek ballpark that featured William Mitch, District 20 United Mine Workers of America president as keynote speaker. William Collins, president of Hill Creek Miners' Union, was master of ceremonies. Early in the day a parade led by the West Blocton High School Band (all six members) riding on a truck bed wound through town. The band later performed at the ballpark. Following speeches, there was a picnic lunch available at a canteen set up on the field. Proceeds were given to aid sick and disabled workmen. Boothton defeated the local team in a doubleheader baseball game. The day's celebration ended with a performance in the high school auditorium by country music star Happy Hal Burns and his troupe of popular radio stars.

On Armistice Day, the (African-American) West Blocton Junior High School, led by principal Elmore Suttle, paraded through town singing "America." The *Press* of November 26 gave details saying 160 children and five teachers and elementary school students arrived at the West Blocton Theater at 11 a.m. The program included a flag raising, songs by the Junior High Chorus, patriotic recitations, and a speech by Dr. George W. Randall. Others on the program were Rev. W. W. Brown, Bernice Heard, Adell Allen, Willie Mae Brown, Athlete Calloway, and Mable Brown.

Though the times were improving, the community was still economically depressed. The 1936 Christmas season found the Elks Lodge and the American Legion delivering "mission of charity groceries and other items to the needy."

Finnen Filling Station. (Courtesy of Libby Fitts, West Blocton, Alabama.)

J. F. Wood Motors' employees could repair anything with wheels. Pictured are Emmett Carroll, Boots Carroll, Leonard Carroll, and Johnny Leonard. (Courtesy of Bettye Kitchen Hughes.)

WBHS Class of 1936. The small class of 1936 produced a high percentage of successful men and women who served their country in war and peace all over the world. Glenn O. Perry (third from left, top row) was the most widely traveled. An alumnus of the University of Alabama, he held a number of corporate positions until World War II when he joined the U. S. Government Civil Service. Much of his career was in the Diplomatic Corps in consulates and embassies in Latin America and the Middle and Far East. During his China tenure, as Vice Counsel in Shanghai, the Communist government detained him for several months. His last government post was Assistant Chief of the Audit Branch, State Department. He retired in 1980 as Executive Auditor, Fluor Corporation, in California. He resides in Olympia, Washington. (Courtesy of Elmo Israel Ellis.)

West Blocton Methodist Church. (Courtesy of Cahaba Trace Commission.)

West Blocton Elementary School. (Courtesy of Nancy Ray Sansing.)

Still, things were looking a bit better. A new elementary school building, part of a reported $100,000 building program for the county, had been constructed as had a new Methodist church; a new school on Smith Hill; and street paving. Two large projects, a new water tank and new water lines, were in the offing. To supplement federal Works Progress Administration funding, West Blocton sold $13,000 in water works bonds at 6% interest. In December, the town announced the installation of a new water system which included a large electric pump at the spring; about three miles of water mains; a 100,000 gallon, fifty four foot high steel tank; and 275 meters and meter boxes at a total cost of $22,000. Part of the cost was a grant of $9,900 from the Public Works Administration. The water works project employed thirty-five men. W. I. Coston and Sons of Bessemer and the Chicago Bridge and Iron Works of Birmingham were the successful bidders.

Still a major factor in the nation's economic recovery, the WPA spent $44,312 in Bibb County in 1937. A sizable amount of that total was applied to West Blocton projects. Early in 1938, three-quarters of a mile of Main Street through West End was paved. No longer were residents along that high-traffic street tempted to hose down summertime dust. Water pressure ceased to be a problem elsewhere in town on hot, dry summer afternoons. It took courage and foresight on the part of Mayor W. R. Ray and the town council to obtain funding for this water system project and to see it through to completion.

A national fund-raiser for the Warm Springs Foundation, a crippled children's program promoted by President Roosevelt, himself a victim of infantile paralysis, was supported in West Blocton in a social event at the Elks Hall. The special dance and program was called President Roosevelt's Birthday Ball. The county chairman in January 1937 was Tulon Desmond. Eddie Mahaffey and his "Bama Skippers" played for the dance. Showing their interest in national political affairs, local residents Mrs. Ed Gillespie and Mrs. Frank Morrison went to Birmingham in April to attend a speech by Mrs. Roosevelt.

While there were opportunities for inexpensive or free amusement, West Blocton's "younger set" during the mid-1930s often had to create their own entertainment. Local talent helped provide the music for social events at the Elks Hall. Notable was a small dance orchestra that called themselves the Melodeers. Its musicians were Floyd Jones, Malcolm

The Blocton Elks Ritual Team won the state championship in 1935. Front row (left to right) Dave Israel, Tulon Desmond, Ed Gillespie, and Dr. William McCarty. Back row - John F. Golston, Dave Klotzman, Frank S. Israel, Samuel B. Israel, and Willcox Randle. (Courtesy of Blocton Lodge 710, B.P.O.E.)

Dailey, Elmo Israel, Robert Davie, Harry Burton, Billy Ray, and Leno Rossi. The old Blocton Home Band's bass drum was the basic part of the percussion section.

West Blocton miners joined the strikes of 1934–35 that brought significant conflict to other mining towns in the county, principally Piper and Coleanor, where serious and threatening confrontation brought the state militia to keep order. James W. Walker's Piper history, *The Struggle and the Joy*, recounted the details of this event. The merchants of West Blocton, already hard hit by a deepening depression, were particularly affected by this strike. Reporting on another strike in April 1937, the *Press* stated, "The coal mine strike is still on. Everything is very quiet and everybody seems to be in good humor hoping for a better day." In May the strike ended with the commercial operators agreeing to pay the strikers better wages. Editor Oakley deplored the strike saying, "Strikes in the future should be averted, and some means should be worked out whereby negotiations for higher wages and other considerations can be agreed to without the necessity of a general strike."

Tip McCool and Lewis Deerman. Even during the depths of the Depression, low paid and often out of work miners would splurge on something frivolous. Here Tip McCool and Lewis Deerman recorded their friendship for posterity following a shift at a local mine. (Courtesy of Sandra McCool Everse.)

Meanwhile the big development in 1937 at the high school was the construction of a lighted athletic field. Again, WPA construction workers joined county commission workers, using county machinery to excavate a hillside in front of the school building for the facility. The lighting was financed by members of the local Board of Education with the understanding they would be repaid from a portion of the gate receipts. The 1937 WBHS football team was an above average one. The game with Bibb, always close, always hard fought, was won under the lights by the Tigers, 12–0, the first win since 1931. There were no wins for WBHS for another decade in this intra-county gridiron battle.

The town was shocked by the suicide on August 18 of its chief of police, R. E. McSpadden, by a bullet wound to his head. The county sheriff said no motive was determined. Tulon Desmond and others at his service station had seen the chief walking briskly toward his home with a sack in his hand a few minutes earlier. Desmond had asked Sheriff "Mac" where he was going in such a hurry and what did he have in his sack? He replied he had a gun and was going home to kill himself. The questioner took that as a joke and he and the other bystanders laughed. A few minutes later a boy running up Main Street shouted that Mr. McSpadden was dead.

The Red Men's Nebe lodge was evidently still active in 1937. The weekly paper reported that the lodge would sponsor a Labor Day picnic. Competing for picnickers for the holiday were a "Dixie Frolic Show" at the high school featuring radio favorite Happy Hal Burns in a repeat performance from the previous year and a dance at Huey's Place featuring a Birmingham orchestra, "Cliff Nation and His Clarinet."

Having come into widespread national popularity in the late 1920s and early 1930s, radio in mid-decade was the big communication and home entertainment medium. Almost every home had a radio and a list of favorite programs. *Lum and Abner*, *Amos and Andy*, *Fibber McGee and Molly*, various news commentators, sports broadcasts, and a great variety of programs were listed in daily newspaper radio logs. West Blocton took special pride in pointing to the broadcast career of native Frank Israel who was briefly an announcer at WJRD in Tuscaloosa, and in December was promoted to WAPI Birmingham. The *Press* said, "Mr. Israel was forced to change his name to Don Frank for *Euphony's* sake." Another local

"The Mighty Four, 1937" (left to right) James Moore, Dan Moon, Fred Liles, and Frank Ferrire. (Courtesy of James D. "Dan" Moon, Stockton, California.)

student of radio technology, Marshall Davie, Jr., was a short-term manager of Station WBH-Huntsville before entering Georgia Tech. While a student at Tech, he had a part time job at station WSB. In November 1939, he was honored as the outstanding student of the previous year at that prestigious institution. West Blocton of the 1930s took great pride in the achievements of its successful college students.

The town lost two prominent citizens in 1938. Dr. George W. Randall, age 60, who had been prominent in civic affairs at local, county and state levels for almost four decades, died in early April while on vacation in Florida, and in late November, Huey Wood, aged 53, former law enforcement officer and businessman, died at his home. Both had been effective and loyal West Blocton leaders. Wood had been particularly effective in the development of an accredited high school in West Blocton in 1920 while he was serving as Bibb County Sheriff. Also passing quietly at his son's home in West End in October was George Campbell Burton who had lived the last several of his ninety-eight years in West Blocton. According to his grandson, George R. Burton of Shelby, Burton was a native of east Tuscaloosa County and had gone into the Confederate Army in his early twenties. He was serving as messenger to General Albert Sidney Johnston when the general was killed at the battle of Shiloh. Although he was not a long-time resident, he was the last Confederate veteran living in West Blocton. He is buried in the cemetery next to the Methodist Church in Brookwood.

Happy Business Days are Here Again

The big news of early 1938 was the announcement that the Black Diamond Coal Company of Birmingham planned to open the old TCI Number 9 mine. While it had not been reported in the weekly newspaper except in syndicated columns on international news, Europe was on the brink of World War II. Germany's resurgence as a military power and the apparent weakness of her continental neighbors and Great Britain invited conflict over territorial ambitions on the continent and in European overseas colonies. The predictable conflict had its effect on the nation's economy as the need for fossil fuel was anticipated. It had been just under a decade since the closing of Number 10 and the departure of TCI that had so drastically affected the local economy.

The *Press* editor observed in the March 1, 1938, issue:

> THE OPENING OF THE MINE AT WEST BLOCTON
>
> Work on opening of the No. 9 Mine at West Blocton by the Black Diamond Coal Company of Birmingham is progressing nicely. Several months will be required to get the mine operating. It is planned to ship several thousand tons of coal each day from the mines, and about 250 men will be employed when operations are at a peak. There are now several hundred people employed in the mining business in Bibb County. In fact this is the largest industry in the county with a weekly payroll of many thousands of dollars. The mining companies pay a large amount of taxes in the county. For the past several years it looked as though the mining business in Bibb was "shot," but it is hoped it will pick up and continue to operate at a just profit.

The 1939 *Industrial Relations Report* identifying mines, management, and number of employees listed J. P. Upchurch as superintendent and John Samsel as mine foreman, and there were 152 employees in the Black Diamond, Blocton #9 mine. In the same report Hill Creek mine employed 232, with A. A. Ashworth superintendent and Vance Tatum foreman. Blocton Mining Company's Klondyke mine employed 165 with R. G. Moore superintendent and J. W. Fancher foreman.

When the Black Diamond Company achieved full operation, it opened a company store that offered groceries, dry goods, and furniture. Immediately it was the largest and busiest store on Main Street, occupying the ground floor of the Masonic Hall, and later the old Joe Israel Store building next door while featuring the town's first neon sign. Following an old-fashioned company policy, daily miners' checks could be cashed in the company office and company scrip, called "clacker," could be spent in the store. Those cashing checks before payday were obliged to pay the company a 10% fee. The abandoned Presbyterian Church was remodeled and served as a furniture warehouse for the store, and a company restaurant occupied the old metal-sided Alabama Power building just to the east of the main store. This was one of the last pre-1927 structures and was removed in the 1970s to enlarge the Southern Bell Telephone building. Eventually, the company restaurant was moved to a building next door to the company store and another building housed a furniture department.

While wages were still low, mine operations had picked up in the other mining towns as well as in West Blocton. Because of more area mining jobs being created and mines operating at full capacity, retail sales increased and within a short time all the old store fronts on Main Street were occupied. Other businesses opening in the late 1930s were Pierson's Hardware, occupying the former Klotzman building. The manager was W. R. White, who moved his family to town from Centreville. Earlier Clara Sue Holifield had opened a beauty shop in the Peacock Building (the drug store), though this business was short-lived. Following the beauty shop in that space was the Jackson Clothing Company, a ladies ready-to-wear shop with W. D. Harris manager.

The clerical workforce in town grew enough to attract the attention of union organizers. By early 1939, leadership from the mine locals sponsored action to unionize store clerks. At full employment times there were at least three dozen employees on Main Street qualified for membership. One of those involved in assisting with early organizational efforts was Percy Riley, prominent mine local union officer. Another active clerks' union officer in the mid-1940s was G. Odette Slaton. The organization was active, but according to recollections of its members, accomplished little for their welfare. Clerks eventually relied on the wage and hour law for the small wages received. As late as 1946–47, some thirty-six stores on Main Street and on Highway 5 agreed with Clerks' Union No. 12578 to close on Thursday afternoons, except the three Thursdays before Christmas.

As mining jobs and population increased, traffic in the narrow street of the business district, especially on Saturdays, endangered pedestrians and caused Main Street traffic problems in the center of town. To remedy the problem, the city council elected to purchase a traffic light for the Main and Florida Street intersection. Some claim it to have been the first

First brick house in West Blocton built by Ed and Mary Etta Gillespie, *circa* 1938. (Courtesy of Mary Jo Jennett Tibbs.)

traffic light in the county. Others say that Centreville had installed a light near Courthouse Square, and West Blocton, still the largest town in the county, was not to be bested on civic improvements. In any event, the light often was side-swiped by high-loaded trucks resulting in its being knocked askew. The town often called on the local Alabama Power office manager to effect the light's proper realignment with the streets. A standing joke by an element that thought the light unneeded and frivolous said the black and white races used the red and green to cross the street separately while cows used the yellow. West Blocton was on the rebound and able to "laugh at a fall, and baffled get up and begin again" as predicted in the *Birmingham News* editorial after the 1927 fire.

Several major fires plagued the community in the 1930s after the town lost its 1929 fire truck in the early, desperate days of the Depression. In late November 1937, during an early season hard freeze, the Kenneth D. Young home burned to the ground. Defective fireplugs (some said the nearest one to the residence was frozen solid) and a lack of fire-fighting equipment were blamed for the loss. Little was saved in the late Sunday night blaze. Two fires struck the M. C. Davie family. The Main Street store that had been rebuilt after the 1927 town fire was a total loss in October 1932, and within a few weeks, their home on Tin Top Street was also consumed. Both fires could have been extinguished, the family claimed, if the town had still owned a fire truck. In the early morning of January 19, 1939, the home of the J. A. Johnsons burned on School Street with their fifth-grader son and Mrs. Johnson's mother barely escaping with their lives. The home was a complete loss. Only a weak, single stream of water from the volunteers' fire hose was available. In early February, the home of Dr. L. E. Peacock was badly damaged by an upstairs fire. It was detected early and luckily controlled by the prompt arrival of fire fighters. The town government was urged by some of its most prominent citizens to make the place safer from the threat of fire.

Huey Frederick and his wrecker were a common sight around Blocton. (Courtesy of Huell Lane Frederick Martin.)

Huey Frederick could tow your automobile to his café and sell you a hearty plate of bar-b-que.
(Courtesy of Huell Lane Frederick Martin.)

The April 6, 1939, *Press* announced in the West Blocton personals news column that, "The Town of West Blocton has purchased a fire truck. We are sure every citizen is proud of this movement." After being handicapped for seven years, the local volunteer fire department again had a truck with a pump to fight fires. L. P. Parker, long-time member of the volunteer fire department, believed the truck came from Elba, Alabama, and that it had been submerged in one of that community's many floods. This, if true, would account for the vehicle having continuous motor and gear problems. City records are missing from several years in the 1930s thus making it impossible to verify the actual supplier. It was big, shiny, newly painted red, and when cranked could be heard all over town. Apparently, the muffler was either missing or defective. West Blocton Fire Department was printed in bold, gold letters on its hood. Distracted by its new paint job and generally impressive appearance, the casual observer hardly noticed the six-inch square welded patch on its massive radiator. Regardless of its imperfections, the water pump worked handily allowing at least two strong, high flow streams of water to be put on threatening flames. Many remember it as a great addition to football pep rally parades with megaphone-equipped red and white clad cheerleaders riding proudly in the hose compartment hoping with the driver that the truck would make it up the next steep hill on the parade route.

The old workhorse truck had to last longer than the town fathers envisioned. This was a stopgap, affordable purchase to be used until a new truck could be considered. The mayor, remembering the fate of the new, expensive apparatus bought in 1929, was not about to over-step uncertain revenue for a new truck. The Depression was not officially over and seemed to linger in West Blocton longer than other places. Then World War II intervened giving vehicle production priority to the military. Small Alabama towns were near the bottom of entities qualifying for a fire truck. Before the old truck's last days, its advanced degree of disrepair was painfully evident. The aging truck's gear mechanism eventually gave way to an irreparable state making it impossible to be driven. The pump, however, still could do its duty as well as ever. So, in lieu of buying a different unit, new or used, the thrifty town council compromised by having the vehicle towed to fires to make use of the pump. The fire chief, Huey Frederick, also operated a cafe, dance hall and auto repair shop, called "Ugly But Honest Huey's Place," a slogan painted on the doors of his small wrecker. This service truck was given the standing assignment of pulling the fire truck to fires, its front wheels elevated just enough to clear the roadway. A 1940s visitor from Tuscaloosa witnessed this extraordinary duo-unit go into action on the occasion of a mid-summer's fire alarm and, according to his host, fell to the ground in a spasm of laughter. The witnessing locals did not take kindly to his apparent contemptuous behavior. The old fire truck ended its days serving as a giant toy on a town playground to be climbed on by juvenile firefighters. Finally, the semi-antique derelict was declared a hazard and hauled off as scrap.

CHAPTER 8

Fighting Hitler and Tojo

In 1940, sixth-grader Turner Crowe had one of the most impressive scrap iron piles in his West Blocton neighborhood. He was among the town's boy scavengers who made frequent scrap metal gathering forays with their wagons seeking discarded steel, iron, copper, and aluminum. Copper wire brought the best price, a penny a pound. Generally, the "scrap iron man" would give 30c a hundred pounds for most metal pieces on his periodic visits. While other boys quickly sold off their collections for pocket change, Turner held back. The scrap piles were destined for Japan. He told friends and classmates the scrap iron they sold would be converted into bullets by the Japs to shoot at American soldiers. He vowed he would hold out until his scrap could be used by his own country.

Copper wire brought the best price, a penny a pound. Generally, the "scrap iron man" would give 30c a hundred pounds for most metal pieces on his periodic visits.

One summer day in the late 1930s, thirsty playmates had followed Turner into his grandparents' kitchen for a drink of water. On a shelf of the sleeping porch was Grandfather Turner's old Atwater Kent radio with an external, round speaker atop its metal case. It was tuned in to a rebroadcast of a short-wave recording. The eerie sound being aired was of a frantically excited foreign-speaking voice that faded in and out of background static obviously delivering an emotional oration. Periodically, when the speaker paused, there were odd, reverberating crowd responses of, "Heil Hitler! Heil Hitler!" One of the boys asked, "Who's that?" Papa Turner answered, "That's Adolph Hitler, the Dictator of Germany. We'll soon have to fight him."

The first pre-World War II newspaper story telling of a West Blocton boy enlisting in the service appeared in the August 17, 1939, *Centreville Press.* "Mr. Billy Harris joined the Army Air Corps at Ft. McClellan, he will train at Charleston, S. C. and then sail for Hawaii where he will be stationed for two years."

Hitler launched his attack on Poland on September 1, 1939. Two days later the familiar voice of news commentator Elmer R. Davis, popular radio newsman, announced to his evening American audience including West Blocton households, "Great Britain and France declared war on Germany today, twenty five years after the start of the last war." On September 7, 1939, a week after the war's declaration in Europe, *Press* editor Oakley wrote, "Everyone agrees that Hitler and Germany are to blame for the current conflict." While the reason was not fully understood for several years, letters from Jewish relatives in Poland to the Joe Israel family in West Blocton slowed to a trickle and then suddenly ceased. Annual fall High Holy Days greetings and letters sent to Poland by the Israels went unanswered. Adolph Hitler's Holocaust thus reached into the homes and hearts of the last Jewish families in this small, southern town.

The West Blocton correspondent to the *Centreville Press* during the times was Mrs. Glen Seals. She made every effort to keep her readers aware of the community servicemen's stations and the achievements of the college set. Another reporter during the war years was Mrs. Bill Keyes. Sometimes international news took a back seat to local events.

In late January 1940, a ten-inch snow covered the area accompanied by sub-zero weather. The cold siege lasted long enough to give school children a two-week vacation. Water pipes froze all over town. The snowed-in families melted the white stuff in washtubs on kitchen stoves to supply water for household use. Family pets would not venture out into the elements and chickens remained on their roosts. Youngsters braving the elements hiking to Caffee Creek found it frozen over and the ice thick enough to withstand a cautious walk on its surface. Old-timers likened it to the winter of 1899 when a similar severe cold spell spread over the deep South. Some remembered that during the 1899 freeze, an owner of ice skates from his days up north actually skated on the Cahaba River. Joseph Squire wrote in his autobiography that the thermometer read 10 below zero on February 13, 1899, at Winfield in northwest Alabama. The same reading likely visited the Cahaba River Valley in Bibb County.

The people of West Blocton had always been interested in state and local politics. In early evening on April 30, a Main Street political rally sponsored by the West Blocton Lions Club attracted local and state office seekers. Tulon Desmond arranged a parade and performance of the American Legion Drum and Bugle Corps of Birmingham. With the music, fireworks and political speeches, it was like old times, reflecting the return of an element of civic energy after a long quiet spell nurtured by the Depression.

West Blocton had been aided significantly by a number of government-sponsored projects, most resulting in water works improvement and school buildings. In May 1940 the Works Progress Administration (WPA) was still making its impact felt in the community. A banquet for the professional and service division of the agency was held in the Methodist Church basement to recognize their achievements. L. S. Moore, H. H. Farrar, and Russell Thomas spoke about significant projects. Russell's subject was "What Beautification Has Done for the Schools of Bibb County." A news story said that over 2,000 plantings had been made in the county around schools and other public buildings. If one knows where to look, evidence of tree plantings during the WPA era are still in place. A row of water oaks planted by a WPA work crew near the Johntown Baptist Church marks the former location of the community's three-room school building. Trees and shrubbery were also planted on the 1935 West Blocton High School campus.

The county weekly announced that Elmo Israel, who would graduate in May at the Capstone, was named the most outstanding male student for the last four years at the University. He earned a 2.96 quality point average out of a possible 3.00; was elected to Phi Beta Kappa; was editor of the *Rammer Jammer*, the school's humor magazine; was associate editor of the *Corolla*, the yearbook; and recipient of many other honors. Also at the University of Alabama, James Coleman Parker was awarded ROTC rank of Second Lieutenant, as was Milton Fancher. Jack Lawrence was awarded

WPA Worker Crew. West Blocton, 1940. (Author's collection.)

Elmo Israel Ellis, college photo. (Courtesy of Elmo Israel Ellis.)

a degree in textile engineering at Auburn; Henry Clay Randall graduated from prep military school at Columbia, Tennessee; and Billy Harris was stationed at Hickam Field in Hawaii. In October, Bill Carroll and Henry C. Randall were listed as fraternity pledges at the University. Carroll had been a member of the freshman basketball team, was active in intramural sports, and held membership in the Pershing Rifles, an honorary ROTC organization.

In November, in an unprecedented run for a third term, President Roosevelt won over Wendell Willkie ten to one in Bibb. In Beats 10 and 11, Roosevelt won 312 to 38. Also in November the town learned of the death of banker J. B. Wadsworth, aged 76, of Gadsden. He had been a citizen of West Blocton in the 1890s and was known locally as the founder of the Blocton Savings Bank in 1892, the county's first bank, which he operated for ten years. He also built the bank and hotel building in Centreville where he founded the Bibb County Banking and Trust Company in 1897. He once owned and operated the coal company at Coleanor while a resident of Bibb.

The Selective Service draft started in September 1940. Governor Dixon appointed W. E. Nicholson and N. W. Hunnicut of Centreville and Tulon Desmond of West Blocton as members of the Bibb County Draft Board. By October 31, the agency had begun its work. Edward Russell Smelly's number was called making him the first man from West Blocton eligible for the draft. However, Sammy George and Rufus Cristo Lagrone had the distinction of being the first to leave for basic training having enlisted before being drafted. In early December, they departed for Fort Barnacus, Florida. Other early departing boys were Raymond "Chink" McKinney, Jack George, and brothers Herschel and George Cunningham. Luther Elmer Johnson, Loyd Palmer Marchant, and Elija Nelson, men from West Blocton Routes 1 and 2, were also listed in the early draft.

Other events made the December pages of the *Press*. W. W. Perry of West Blocton was honored in Birmingham as one of the 100 oldest Shriners in the state. In 1940 for the first time, local merchants and the town sponsored Christmas lights strung across downtown Main Street. The festive decorations brought an element of cheer to a community painfully aware of a world threatened by a spreading and worsening European and Far Eastern war. In mid-year the results of the 1940 census showed that West Blocton had 1,325 residents, a gain of 255 over the 1930 count. Centreville had 891 and Brent, 828. The West Blocton count included the city limits only so the numbers for beats 10 and 11 combined would have shown a more accurate number since the 1930 census had shown a count for Blocton and West Blocton, making the 1930 and 1940 numbers hard to compare. The population increase was a result of the opening of the Number 9 mine by the Black Diamond Company. Main Street businesses were still on the increase. Sumter Farm and Stock Company that had operated Yellow Front Stores in Centreville and Brent for several years, added another in West Blocton to their chain by opening on March 15 in the old Allen Furniture and Undertaking building. This general store offered groceries, dry goods, as well as domestic animal feeds.

The 1940-1941 school year saw the addition of a band director for the two county high schools. He spent half time each in Centreville and West Blocton. New director James Hoit was a talented musician and teacher who brought his two bands along at a rapid rate, both groups were playing elementary marches and concert band pieces by Christmas. Several WBHS band members serenaded on Main Street on Christmas Eve 1940. These members of the band knew one tune, "Rose Bud," a beginner-band staple, and according to Tulon Desmond it was played at every stop. Despite the limited repertoire, many household members came to their porches and applauded. Among the WBHS musicians were several with special talent who formed a small jazz ensemble playing contemporary pieces. Among the members were: Nell

Ambrose Place, 1940. (Courtesy of Geraldine Ambrose Pate, Bessemer, Alabama.)

Finnen, piano; Tom Finnen, trumpet; James Harold "Bozo" Finnen, trombone; Jimmy Roycroft, saxophone; Brooks Young, clarinet; Tulon Desmond, trumpet and piano; and Buddy West, drums. Their selections included a special arrangement of "Tuxedo Junction"; other pieces were "In the Mood," "Blues in the Night," "Perfidia," and "Basin Street Blues."

The earliest uniforms of the marching and concert band were simply navy blue coats and white trousers. Later, WBHS uniforms were red wool capes with white satin linings made by the home economics sewing classes under the direction of teacher Effie Ozley Ray. The school bought caps and individual band members supplied black ties, white shirts, white duck trousers, and white shoes. It was a sharp looking aggregation, and in West Blocton occasional parades and concerts gave a special boost to school pride and community spirit. The weekly newspaper reported the band in concert on February 9, 1942, boasting forty-five members. Band director Hoit, along with other male teachers, were soon called to the military. He was not on the faculty in 1943-44.

The organized labor movement was always strong in the West Blocton area and the locals were disciplined to adhere to the national headquarters in walkout decisions. Such was the case in April 1941, when area miners went on strike over contract disagreements. In early May the Klondike Mine was destroyed by fire. It was reported in the *Centreville Press* as follows:

> KLONDIKE MINES DAMAGED BY FIRE
>
> Fire of undetermined origin destroyed the tipple, washer and two Southern Railway cars at the Klondike mine last Thursday night.
>
> The fire was discovered about midnight but had gained such headway that it was impossible to bring it under control. The washer had been recently completed and had been in operation about two or three months.
>
> Fire fighting equipment was removed from the mines about 3:00 o'clock when it believed all danger was passed, but was called back about 6:00 o'clock when fire broke out in the shaft of the mine caused when burning timbers from the tipple fell into the shaft.

WBHS Band near the Drugstore, 1942. The town's first and only traffic light hung solemnly above the occasion. (Author's collection.)

Reminders of a world at war were everywhere. Herman Fancher and Melba Shamblin posed before a billboard asking citizens to pay income taxes early, thereby helping the country's war effort. (Courtesy of Martha Fancher Kelley.)

Claude Holland. (Courtesy of Roselle Brewer, Lowerytown, Alabama.)

The mine never reopened because of inadequate insurance. Approximately 185 men employed at Klondike had been on strike since April 1 due to the national coal strike. Some said the fire was deliberately set, a thought harbored for many years by members of the Young and Lawrence families, the mine owners. Union leaders countered the sabotage rumor with their opinion that the mine was losing money and the owners wanted out of the operation. However, investigation turned up no evidence of arson, and the fire was listed officially as accidental. Where the Klondyke workers found employment is not clear, however, Belle Ellen, Hill Creek, and Black Diamond mines were in full operation once the strike was settled. Dave Evans and George Young also had truck mine operations employing small crews.

Defense and energy conservation had become necessary considerations by the nation with the threat of war reaching across the Atlantic and Pacific. Local evidence of this occurred in November when the football game between the Tigers and Choctaws was played in the afternoon to save energy. The Office of Production Management prohibited the use of lights for nighttime play. Bibb won the game. According to William J. Murphy, early departure of many boys for military service left but 15 players on the Tiger squad. There were no football teams fielded by WBHS in 1943 and 1944. Senior classes also suspended the publication of annuals those years.

Many of the high school's male graduates in 1940 and 1941 joined the armed forces either by Selective Service draft or volunteering. The local newspaper writers dutifully identified their training and base assignments. Many thought the mandatory year's military service seemed little enough for boys to do for their country before starting careers. But history changed those short-term plans.

Word reached West Blocton's radio listeners on Sunday afternoon, December 7, 1941, that the Japanese had attacked Pearl Harbor in Hawaii. Five West Blocton-born servicemen were stationed on that base. The following Monday, the United States Congress declared a state of war with Japan. On December 11, Germany and Italy, because of their pact with Japan, declared war on the United States, and the Congress reciprocated on the same day. The long expected war had come.

Claude Holland, nineteen years old, from Lowerytown, former WBHS student, had joined the Navy on October 4, 1940. He was a member of the crew of the battleship *Arizona*, which capsized after taking a direct bomb hit during the air attack on Pearl Harbor. His family was officially notified of his loss a few days before Christmas. Seaman Holland was the first World War II casualty from Bibb County. His body and those of his shipmates remain entombed in the *Arizona*, now a shrine to those who lost their lives in the Pearl Harbor attack.

W. D. "Billy" Harris had joined the Army Air Corps in 1939, and was assigned to Hickham Field on Oahu. His parents received a terse telegram from him ten days after the attack that said, "Am O.K." The Harris family reported to the *Press* that they "greatly appreciated this short message at so serious a time for our boys in Uncle Sam's service in Hawaii." Harris remained in the service as a career officer. He was assigned to the Pentagon and retired as a Lieutenant Colonel.

Remarkably, Dan Moon (James D., Jr.), survived the Pearl Harbor attack on his ship, the light cruiser *St. Louis*. He had joined the navy in July 1940 and was assigned to his ship in November at Hampton Roads, Virginia. The following is from his "Wartime and Personal History" written in 1994 for inclusion in a special "Time Capsule" at the Lodi Memorial Cemetery in California:

> On December 7, 1941, the *USS ST. LOUIS* was in the naval shipyard at Pearl Harbor being outfitted with a new radar when the Japanese attacked. The *St. Louis* was outboard of the *USS Honolulu* which was alongside the dock. When the alarm sounded I was asleep in my bunk and I jumped out of it and ran all the way to my battle station in my shorts thinking it was a drill. When I got to my battle station which was in the lower handling room of number 3 and 4 AA [anti aircraft] mounts, there was no one there and there was no electricity to run the hoists. Later people began to show up and said we were at war with the Japanese. Being young and naive, I did not know we were even close to being at war with anyone. An enemy aircraft raked the ship with machine gun bullets and a bomb dropped through the dock against which the Honolulu was nestled. The concussion was so tremendous it not only damaged the *Honolulu* but shook the *St. Louis* to the extent that we were certain she had sustained a direct hit. About an hour after the shooting started our AA mounts started and we went out of the channel at 25 knots. On the way out we sank a two-man sub and later downed three aircraft. We were the only large warship to get out of the harbor during the battle. After a three day search for the enemy we returned to Pearl Harbor. The water was black with oil and debris. At Alea Landing caskets of deceased personnel were piled on high as far as you could see. We lost over 2,000 men that day and one thousand are still entombed in the *USS Arizona*. A friend I went to school with, Claude Holland, is one of those men.

Moon spent the rest of the war in the Pacific, his ship taking part in action at Kiska in the Aleutian Islands, and also at New Georgia Islands, where his ship was damaged by a torpedo attack. He was injured during that action and received the Purple Heart. He saw further action at Bougainville, Guam, and Leyte Gulf in the Philippines. At Leyte, his ship was hit by Kamikazes, and for his efficiency and bravery in that action he was awarded a commendation by the commanding officer. His ship was involved in attacks on the Japanese home island of Kyushu and later at Okinawa. Moon remained in the Navy for just over 22 years. Later, he worked with the Civil Service, finally retiring in Stockton, California.

James D. "Dan" Moon. (Courtesy of James D. Moon.)

Ben Jennett, Jr. (Courtesy of Elizabeth Jennett Brannon.)

A West Blocton-born Marine, Bernice S. Swindal, brother of Col. James B. Swindal, was a gunnery sergeant stationed at Pearl Harbor on the day of the attack and was able to get off some return fire when enemy planes strafed his barracks. He had a number of anxious relatives in West Blocton. He served in major Pacific actions the remainder of the war and later saw action in Korea. The Swindal brothers were born in the same West End, Main Street house. Another West Blocton vicinity serviceman at Pearl Harbor, Eugene Fountain, survived the attack on the islands and made a career of the military.

Another of West Blocton's sons was in the midst of the Pacific war. Ben Jennett, Jr., enlisted in the navy when he turned seventeen, and in early December 1941, he was assigned to the recently commissioned aircraft carrier *Hornet* in Newport News, Virginia. After the enemy attack in the Pacific, preparations were soon underway to strike back at Japan. The *Hornet*, a part of those plans, soon departed for California where it took aboard sixteen B-25 Bombers. It was from Jennett's ship that the famous April 1942 raid on Tokyo was launched under command of Col. Jimmy Doolittle. President Roosevelt humorously said the planes were from a base called "Shangri-la." Jennett was also in the important battle of Midway the following June. In late October 1942, while Jennett was still aboard, the *Hornet* was sunk in the battle of Santa Cruz Islands while defending American positions on Guadacanal Island. Jennett survived the sinking to fight throughout the rest of the Pacific war. At one point in his career, Jennett was the youngest petty officer in the Navy. He earned officer rank and remained a career officer, retiring with the rank of lieutenant commander. Jennett, an avid and talented golfer, pursued a second career as a golf pro. Again retired, he is a resident of Memphis, Tennessee.

These men are worthy of remembrance for their being in early major war battles of our country in World War II. They were West Blocton youth in harm's way. Considering the town's small population, it is remarkable that five native sons were assigned to Pearl Harbor on that fateful, historic day and that another was aboard the famous carrier *Hornet* for a part in the first air-strike on Tokyo.

School children in West Blocton were made aware of the urgency of the country being at war by air raid drills. The usual fire drill was a series of short school bell rings. New instructions issued in January 1942 added air raid drill signals, long rings of 15 seconds and each class appointed an air raid warden with assigned duties. School reporter Jo Ann Robinson wrote, "We hope to have our group prepared if the enemy planes should enter our territory."

Rationing of strategic materials was begun in earnest in Bibb County in the early part of 1942. Rubber was one of the scarce resources. Details appeared in the newspaper announcing regulations governing issuance of certificates to allow citizens to buy tires and tubes for automobiles and trucks. One of the most difficult rationing problems had to do with gasoline. All over the country there were cases of counterfeited coupons, ticket stealing, as well as the stealing of gasoline from industrial supply tanks. In West Blocton, watchman Len Thompson at the Red Dot Coal Company, owned by George Young, shot an intruder that had attempted to steal gasoline. The wounded man was caught and taken to jail. The *Press* commented, "Gas thievery has been a very common practice in Bibb County since the rationing of gasoline started several months ago, but a few well placed shots should help stop [it]."

The mid-February paper stated that 1,048 Bibb men had registered for the draft in the first wartime call. In mid-March West Blocton was made very much aware of the war by the county's first "blackout." It was termed a success and reported as follows in the *Press*:

> WEST BLOCTON TEST BLACKOUT
>
> West Blocton March 17—West Blocton successfully observed its first blackout Tuesday night. The blackout period began at 8:55 and lasted until 9:15. [The town fire siren sounded to warn of the simulated attack and to signal the all-clear].
>
> A member of the civilian defense council of West Blocton said he had never seen the town as "black." Not only West Blocton participated but also the surrounding communities of Smith Hill, Hill Creek, Primitive Ridge, and Lowerytown.
>
> There were 125 officials on duty throughout the district. After the blackout, they reported to L. S. Vanderford, West Blocton Defense Chief.

> Dr. H. M. Fuller, Chairman of the Bibb County Defense Council stated this week that plans are now being made to have a test blackout for Centreville and Brent in the near future.

Mimeographed instructions had been delivered to every residence and business in the area prior to the exercise. The blackouts continued at various times during the war. A letter dated September 26, 1943, from Alice Quinn to her future husband, Willis Pope, who was in the Navy stationed in San Francisco, contained a blackout reference, "We have just come out from under a trial blackout-it lasted an hour—while it was going on I tried to picture the service boys that work altogether in the dark—I'm sure my picture was very mild compared to the real thing."

Willis Pope. (Courtesy of Wanda Pope Miller.)

In a February 1942 *Press* issue the organization of the county Sugar Rationing Board was announced. First registration was scheduled at the high school for April 28-29. Earlier, a local Rationing Board Office was opened on April 19 concerned with food, tires, and gasoline. Its offices were in the old West Blocton Savings Bank building. All citizens were required to register for ration books.

Other local developments in early 1942 included the appointment of M. C. Davie to the county draft board in place of seriously ill Tulon Desmond. The Bibb County Health Department opened a North Bibb Health Office in the rear half of city hall to serve expectant mothers, babies, and venereal disease patients. Midwives were offered monthly training meetings.

In March, James Donald McCool joined the Navy, one of the youngest local boys to volunteer for military service. Those who knew James Donald believed that he was not old enough but his age obviously was not closely checked. Another very young man, James Harold "Bozo" Finnen, joined the Navy in the summer of 1942. Many believed both these boys in their zeal to serve the country, had given incorrect birth dates for enlistment. An often-heard explanation about youthful enlistees was, "his parents signed the papers."

Wartime did not cancel the annual high school event of Stunt Night, Reds versus Whites, which was a part of local school events from 1940 through 1947 and patterned after College Night at Alabama College in Montevallo. The evenly divided student body had king and queen contests, athletic and poster competitions, and a parade that preceded the big night. It featured student entrances by each side, songs, yells, and a short dramatic "stunt." Bibb County High had held a similar program from the late 1930s. During the war, intensely patriotic Principal L. S. Vanderford gave the entire proceeds from the event to the Red Cross, foregoing the purchase of badly needed school supplies.

In April 1942, the community was saddened by the death of Tulon Desmond, local successful businessman and civic leader for nearly three decades. Desmond was active in county and state politics. As a veteran of World War I, he was an organizer of the local American Legion Post and also served several years as West Blocton's city clerk.

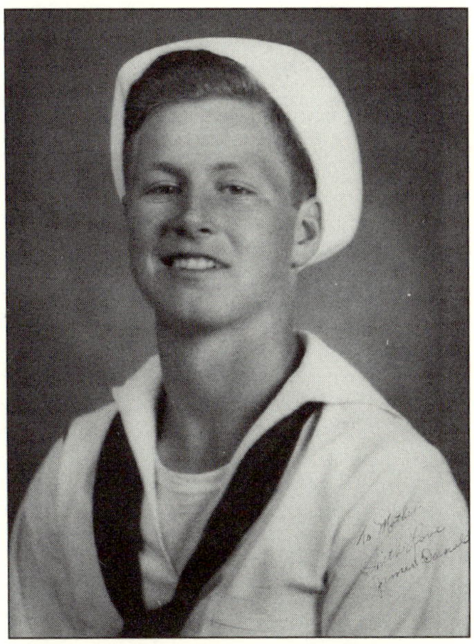

James Donald McCool, fourteen years old. (Courtesy of James Donald McCool.)

An interesting front page *Blocton Enterprise* story of April 16 carried the picture and story of Mrs. A. W. Davenport of Blocton, living with a daughter in Pasadena, California, who was honored for having five sons in the navy. Mrs. Davenport was invited by the Secretary of the Navy to sponsor the launching of a ship in San Francisco in August. It is believed the Davenport family lived in the community only a short time, but long enough to call it home. Closer to home and well-known was Mr. and Mrs. C. P. Parks who eventually saw five sons in service: John, Comer, Carey, Robert, and Frank. A daughter-in-law, Comer's wife (Catherine Blake Parks), also was in service as was a Parks' foster son, Herbert Pope, known as "Little Pope." Mrs. Parks was remembered as a "Five Star" mother in the November 1998 Mt. Carmel Baptist Church's veterans program.

Zebulon V. and Maude Prather's four sons were in the service: Charles H. and Howard "Booty" in the Navy and Hollis Edwin and Herbert H. in the Army. Herbert was wounded and awarded the Purple Heart, and Hollis earned a Silver Star for bravery during Northern Solomon Islands fighting in the Pacific.

Mr. and Mrs. James D. Moon, Sr. also had four sons in the armed services. James D., Jr. (Dan) and Howard are subjects of this chapter. Also serving were Charles Roland and John Thomas (J. T. or Jake).

Attention was focused on the Pacific theater during the early months of the war, but the main fight as agreed by Allied leaders was in Europe. Stopping Germany, conquerer of the continent and threatening invasion of Great Britain, took priority. In all serious dramas, there is a need for comic relief. A bit of that element appeared in a syndicated photo and its caption that was spread from coast to coast in May 1942. It showed 28-year-old Private Frankie Farnetti of West Blocton teaching Irish girls to dance the jitterbug. The family and local postmaster received copies of the daily newspaper photo from as far away as New York and Philadelphia. Farnetti, son of an Italian immigrant, had joined the army in April 1941. After basic training in New Jersey and Louisiana, he had been sent to Northern Ireland. The local newspaper erred in identifying the serviceman as Frank Ferrire also from West Blocton. West Blocton's Frank Ferrire was 21 years old, and in July 1942, was stationed at Ft. Sill, Oklahoma. Ferrire spent his fighting time in the Pacific where, in the jungles of Luzon, Philippine Islands, in May 1945, he was awarded the Bronze Star for wiping out a nest of Japanese snipers. After the war he was a businessman in clothing and florist services in West Blocton. His tenure as the town's mayor is covered in later pages.

Frankie Farnetti. (Courtesy of Frankie Farnetti, Santa Monica, California.)

Frank Ferrire and Elizabeth "Baby" Wood, Birmingham Airport, 1943. (Courtesy of Bettye Kitchen Hughes.)

As had been the record of the Blocton-West Blocton area in World War I, response by the home front was always in the "over-the-top" category. Early in the war, sale of war bonds in north Bibb was lead by West Blocton, where $65,023 was sold. The area had pledged to buy $123,625 during 1942. The county pledged $189,825 with the north end of the county accounting for 65% of the total. Coal mine employees at Black Diamond and Hill Creek, showing their patriotic fervor, pledged $35,500 in purchases.

Scrap drives for critical material began by mid-year. As chairman of the Bibb County Petroleum Committee, George E. Dailey of West Blocton was county chief of the scrap rubber drive. All filling stations agreed to be collection points. To meet the county quota, five pounds of rubber from every man, woman, and child was requested. Schools sponsored scrap drives to collect discarded metal and rubber. From time to time the amount of "salvage" each school had collected was printed in the weekly paper. As late in the war as January 1945, Boy Scout Troop 35, directed by scoutmaster Francis Leonard, was actively engaged in the collection of scrap paper.

Kent West. (Courtesy of Eloise West Wintter, Bessemer, Alabama.)

The American entry into the war had begun to make a difference in Europe by late summer 1942. That factor became profoundly evident to West Blocton citizens when word came to Mr. and Mrs. George H. West that their son, Kent, an army aerial gunner, had been credited with being the first American to shoot down a German pursuit plane. His plane, a B-17, was nicknamed the *Birmingham Blitzkrieg,* and West was the "belly gunner." It was the first all-American raid on German rail yards in occupied Rouen, France. The *Birmingham News* carried West's photograph on its front page and said, "His home is at West Blocton but chances are Sgt. Kent R. West will become known as the 'Birmingham Blitz'." When the Movietone Newsreel showing Kent's return to England after the successful raid was shown locally, Earl Philpot, manager-projectionist of the Strand Theater, stopped the film so all could clearly see a still frame view of Kent astride a bicycle about to ride away from his plane. The cheers and applause by the audience were deafening.

Tragically, Kent was killed when his aircraft was downed by enemy fire on December 26, 1942, in North Africa. He was first listed as missing in action when the plane failed to return to its base. His death was not confirmed until February 1944. On March 5, a memorial service was held in a packed First Baptist Church of West Blocton. L. S. Vanderford, Kent's high school principal, read the eulogy. Among West's honors noted was the recognition of his bravery by the British people by having a life sized portrait of him in the Hall of Fame in London with the story of his life beneath. Earlier the family had been notified that his portrait and personal history were displayed in the British Royal Air Force headquarters. It is not known if this portrait is still in place in London. Principal Vanderford also identified the awarding of the Purple Heart with One Oak Leaf Cluster by the U. S. government. In the accounting of his life history, Kent's playing on the WBHS football team was mentioned. West was also remembered in the November 1997 Veterans' Day recognition at Mt. Carmel Baptist Church.

The first West Blocton soldier in World War II to be wounded seriously enough to be discharged was Edward "Punch" Lynch from the African-American community. His was a leg wound inflicted by enemy fire while he was serving in the South Pacific. After returning home, he was employed by Roland Funeral Home. Later, his son, Fred Lynch, was elected to the West Blocton City Council.

Private Chester Hargrove, an African-American soldier from West Blocton, was first reported missing in action. His sister, Daisy Wesley, received a telegram dated June 13, 1944, reporting that her brother had died in Sicily on March 10. Community historian Willie Jo Parker identified Hargrove as "the first Black soldier from this area to lose his life in a war."

The John W. Fancher family had double sorrow. Their oldest son, Corp. John W. Fancher, Jr., serving in an anti-aircraft unit, was killed by enemy fire in the invasion of North Africa on February 14, 1943. Their next son, Lt. Milton Fancher, University of Alabama graduate in 1942, was piloting a B-17 bomber in a raid over Germany when his plane was shot down on May 8, 1943. He was awarded posthumous honors the following March including the Distinguished Flying Cross for extraordinary achievement while serving as a pilot of a B-17 and also the Air Medal with four clusters for exceptionally meritorious achievement while participating in ten separate combat missions and other heavy bombardment missions over enemy territory. The posthumous awards were presented to his mother in Birmingham in March 1944. The Fancher brothers' bodies were returned to the States in 1949. The family by then had settled in Marion, Illinois, so the burial was in that locale.

West Blocton not only provided men for the war effort, young women had also joined Uncle Sam's ranks. In February 1944, local news items in the *Press* told that Sergeant Frances Kren, a graduate of Tulane University, had spoken over Birmingham radio station WBRC in her capacity as a recruiting officer. She was stationed in the Frank Nelson Building in downtown Birmingham and invited local girls interested in the armed services to visit her there. Other local women in service were Mrs. Comer (Blake) Parks, whose name is the lone female appearing on the West Blocton Methodist Church's honor roll for World War II. The paper also mentioned Private Grayce Parsons as well as Lt. Roselyn Farrar and Lt. Kathleen Stemetz who were army nurses.

The community was in mourning on May 18 when it learned of the death of Navy aerial gunner Loiel Franks in a Norfolk, Virginia plane crash. Loiel was widely known in West Blocton as a bright, friendly youth with remarkable promise. His body was returned to West Blocton for full honor military rites performed by the local American Legion and a naval escort from the Birmingham Recruiting Station.

D-Day, the Allied invasion of Europe, was anticipated over the spring months of 1944. The following appeared in the *Blocton Enterprise* of May 11:

> D-DAY TO BE OBSERVED AT WEST BLOCTON
>
> Official notice has been received here by Mr. L. S. Vanderford, Principal of the Schools, to be on the alert for D-Day, the day on which the invasion of Europe will begin.
>
> When this announcement is received, the fire siren will sound five or six short blasts and when this

Milton Fancher. (Courtesy of Martha Fancher Kelley.)

Loiel Franks. (Courtesy of Nina Franks Smith, Milton, Florida.)

occurs it is asked that each person leave their tasks and gather at one of the churches for an hour of united prayer for all our boys on the battle fronts. Business houses will also be asked to close for that period.

Press and radio announcements of the invasion may be taken as official and if this announcement comes late at night, D-Day will be observed early the following morning.

All people of every faith are urged to observe this prayer period for God alone can give us the help, courage and fortitude that we're all going to need when the invasion begins.

On the morning of June 6, the fire siren sounded its several short blasts. It signaled that Allied Commander, General Eisenhower, had announced that the invasion had begun. All Main Street stores closed their doors, and managers, owners, and clerks filed quickly into their respective churches. The Methodist Church was filled. The author, then a teenage summer clerk at the Yellow Front Store, remembers being deeply touched by the solemnity and awe of the moment. He recalls not the exact words but the devout and sincere nature of the prayer for peace and safe deliverance offered by Episcopalian lay leader, Dr. L. E. Peacock. It was an occasion one remembers for a lifetime.

Harry Cecil Dyer. (Courtesy of Virginia Dyer Grissett, Hueytown, Alabama.)

Fred Marchant. (Courtesy of Mary Marchant Lippeatt, West Blocton, Alabama.)

Harry Cecil Dyer was killed by enemy fire on a Normandy Beach that historic day. Again, a soldier from a far off, small Alabama town had given up his life for his country. The following November, a memorial service was held for Dyer at Mt. Carmel Baptist Church. His body was brought home for burial in Mt. Carmel Cemetery in March 1949. On the second day of the invasion, Fred Marchant, who had attended West Blocton High School, was hit by friendly fire. A nervous G. I. did not hear his password when challenged and shot him mistaking him for a German infantryman. Marchant died from wounds while being cared for in a small French hospital. Marchant's wife Dottie, then a resident of Gastonia, North Carolina, received a telegram that said, "THE SECRETARY OF WAR ASKS THAT I ASSURE YOU OF HIS DEEP SYMPATHY IN THE LOSS OF YOUR HUSBAND SERGEANT FRED B MARCHANT REPORT RECEIVED STATES HE DIED FIFTEEN JUNE IN FRANCE LETTER FOLLOWS DUNLOP ACTING THE ADJUTANT GENERAL"

Marchant was one of thirteen first cousins, the sons of three sisters, Carrie, Cross, and Martha, nee Woods, serving the country in World War II, all from Bibb County. The family's story appeared in the *Tuscaloosa News* on Memorial Day 2000.

George Burt, another West Blocton soldier, member of the 17th Airborne Division, was killed on March 24, 1945, in a parachute jump during the Allied crossing of the Rhine. He was buried in a Dutch cemetery near the village of Margraten. A special memorial service was held at the West Blocton Methodist Church in March 1946. It was reported that he was hit in mid-jump by a sharp-shooter German defender and was dead before he hit the ground. Burt's body was exhumed from its European cemetery in October 1948, and returned home for burial in the family plot at Scottsville Cemetery. Thomas Finnen died in an accident in England prior to D-Day. Tom was among the large mobilization forces in England during the spring of 1944 preparing for the assault on the Channel beaches. In the close temporary quarters occupied by his unit, a soldier in a tent next door accidentally discharged his carbine. The projectile fatally struck Finnen and, again, there was a moving and appropriate memorial service in the West Blocton Methodist Church.

George Burt. (Courtesy of Martha Fancher Kelley.)

Another West Blocton soldier, Sgt. Raymond Lawrence, son of Mr. and Mrs. A. R. Lawrence, died in an airplane crash at Eglin Field, Florida during a training flight on March 17, 1943. The funeral was conducted locally with visiting military service units, the American Legion, and Boy Scout Troop 35 taking part. The May 3, 1945, *Enterprise* reported the last European war death of a West Blocton serviceman when it announced the loss of Cristo Lagrone during a March 1945 Allied drive into Germany.

Henry Thomas Finnen. (Courtesy of Nell Finnen Bleidt, Birmingham, Alabama.)

The Lewis Farmer family was notified that their twenty-year-old son, Marine Lewis Farmer, Jr., had been killed by enemy action on February 1, 1945, during the invasion of the Japanese-held Marshall Islands. Lewis had been a member of Local Union 7578, a miner at the Black Diamond Company, before entering the Marines. Marine Howard Moon was killed in action on February 26 at Iwo Jima, one of the bloodiest island invasions of the war. As was the case with many others killed in action, Moon's body arrived in the spring of 1948 by train in Woodstock with a military honor guard escort, three years after his death. Ironically, Moon had been employed as a Southern Railroad fireman and the paper reported that his remains came via the same train he had worked on before leaving for service. He was buried in West Blocton's Hickory Hill Cemetery with full military honors. Neil Holland, son of Will and Carrie Holland of Smith Hill, lost his life in early 1945 on the Philippine Island of Mindinao when he was caught in American and Japanese crossfire. Another area Navy serviceman, J. C. Goggins, whose father lived on Smith Hill, also lost his life in the Pacific war. At one time, Goggins lived with the Sewell family on Main Street. He is buried at Mt. Carmel Cemetery.

President Roosevelt died on April 12, 1945. West Blocton joined the nation's sorrow in a simple and moving prayer service memorial in local churches from 3:30 to 4:00 p.m. on the day of the funeral. All local stores closed for the observance.

In mid-May 1945, W. R. Ray mayor of the town for twelve years, died after a short illness. Ray had seen the town through its most difficult days of the Depression and most of World War II. The obituary in the May 15 *Centreville Press* said, "He had many fine characteristics and always stood for justice and right and the up-building of the community. During his administration many improvements have been made. Among the major ones were installing new water works, with new tank and meter system, paving the streets, buying new fire fighting equipment and a city hall, the erection of two new school buildings and numerous other things." Ray had been a mining boss at the Number 7 TCI Blocton mine and superintendent at Hill Creek mine. Mine operator Dave Evans was named interim mayor and in September 1948 was elected to the office for the following term, serving until 1949.

When the Germans surrendered on V-E Day, there was no celebration in the local streets as there had been at the end of World War I. The *Centreville Press* simply announced that the following Sunday would be a day of prayer and thanksgiving. Street parades and impromptu celebrations were left to the big cities while the local folks listened to the proceedings on their radios and waited for the images of war's end to be shown in newsreels at the Strand Theater. There were feelings of relief and thankfulness that half the battle was finished. Especially grateful were the families of war prisoners Herman R. Fancher and Lawrence Frederick. These boys had grown up on the same block in West End and though Frederick had moved to Oneonta before he entered service, they still maintained their friendship, taking military leaves at the same time and visiting before going overseas.

Lewis Farmer. (Courtesy of Judy Farmer Richardson, West Blocton, Alabama.)

Howard Moon. (Courtesy of Mary Clyde Moon Owens, Centreville, Alabama.)

Herman Fancher and Lawrence Frederick, back home, 1945. (Courtesy of Martha Fancher Kelley.)

Both served in the Army Air Corps. Frederick was listed as missing in action when the B-24 Liberator, (nicknamed "Fearless Fosdick" from a Dick Tracy comic strip character) on which he was serving as radio operator was shot down over the Brenner Pass in Italy. He was one of three crewmen to survive and was taken prisoner by the German Army.

Fancher was a B-17 engineer whose plane was crippled by enemy fire in a raid over Germany and forced to crash-land. All crewmen survived and were taken prisoners. When Fancher was under escort to be interrogated by a German intelligence officer in the building where prisoners were processed, he spotted Frederick in a room off the corridor. They were later placed together. How ironic and coincidental that boyhood friends from a small Alabama town, both prisoners of war, would meet again under the same roof in far-off Germany. He and Frederick had endured but were very badly treated by their German captors. The families of both men were grateful for their survival and anxiously awaited their safe return home. At war's end, Fancher was released to American forces in Munich and was home by June 1945.

Technical Sergeant Harry J. Adams was drafted into the army from his job in the Number 9 mine in September 1942. After infantry training and state-side duty, he set sail with his unit for Europe aboard the *USS Brazil* from Staten Island in August 1944, ten weeks after D-Day. After a ten-day crossing in sardine-packed quarters, his 26th Yankee Division, assigned to General Patton's Third Army, disembarked at the French port of Cherbourg. He was in the unit's headquarters company, but at one point during training he had been with Company C, 104th Infantry, a company all but wiped out by enemy action in the Alsace-Lorraine area near Nancy. From the arrival date at summer's end, his division raced

Harry Adams (far right) in Europe. (Author's collection.)

across France into Luxembourg and through southern Germany into Austria, and at war's end was in Czechoslovakia. In his reconnaissance outfit, Adams was often in front of General Patton in the thick of battle. He saw action in major battles and was on the southern edge of the Battle of the Bulge in December 1944, winning several battle stars and the Purple Heart. On May 9, 1945, he wrote to a sister:

> Dear Dot and all,
>
> The war here was officially over last night at mid-night. It was not an occasion of wild excitement as I had thought it would be. We have known it was over for quite some time. The war was won west of the Rhine, although there has been quite a bit of fighting since. Most of the Germans realized it was over and they have been surrendering in droves. It created quite a problem of how to get them back to P.W. [prisoner of war] enclosures. About the only ones who have been fighting have been the SS troops who knew they had committed so many crimes against humanity that they would be killed anyway.
>
> Well, I don't see why the censor should care about me telling where I am now. I am in the little Czechoslovakian town of Friedburg near the Austrian border. I don't guess you will be able to find it on a map but it is not very far from Prague, where the very last resistance is being put up. It is not a very big place, about the size of Blocton, I guess. The people are pretty nice and it is nice to be around people you can be friendly with. We were not allowed to associate or even talk to the Germans and Austrians unless it was on official business. It isn't natural for an American to treat people like the Germans should be treated. They are going to have a lot of trouble with the occupation army in that respect unless they use combat troops who know just how they should be treated. They have been almost falling on our necks they were so glad we got there before the Russians.

> The weather is much milder here than it was in the Bavarian Alps. Just a week ago I was in a pretty hard snow storm. I never thought I would see snow on the ground in May. The scenery back there was beautiful though. Got down far enough to see the Austrian Alps in the distance and was glad we turned east there because those things were covered with snow and really looked cold. I am enclosing another photo. I look like h— but that is just natural I guess. I was that dirty most of the time on this drive. Hope you are all O.K. Give Marion [his niece] my love. Wish I could see her.
>
> Love, Harry

Adams was discharged in November 1945 following the war's end in Japan. By mid-winter he was back in his Black Diamond Number 9 job operating a mining machine and pinning roofs.

Ex-servicemen were invited to a special American Legion "Homecoming Meeting" June 15, 1945. Special guests were former post adjutant James W. Bently, then of Billingsley, who had fought both in World War I and World War II; Air Corps Capt. Bill Carroll, who had four years flying service; and Henry D. Riley, who had been interned by the Japanese invaders in the Philippines. All veterans were invited to the meeting so the West Blocton Legion Post could extend the town's appreciation for their efforts during the war.

The arrival of the Henry Riley family in town was an interesting result of the May 1945 liberation of Manila in the Philippine Islands. The former Navy Civil Service official, a West Blocton native, had arrived to be near his brother, local cafe owner Walter Riley. The family had spent the years of the Japanese occupation under very difficult conditions. Riley's wife was Japanese and a Christian missionary in the Phillipines when they married. Most of the time while interned, Henry Riley was separated from his family. Japanese Army representatives asking for her older daughters to marry Japanese officers occasionally approached Mrs. Riley. Each time she won an uneasy delay in these proceedings. There were two boys and five girls in the family.

Walter Riley received a hastily typed note on Red Cross stationary from his brother Henry in February 1945. From the Santo Tomas Internment Camp, Henry wrote:

> We have all pulled through by the skin of our teeth, and are all so thankful to be alive. The boys [U. S. Army] got here in time. They are the swellest bunch of fellas we've ever seen.
>
> It was the Texas division that liberated us. They've brought in enough food to last us for the longest time. The first time we've seen so much food together in years. They got in here just in time, cause we were really starving to death.

The Rileys were residents of West Blocton for a short time before settling in the West. As a matter of concern as to reaction of local citizens to Mrs. Riley's ethnic heritage, the local news column writer said she was "Chinese."

There were more battles to be fought to subdue Japan. Those families with servicemen in Europe now had to face their being sent to fight the Japanese, and in the spring of 1945 that promised to be a long and bloody road. There was not yet a great homecoming of fighting men and women. After atomic bombs were dropped on Japan in August, peace finally came. West Blocton was thankful for the war's end and, as had been the mood in May at the fall of Germany, the community went to its churches to give thanks. In the personals column of the West Blocton news, the writer said, "churches had special thanksgiving services last Tuesday night after the surrender of Japan. The town celebrated in a muted way compared to other places."

West Blocton lost at least nineteen young men to the war from its immediate vicinity, all had lived within a three-mile radius. From its rural routes others were lost, among those were Sgt. Ross F. Gray who died on February 17, 1945. The special bravery of Gray at Iwo Jima led to the posthumous awarding of the Congressional Medal of Honor. Gray was the only Bibb County soldier awarded this honor and one of only a few in the state. Later, a Navy ship was named for him.

Claudie M. Hutchins from the Johntown community on the West Blocton Route also lost his life in the war. He was one of four from his family in service. Another Johntown soldier, Victor Donnis Fountain, son of H. E. and Lola

Ross Gray Funeral. (Courtesy of Alabama Historic Ironworks Commission.)

Victor Donnis Fountain. (Courtesy of J. C. Fountain, Cedar Grove, Alabama.)

Claudie Hutchins. (Courtesy of Anna Dean Hutchins Martin, Woodstock, Alabama.)

WBHS Veterans, 1946. The 1946 WBHS class included a special page in their yearbook to the veterans of the class. "To these our Veterans - Roland Moon, W. A. Cox, Harmon Frederick, Walace McElroy, Joe Price, James Stewart, James Roycroft, Bill Cox, and James Hardemon - Who willingly took time-out of their school days to fight for our American way of life." (Author's collection.)

Fountain, lost his life while serving in a tank unit on a Belgium battlefield in January 1945. He was buried with military honors in Ada's Chapel Cemetery. Donnis attended West Blocton High School before entering the army.

West Blocton probably had a much larger proportion of its men and women in service in World War II than most in Alabama its size with a larger proportion of its population killed, wounded, and missing in action. During the war, J. A. Johnson served as the local AGS depot agent and also Western Union telegraph operator. His job was said to be particularly difficult considering the number and frequency of telegrams reading, "we regret to inform you" he was obliged to deliver to local households.

By Christmas 1945, many having served their country were discharged. Bodies in government-issue, box-like coffins continued to arrive for several years, prompting second funerals and memorials. At least two remain abroad: Claude Holland, whose remains are aboard the battleship *Arizona*, a war memorial at Pearl Harbor, Hawaii; and Cristo Lagrone, whose grave is marked by a government-issue cross in an American cemetery in Luxembourg.

When the warriors returned home, some continued where they left off in high school. Nine former servicemen are pictured on the dedicatory page of the 1946 *Westonian*, the high school annual; most standing in a military "at ease" posture for photographer Tuggle's lens. The WBHS classes of 1946 and 1947 both graduated several veterans. Others, some of whom were coal miners' sons, enrolled at state colleges and universities taking advantage of Uncle Sam's GI Bill, while others went back to local jobs or arranged to commute to positions in surrounding towns and cities.

Returning servicemen found the town much as they had left it. To have a beer, one had to motor to Jefferson County and stop in at the Green Lantern or at another of the familiar watering places along U. S. Highway 11 just across the county line. Cows still roamed at large on the town's streets. A May 1945 town election outlawed hogs in the city limits by a margin of one vote. Cows were allowed to stay, the argument being that to restrict cows would take milk out of the mouths of children. Greyhound Bus drivers, long-time critics of the town's animal traffic hazards, and other vehicle operators still had to vie for right-of-way with bell-adorned quadrupeds. This was the hometown scene after the local boys had seen Paris, London, Berlin, Tokyo, and the great U. S. cities at all points of the compass. Regardless of its countrified, provincial atmosphere, it was good to be home in West Blocton back to civilian status, out of harm's way. Incidentally, cows lost residence in July 1948, when a stock law won approval handily by the residents who chose to join, at last, other municipalities in the county that for years had not permitted town limits to be a free grazing range. Those who believed that a cow in the back yard was a necessity had to consider relocation to the country.

The veterans at WBHS helped Coach Oscar M. Zeanah renew the varsity football program. For lack of enough boys to field teams, football had been suspended for the 1943 and 1944 seasons. Zeanah's 1945 team with only two players who had ever participated in a high school football game lost two games, tied two and won the rest. The two losses were to undefeated Bibb County, 13-0 and 31-0. That was to change in 1946, when West Blocton beat Bibb for the first time since 1937. Remarkably, in the contest on the Centreville field, the Tigers scored five touchdowns but only two of them counted. Both extra points failed, leaving the score 12-0. In 1947, the Tigers beat Bibb 7-6, the first time that WBHS had won two years in succession. It was also West Blocton High School's first undefeated, untied team since a football program began in 1925.

The town lost its Rationing Board in September 1945 to a district office in Tuscaloosa, an indicator of how quickly post war life was getting back to normal. (When the agency moved from its headquarters in the old West Blocton Savings Bank, that facility had yet another tenant, the West Blocton Loan Company which opened in May 1946.) One of the most critical of rationed staples was sugar. The local Coca-Cola plant was forced to cut its output drastically during the last year of rationing, but in June 1946, the owner, W. A. Jones, pledged to maintain the quality of his beverage and would not use sugar substitutes. "We will keep our distribution up on a regular basis although the supply will be curtailed."

Unbeaten West Blocton. The 1946 team beat everybody. Bottom row (left to right) L. B. Pierson, manager; Enos Franks, Jack West, E. H. Gentry, Ken Roycroft, James Gallent, Jack Smith, Clifton Lawrence, John Kish, Clarence Franks, Perry McKenzie, John Reach, assistant manager. Second row - Coach Zeanah, Harold Rogers, Jim Ray, Wallace Hayes, Walter Riley, B. E. Champion, Olvis Lawly, Douglas Jones, Ernest Ayers, assistant coach. Third row - Billy Sansing, Aaron Fancher, Charles Martin, Robert Gleaton, Sam Govie, Phelon Lawley, Oscar Higginbotham, James Dailey, and Harold White. Fourth row – Ed Suell, Ronnie Giles, Robert Dennis, Hayward Robertson, Willie C. Pounds, and Preston Hutchins. (Courtesy of Martha Fancher Kelley.)

CHAPTER 9

Coal Mining's End in West Blocton

The organized labor scene in West Blocton continued to mirror national trends. A wave of strikes occurred in the nation following the war. The leaders of organized labor were resentful of a stabilization program that they thought had deprived workers of deserved wage increases. Soon John L. Lewis demanded wage increases and fringe benefits for his union members. The operators refused. The miners walked out on April 1, 1946, and the government, at President Harry Truman's order, seized the mines. More strikes and trouble ensued to affect the local scene.

The miners walked out on April 1, 1946, and the government, at President Harry Truman's order, seized the mines.

The work stoppage in West Blocton took its toll on workers and merchants alike. The *Centreville Press* editor summed it up in an item printed November 28, 1946.

> John L. Lewis has done much for the coal miners in the past ten or twelve years. He has been a great leader for them, but he is now leading them into poverty and ruination, and along with the miners he will force the bankruptcy of a majority of the coal mines in the United States within a few years, because of their inability to sell coal at a profit in competition to other fuels. Coal is daily losing its place as one of our main fuels.

Editor Oakley was stating the obvious. The technological changeover to diesel oil and natural gas in the country may have been accelerated by the increase in the price of coal, but the transition was inevitable. Already railroads and other energy consumers were converting to other fuels, and diesel engines were replacing steam locomotives. This trend would later dramatically affect West Blocton.

The strike in the spring of 1946 lasted 59 days and others followed, but "ruination" was a bit farther down the pike. The situation calmed, and in 1947 and 1948, there were fewer shutdowns. Oakley repeated his concern about high wages granted to miners in July 1947 and predicted that the increase would cause another depression. He wrote, "We all like to see the miners make everything they can, but the price of coal has about reached the place where it is entirely out of the reach of the common man."

Young men finishing school in 1945 were still obligated to serve in the military. The draft had not ended. Some who enlisted or were drafted into military service just at war's end, found interesting assignments. Edwin Smith, son of John and Lucille, had joined the navy in August 1945. By February 1946, he was in Washington, D. C. assigned aboard President Truman's yacht *Williamsburg*.

L. S. Vanderford resigned as WBHS principal in June 1947. He and his family had been in the community for fourteen years, arriving in 1934 to teach history, spending a year at West End High School in Birmingham, and returning to serve as principal for thirteen years. Vanderford was a community leader as well as a professional educator. He was a Sunday School teacher and superintendent in the Methodist Church. During the war years he worked with the Red Cross, served as Civil Defense Chairman, was a member of the local Office of Price Administration, and was a leader in war bond campaigns. Vanderford served several terms as president of the Bibb County Teachers Association. Mrs. Vanderford was

Hill Creek Mine Tipple. (Author's collection.)

a leader in church and PTA activities and taught high school commercial subjects. A newspaper columnist reported several farewell entertainments for the family with presentation of gifts of appreciation, and closed with, "they will be sorely missed here, we wish them success and happiness." Vanderford moved to Montgomery where he served on the administrative staff and faculty of Starke Military Academy, then as principal of Brundidge High School, and later headed the student guidance and counseling department at Troy State University.

The new principal in West Blocton was C. K. Elrod from Etowah County High School. Elrod was to serve only a few months, however. In November he was diagnosed with leukemia and did not return to school. He died the following April, at the young age of 29 years. Oscar M. Zeanah, after serving as acting principal, was named to the position. He was to remain with the school until 1957. In March 1947, the town also welcomed a new doctor, P. H. Warren, whose tenure was but three months before he entered the army. In July, Dr. Percy Nolen began a practice that lasted several years.

February 1947 saw the installation of a heating and cooling system in the First Baptist Church, the first reported "air conditioned" public building in town. Governor "Big Jim" Folsom appointed Charles Irwin to the Board of Mine Inspectors. In August, the Black Diamond Store expanded next door into the Dave Israel store and held a special bargain sale of Israel's stock. Local baseball star Tucker Brewer signed with the Cincinnati Reds in August, and Henry Clay Randall left to study at England's Cambridge University.

The Birmingham newspapers and the county weekly reported glowingly in mid-August of the choice of Nell Finnen to be in the 1947 Miss Alabama finals. Her college freshman year had been completed at Wayne State University in Detroit, and she was enrolled at the University of Alabama for her sophomore studies in music. At the competition on August 20, she was awarded fourth place, quite an accomplishment for a West Blocton hometown girl.

C. J. Patton of Bessemer opened Bibb County's first florist business, Ann's Flower Shop, on Main Street in November. Also in November, while en route to Florida, Mel Allen, "The Voice of the Yankees," accompanied by the University of

Alabama's famed coach, Frank Thomas, stopped by West Blocton briefly to visit Sam Israel, Mel's uncle. In December Mrs. Boyd Hartley, nee Margaret Lane Edmonds, became the local reporter for the *Centreville Press* and the *Birmingham News*. A state Bureau of Mines inspector's report on the Black Diamond mine in November 1947 indicated there were 246 men employed there in the production of a daily average of 620 tons of coal.

The Birmingham papers' front-page headlines told of the big, destructive fire on West Blocton's Main Street in January 1948. It was the largest and most threatening blaze since the town fire of July 1927. Frozen hydrants on a 22-degree night had to be thawed by acetylene torches before the first streams of water could be aimed at the fire. Fire Chief Leonard Carroll and Police Chief A. B. Little were both overcome by smoke before the fire was brought under control. They were replaced by A. N. Kitchen and Maurice Jones, both veteran firemen. These men directed the volunteer firemen in a three-hour battle that included help from Centreville and Tuscaloosa. Bessemer was also asked for help but because of restrictive municipal ordinances was unable to come to the town's aid as they had in 1927. The fire destroyed two buildings housing ladies wear and furniture. Davie's 5 & 10, next door, caught fire but was extinguished. At one point the fire threatened the Black Diamond men's store and the main building housing the grocery, meat market, and company office as well as the Masonic lodge hall above. All were damaged by smoke and water. Loss was estimated at $75,000 to $100,000 in stock and buildings. Those business owners on Main Street who had been through the 1927 business district fire believed they were in for a repeat disaster. This appeared especially evident at the fire's worst when high flames leaped skyward as roofs collapsed. However, a pump-equipped fire truck and a better downtown fireplug system as well as having early help from neighboring towns kept this 1948 fire from spreading.

L. S. Vanderford family. L. S. Vanderford served as WBHS principal through the difficult years of the Depression and World War II. Pictured are daughter June, Levi Stanford Vanderford, and Dorothy in May 1946. (Author's collection.)

By April the burned buildings were almost ready for re-occupancy. Carpenters, roofing crews, and brick masons had been busy at work all of March. Grand opening sales were announced in early fall after all new stock was in place. There were other new businesses. Floyd Morrison had built a new structure on Main Street to house the Alabama Power Company, and A. B. Little had remodeled the West Blocton Supply Company to be used as a restaurant. Unfortunately, the merchants were again hampered in the spring by another miners' work stoppage.

Encouraged by the rebuilding of the company stores, Mary C. Street, who had lived in West Blocton off and on for many years, announced the opening of a Western Auto Store. The opening was properly staged with Mayor Evans cutting the "Welcome Cord" at 10 a. m., Wednesday, January 29. Nell Finnen, "Miss West Blocton of 1948," and official hostess for opening day, gave each visitor a ticket for a prize drawing to be held at 6 p. m.

The politics of 1948 were complex. President Truman had won a surprise victory over Republican challenger Thomas E. Dewey. He had done so without the "Solid South" that had left the old Democratic fold over the issue of civil rights. Neither Bibb Countians, nor any Alabamian, could vote for Truman in 1948, since the Democratic Party's delegate slate was not on the ballot. The segregationist Dixiecrat Party had replaced the Democrats in most of the states of the old Confederacy. The county newspaper editor did not have many favorable things to say about Truman.

In March 1949, the *Press* editor commented that Ku Klux Klan organizers from Jefferson County had been in the

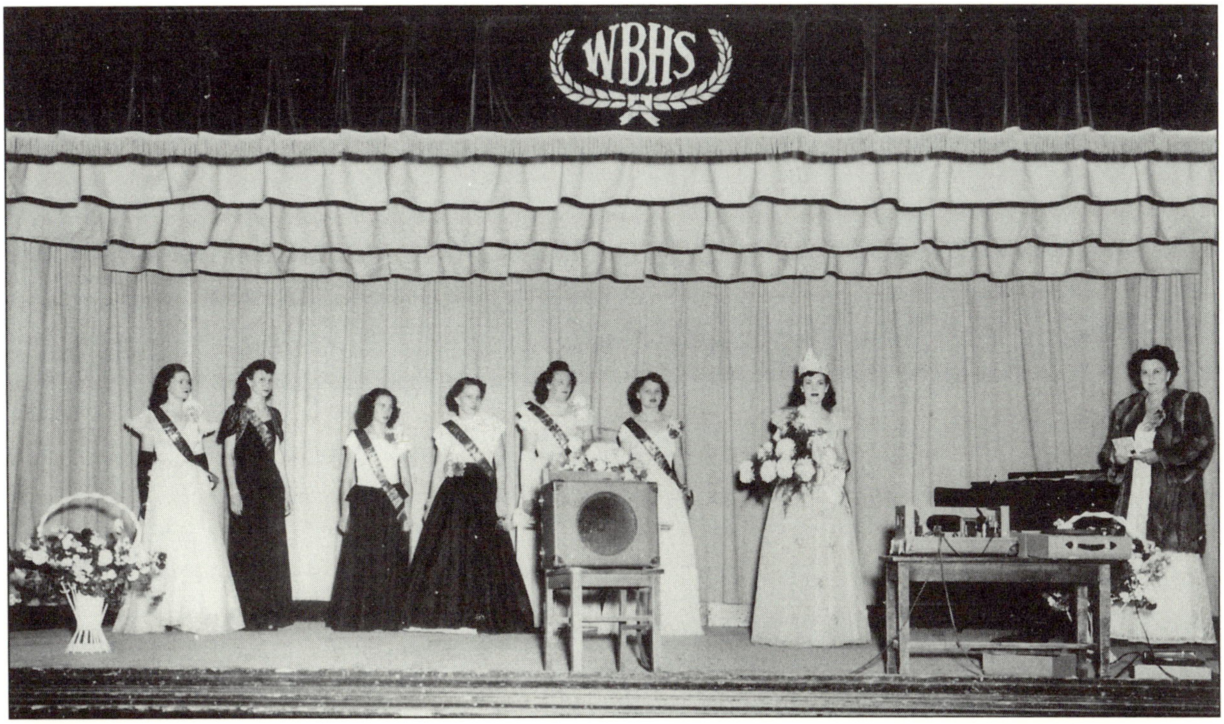

WBHS Pageant. Nell Finnen was the reigning West Blocton beauty in 1948. Contestants at a WBHS pageant were (left to right) Annette West, Sue Pickett Hyde, Ramona Langston, Betty Pratt, Martha Fancher, Juanita Carr, and Nell Finnen. Mary Etta Gillespie served as mistress of ceremonies. (Courtesy of Martha Fancher Kelley.)

county making an effort to start a unit of the Klan. It was his opinion that the Klan would only stir up strife and turmoil serving no benefit. Two years previous, he wrote that the matter had been broached by citizens from another county. The position then had been that Bibb County citizens were living peaceful, normal lives and that an organization creating disturbances was not welcome. This late 1940s stirring would grow into a bigger threat to domestic tranquility in the 1950s and 1960s when those years saw a return of the Klan.

With racial turmoil dominating the political scene, it is remarkable to read in the pages of the November 24, 1949, *Centreville Press* that the West Blocton Methodist minister, T. H. Wilson, preached in the African-American Methodist Church with his youth choir presenting a part on the program. The occasion was the end of the conference year in the church. The following week a note of appreciation from the African Methodist Church was printed in the *Press*.

In May 1949, an official notice of a bank incorporation in West Blocton was printed in the *Centreville Press*. The institution was to be named the West Blocton Savings Bank, the same name of the earlier failed bank. The paper said the state had given approval, and that $35,000 in capital stock had been sold. The bank opening now depended on approval by the Federal Deposit Insurance Corporation. The lack of a bank in the town for several years (since 1931) was noted. Stockholders were listed as S. B. Israel, W. A. Jones, H. M. Roland (West Blocton businessmen), R. S. Ward, E. L. Patridge, H. B. McLauren (Brent Bank officers), and J. D. Pierson (hardware store owner in Centreville and West Blocton). The old bank building was available to house the institution. But the bank was not to be. The July 29 paper reported that the State Banking Committee surprisingly had turned down the proposed West Blocton bank citing that two banks in the county were sufficient. The paper reported that re-application would be made at a later date. Reliable sources indicated that opposition came from board members of the People's Bank of Centreville who effectively influenced the state committee. The remainder of 1949 was beset with the first major post-war recession which may partially explain why re-application for the bank was not made. West Blocton was without a bank for fifteen more years. Subsequent effort to open a branch of the Brent bank in West Blocton was thwarted by what was termed "local politics," stopping another attempt to have local banking services. The lack of a bank was to prove a major handicap in the intervening years in efforts to obtain industry to replace the disappearing coal business. Finally in May 1965, the Bank of West Blocton opened despite reported opposition again from Centreville bank directors. Intra-county sectionalism was still a factor in West Blocton's economy.

Hampered by a strike and a diminished market, George Young closed his Red Dot Coal Company truck mine on the Primitive Ridge Road in the mid 1940s and opened another in Jefferson County. By January 1949, he had given up on coal mining, sold his home and moved his family to Corpus Christi, Texas. The closing of the Young mine signaled the beginning of departures from the area by small mine operators. In February 1948, Dave Evans also departed, resigning as mayor (Wilcox Randle was appointed to fill his remaining term). Evans' mine near Hebron closed as a result of the 1949 recession. In June 1949, the report of the state mine inspector showed only 65 men employed at Hill Creek mine that during the war years had employed 175. When inspected in October, Black Diamond mine employed 240 workers in the mine operations. In March, *Press* editor Oakley summed up the economic situation in Bibb: "Business conditions are still on the decline. John L. Lewis has shut down the coal mines for two weeks, but that will have little effect in Bibb County, because our mines were already down. The W. E. Belcher Lumber Company was forced to close down their veneer mill last week due to lack of orders. The lumber business is also slow."

One of the mine strikes ended in late March 1949. After commenting that coal mining was the biggest local revenue source, Oakley wrote, "If we lose our coal mining, then several hundred families in Bibb County must look elsewhere for something to do, and the entire county will suffer." The editor's prophesy was soon stark reality.

In June when the mines again struck, the editor continued his thoughts about the coal situation. He identified unemployment statistics, freight loss tonnage, and blamed John L. Lewis who he said, "continues to make coal an uncertain fuel." These stoppages, he said, "in the long run mean less employment for the miners. Coal has lost much of its popularity as a fuel because it is no longer dependable. Lewis can call a strike anytime he desires, regardless of its consequences of plant shut down and mass unemployment."

It should be observed, however, that a majority of the organized miners in the county believed in the tenacious union

Red Dot Mine. (Courtesy of J. Turner Crowe.)

leader's promise of better wages and working conditions. The strike was the most potent weapon to win these objectives. The Centreville editor could have his say and opinion, but the miners remained loyal to Lewis. There is no doubt that the increased cost of coal production caused profit margin problems for the operators. It was a time of social, workplace, and technological change. Within a decade, the coal mining era in Bibb County would come to an end partially because of these unstoppable elements. Had the local miners voiced no safer workplace demands, left the union movement, and taken reduced wages, the result likely would have been the same. The frequency of strikes in the Number 9 Black Diamond mine inspired a local joke that claimed if a man accidentally spilled the water from his lunch bucket, the rest of the crew would pour out theirs in sympathy and walk off the job.

Major fire problems continued in West Blocton early in 1949. On Wednesday morning, January 5, the First Baptist Church was heavily damaged by a blaze that started just above the furnace room while a county meeting of the Women's Missionary Union was in progress. Volunteer firemen fought the fire for two hours and were able to confine damage to the church office, Sunday school rooms, and the pulpit. Help came from the Brent volunteers whose truck aided the local department. Damage was estimated at $10,000, fortunately covered by insurance.

Accidental fires were bad enough but in August 1949 the small frame doctor's office located on a rise between Gillespie and Kren's stores was burned. A trace of gasoline at the site indicated arson. It was the office of Dr. George Malouf who had come to the town at the invitation of the Black Diamond Labor Union's doctor search committee. Town businesses and citizens raised a reward of $590 for the arrest and conviction of the person or persons responsible, but the case remained unsolved. The following October, Dr. Malouf relocated to Bridgeport, Alabama. Dr. Crabtree from the Seale Harris Clinic in Birmingham came to be the town doctor. The era of long-term doctors like Peacock and Jones had passed.

There had been a lapse in the development of a high school band. First there were the war years, then change of school administrations and a scarcity of band teachers. For the 1948-1949 school year, Principal Zeanah, working with the principal at Centreville and the county school superintendent's office, arranged to have band director Winston White serve both county high schools. White had a good program and served his students well. Both high school bands were playing at football games, other school functions, and community events in record time. New WBHS uniforms were available by the fall of 1949, said by the school reporter to be of the same style and quality as those worn by the Million Dollar University of Alabama band. By November of 1949, the band was good enough to be invited to several Christmas parades, the premier event being the Birmingham Christmas gala and later the Demopolis Christmas parade. The band, reported the weekly paper, made the town's citizens proud. One bandsman, tuba player Eugene Rotenbury, was featured in an article on August 25, 1949, illustrated with him holding the big horn. The story extolled his tenacity in walking miles from his home in the Scottsville community to his music lessons and rehearsals with the heavy bass horn draped over his shoulder. A more dedicated musician, said his band director, is rare. It was Rotenbury's ambition to join an army band after high school.

Community spirit, both in hard times and good, was shown in a number of ways in this economically fragile town. The largest cemetery in the county (some eight or nine acres) was Mt. Carmel, taking its name from the nearby Baptist church. The cemetery has been in place since 1881. From time to time efforts to upgrade the burying ground had been attempted with only sporadic success. Practically every decade of its history saw an attempt to raise funds to make improvements, to control undergrowth, and in the days of free ranging cattle, to keep up a fence and provide gates. Veterans from all wars since the Civil War as well as area civic leaders are reposed in this hallowed ground. Catholic and Protestant alike are buried there as well as immigrants who came from far off places, their tombs inscribed in the language of their countries of origin. Even now, bodies of former residents are returned for burial in their hometown cemetery. Lavada Lawley Bowen, a California resident for decades, in July 1998, was buried there in the family plot and her son, Charles Hugh Jr., also of California, was interred in September 1999.

In December 1949, Bessie Robinson undertook the job of organizing a committee dedicated to continuous upkeep of the grounds. It was reported that more than 2,500 graves were overgrown with weeds and vines. A series of meetings

Early Mt. Carmel Grave. Fourteen month old Charity Holland is the first known burial at the Mt. Carmel Cemetery. (Courtesy of Cahaba Trace Commission.)

was held to organize a committee and to raise funds. More than 300 responded to a meeting called in April 1950. A request for volunteers to clean the cemetery brought out more than 50 miners who cleared the several acres in three weeks. A fund of more than $500 was raised to keep the grounds cleaned.

Robinson initially chaired the group. She was treasurer of the Mt. Carmel Cemetery Committee for forty years. After Miss Bessie's passing in 1990, the committee continued its work and in the years since has been very effective in keeping the cemetery in good condition. In 1997, the West Blocton Improvement Committee dedicated a historical marker at the southeast corner of the grounds.

The late 1940s and early 1950s saw a drive in Centreville for a badly needed new bridge over the Cahaba River. The state, after much delay and excuse making, finally came through and the bridge opened in late 1950. Smaller objectives were on the agenda for West Blocton. There had been a two-year campaign for a lunchroom at the elementary school and at last, in late February 1950, the cornerstone was laid. A parade down Main Street led by the West Blocton High School Band with town and county officials in cars made its way to the school grounds where an audience gathered around a makeshift speaker's platform for the dedication address by Judge L. S. Moore of Centreville. This got the lunchroom corner stone laid with all the ceremony the community could muster. Now all that was needed was a lunchroom. It was to come, but not overnight. In January 1951, the building was completed. All that was needed at that point was equipment. It was as hard to build a lunchroom in West Blocton as it was to build a bridge in Centreville. The West Blocton folks, however, had no state or Federal aid in their project, delay was nobody's fault except theirs.

The use of a fire truck that had to be towed to fires came to an end in September 1950 when finally West Blocton could boast of a new fire unit. Fire Chief Maurice Jones and Councilman Ben Stanley drove the new vehicle to Centreville for viewing causing the *Press* editor to write that all three municipalities now had reliable fire protection. Centreville had acquired its first real pump-equipped fire truck the year before after much promotion by the local paper. Today that beyond-repair unit, a fine American-La France model, lies under a coat of dust in the town garage. It carries a brass plate identifying town officials' names from the date of its acquisition.

On March 9, the *Press* editor commented on a continuing snag in the local economy. Another coal strike had ended, but he foresaw more rather than less of the same. It was his observation that the coal mining industry was a "sick business." The handwriting was on the wall for Bibb County's Cahaba Coal Field. By late spring 1950, the Bessemer Coal, Iron and Land Company closed its mine in Belle Ellen, affecting 150 employees. In early June, Marvel Number 3 mine closed, taking 150 to 200 jobs. There were only 50 jobs at Hill Creek, and it also closed in 1950. A 30% increase in freight rates caused TCI to cancel its contract with the Boothton mine (just over the Shelby County line from Marvel) and with Black Diamond in West Blocton. Boothton scheduled closing for October 31. Subsequent full-page advertisements listed houses, warehouses, mining equipment, and church structures for sale in Boothton. The community was erased from the landscape.

The 1950 census report indicated how the decline in mining had affected the county's population. The high point

Black Diamond Number 9 Tipple. (Courtesy of Billy and Nina Smith.)

was recorded in the 1920 census with 23,144 Bibb residents. In 1930 and 1940, the count remained around 20,500, but 1950 reflected a sharp fall-off, with a population of 17,955. The 1940 census counted 1,317 in West Blocton and 280 in Smith Hill. It is uncertain if Blocton Number 1 residents were included. The county number was to go much lower in the next two decades. In 1949, state figures showed 1,018 men employed in mining in the county. Although forest products industries came in a close second, coal mining was still the county's biggest industry, but it was fast disappearing and the population with it. When the University of Alabama's basketball arena, accommodating around 14,000, was finished in the mid 1960s, Bibb County's population would have fitted into its seats, no standing room needed.

Discontent and intrigue had remained after the end of World War II and was heightened during the Berlin Blockade. The "Iron Curtain" and subsequent "Cold War" created a complex international division of powers, Communism versus Democracy, which endangered a fragile peace.

A shaky peace gave way to war on the Korean peninsula when the North Korean army crossed the dividing geographic parallel into South Korea on June 24, 1950. The major military role assumed by the United States in what was called a "police action" caused those who had chosen to be reserve military professionals and those who had not been old enough for World War II to be called to arms. The military draft was resumed. West Blocton read the county paper of July 20 on the subject:

> The Bibb County Draft Board reopened last week on a full time basis. Board members are Mr. Ed Patridge of Brent and Mr. Floyd Morrison of West Blocton. Mr. George Herring of Centreville, chairman, resigned last week. His successor has not been named. The draft quota for September for Bibb County is three men. 15 men will be called for examination.

Things did not go well for the United Nations forces in the war's early days. Editor Oakley, unhappy with the Civil Rights movement as were most of his subscribers, wrote that had the Democratic Party spent as much time and effort getting ready for war as they had on trying to reform the South the country would have been better prepared for war. Truman was not a favorite president in most of Bibb. It was a complicated and unpopular war and difficult, turbulent times in the South.

As an increasing number of television sets in local households reported the war (the first sets had arrived in small numbers in early 1949), other community events were making news. The old St. Francis of Assisi Catholic Church on Gunlock Hill in old Blocton had given way to a new church in West Blocton. St. Thomas Aquinas Church was dedicated and consecrated by the Most Reverend Bishop Toolen of Mobile on Sunday, September 3, 1950. Many of the state's Roman Catholic hierarchy were present as were many former members of the congregation. There were 50 members in the parish. The diocese elected to close the church twenty-two years later when the membership dropped to twenty communicants.

Long lists of names appeared in the county paper's columns identifying men called for armed forces pre-induction exams. The town took pride in the December announcement that Mary Elizabeth Fitts, daughter of Reverend and Mrs. Lester Fitts, had been chosen "Miss Livingston University" to win a prominent place in the school's yearbook.

The war came home in all its reality with the announcements that Charles Albert Williams of West Blocton had been injured in Korea, and Paul E. Russell, who had been wounded near Pusan, Korea in September 1950, was in Walter Reed Hospital for a rare shoulder artery replacement operation. In December, Lester Pounds, with West Blocton family connections, was killed in action in Korea.

Local transportation was to be improved in 1950 by the construction of a farm-to-market road from West Blocton through Lowerytown to Cedar Grove. Announced earlier was the contract on a surfaced road to Highway 25 to the east including a bridge over the Little Cahaba.

The year got off to an uneasy, scary start in January 1951, with the bold robbery of the Ferrire Store in Little Italy.

St. Thomas Aquinas Catholic Church Consecration. (Courtesy of Ferrire Collection.)

Mary Elizabeth Fitts, "Miss Livingston University." Courtesy of Beth Fitts Frederick, West Blocton.)

Three masked, armed bandits forced Mr. and Mrs. Ferrire into a feed house where they were locked in and son Frank was tied up. The bold bandits took $15,000 in bonds and $50 in cash. The car was said to have had a Missouri tag. Local, county, and state patrolmen were alerted. No report of the criminals' apprehension was recorded in the local paper.

The high school basketball team had a good season in 1951 and a local committee was formed to raise funds for a $20,000 gymnasium. Just under $3,000 was pledged in the early going but a one-cent sales tax to support the project and other community improvements was voted down in April by a count of 134 to 69. The plan for the gymnasium died. Business projections did not indicate it was time to improve school facilities. History was in a repeat mode in the community. In 1948, when times had been better, high school Principal Oscar Zeanah and the elementary school principal had been able to convince the board that modern rest rooms were a necessity if the schools' accreditations were to be validated, so that matter had been resolved. The 1935 wood-frame WPA school building had space between the auditorium and the classroom wings where rest rooms could be built, so in mid-century the local high school had acquired indoor facilities. Local contractor, Floyd Morrison, won the contracts for these school additions. Restrooms were also added to the 1937 elementary school.

While other areas had begun to pull out of the postwar recession that started in 1949, the spring of 1951 saw no relief locally from economic decline. Mayor Victor Y. Foster and a delegation of city councilmen, civic leaders, and businessmen, members of an Industrial Development Committee, made an exploratory trip to Montgomery in late March to seek advice on how to obtain new industry for the town. That meeting was attended by Charlie Haynes, R. W. McCulley, W. D. Deerman, Joe Stewart, Pete Bearden, Floyd Morrison, Ed Gillespie, Claude Hayes, Ed Finnen, Maurice Jones, Oscar Zeanah, Emery Lippeatt, J. M. Langston, D. C. Moore, Dover Pierson, Floyd Ambrose, G. E. Dailey, and J. W. Randall. The news item said, "Every effort is being made to improve the present business conditions in West Blocton." The town's long-standing reputation of organized labor troubles was a major deterrence to attracting new industry. Other towns competing for industries often used this characterization of the labor atmosphere in West Blocton in their own quests, making it doubly difficult for the town to be competitive in the lean market of the 1950s. Unlike the national need for fossil fuel in previous wars, the Korean conflict did not revive the local coal industry. Reliance was now on fuel oil, diesel oil, and natural gas. The coal mining era in Bibb County was all but over.

The continuing decline of the local coal industry was evident in May 1951, when the Black Diamond mine cut 100 men for lack of orders. By September, the Roden Coal Company of Marvel announced it would close; over 250 men were affected. A subsequent announcement in October said the Marvel mine was leased to another firm and would continue. That arrangement was short-lived.

The tenure of young doctors in the community seemed to be shorter and shorter. Dr. Crabtree moved on and then came Dr. Jenkins. When the James Woods family left in 1951 to settle in Reform, Dr. John Crowder and wife, Dr. Ethel Montgomery, previously at Belle Ellen, bought and moved into the Woods' home. Dr. Crowder served the community until incapacitated by a stroke. Dr. Montgomery, a talented diagnostician who had trained at Johns-Hopkins Clinic, served as the town's physician for the rest of her professional life. In those years she used her home's ground floor as her office.

In November 1951, Wilcox Randle, who had served as mayor in 1948-1949, was given a regional promotion to District Manager of Standard Oil that required the move of his family to Bessemer. He and his wife had been church and

Drs. John Crowder and Ethel Montgomery seated on the porch of their Belle Ellen home. Flanking the doctors are a young John Crowder and Wendall Wade Crowder (right). The family's first grandchild, Carol Zoe, is on Dr. Crowder's lap. (Courtesy of John Crowder, West Blocton, Alabama.)

Armistice Day Parade. The WBHS majorettes were the pride of the Armistice Day parade in November. Posing left to right were Willodean Shaw, Patsy Slaton, Billie Ruth Brown, Eunice Steele, Norma Caldwell, and head majorette, Shirley McCool. The Black Diamond Women's Store, Davie's and the I&N Quality Store are pictured behind the prancing beauties. (Courtesy of Shirley McCool Stamps, Hueytown, Alabama.)

civic leaders for more than a decade. Mrs. Randle, the former Eleanor Dudley, was the high school librarian. The family moved in late March 1952.

War years always aroused patriotic fervor in West Blocton, and the Korean conflict years were no different. On November 11, Armistice Day, an American Legion unit of World War I and World War II veterans along with the high school band, its majorettes and flag carriers, elementary school children, and several cars of citizens led a parade from downtown to the elementary school and on to the high school. The reporter wrote, "In the hearts of those who took part and those who turned out to watch, the rhythm of marching feet was the reminder of what war has meant, of loved ones who gave their lives and of those who are still fighting and dying to preserve peace."

The Bibb County High Choctaws continued their dominance over the West Blocton Tigers in November 1952, winning 7 to 6. The Tigers had tasted victory in 1946, 1947, and 1948. There was a tie in 1949, and Bibb won in 1950. Helping to relieve this pain, the Tigers won over Bibb in a spring 1952 double-header to win the county baseball championship. The WBHS band building and athletic field house burned in January 1953. Band instruments were saved, but the fire had started from a defective flue in the athletic dressing room and all football, basketball, and baseball equipment as well as personal clothing and possessions of basketball players at practice were lost. Plans were made immediately to replace the structure. By the start of the 1952 fall school term, a steel frame building was built east of the auditorium to provide a larger equipment and dressing room and adjacent band practice room. Meanwhile, the band occupied temporary quarters in the home economics building.

Immigrant, pioneer merchant John Kren died in November 1951, and in March 1952, the town lost two other pioneer citizens, John Golson, 80, and P. P. Pritchett, 86, another long-time merchant and civic and church leader.

The February 28 *Press* announced that Victor Y. Foster had been appointed mayor on February 19 by the town council to fill out the term of Wilcox Randle. Foster, a Navy veteran of World War II, was a partner in the Finnen and Foster Grocery Store. He was serving as the North Bibb Red Cross Chairman and was the Exalted Ruler of the local Elks Lodge. By April, Mayor Foster had led an industrial development committee to confer with Birmingham's Committee of 100 to seek industry to replace the waning coal mining operations in the area. The committee focused on locating a plant to manufacture hand and garden tools. The quest was unsuccessful.

Irene Lawrence served as the local news reporter to the county newspaper for a number of years. When appropriate, she included other than strictly personal items. On May 22, she wrote:

> Mrs. J. B. Stokes of the Garden Club of Tuscaloosa and other members of the club came through West Blocton last Saturday on their way to Lily Shoals, which is located east of West Blocton near the Piper River Bridge [the old bridge, up-stream from the present structure]. Many people are interested in these water lilies found in profusion on [the] Cahaba River.
>
> The Standard Oil Company had a picture of the lilies in "Scenic South," in June 1944. It was photographed by the late Mr. T. C. Tuggle. The flowers are a rare species of lilies and will bloom two or three weeks if not affected by high water.
>
> Bibb County is fortunate in having such a beauty spot near by which is attracting visitors from other sections. As Mrs. Street is entertaining her garden club in Birmingham, she is taking some of these lilies to show what a rare flower blooms in Bibb County.

Folk tales surrounded the lilies. Subsequent columns proclaimed that the lilies grew on the Nile in Egypt, and were brought from there by Mrs. Roden of Marvel who planted them in the Cahaba. Another story said they were only found elsewhere in the Seine River in France and in the Amazon in South America. Another said they grew only in one other place, Piper, England. Piper was named for O. H. R. Piper, a Memphis, Tennessee mine investor-developer, later of Birmingham. Wildflower experts were aware that the species grew at stream fall lines throughout the Southeast.

In the September municipal election, Ben O. Stanley, long-time merchant, was elected mayor. A son-in-law of Mr. and Mrs. W. T. Morrison, Pelham J. Merrill, a Cleburne County attorney, was elected to the State Supreme Court in the

Cahaba Lily Postcard. The Little Garden Club, Zone VIII, of the Garden Clubs of America produced this postcard and noted the flower as "a rare and endangered species found among clean, cool river shoals in North Central Alabama. (Author's collection.)

Democratic May Primary. Judge Merrill had been a teacher and coach in the local high school in the late 1920s and early 1930s. Mrs. O. D. Street was serving as state director of the Tuberculosis Association, and in November her picture appeared in the *Birmingham News* as one of "Alabama's Gracious Ladies." Other news of achievement came in November when local school graduate, Mary Bell Reach, was named a secretary for the Voice of America and was assigned to the American Consulate in Salamonica, Greece. Serving in another prominent position during the summer's Democratic Party Convention was Elmo Israel Ellis, another West Bloctonite, who was appointed as an assistant to the director of television coverage. At that convention, the Democrats nominated Governor Adlai Stevenson of Illinois for president and Senator John Sparkman of Alabama for vice president to face Republican Gen. Dwight David Eisenhower and Richard Nixon in the 1952 election.

In Bibb, Editor Oakley, long a critic of President Truman and the Democrats, backed Eisenhower. He was joined by other prominent county citizens including Dr. Ben H. Jenkins then practicing in West Blocton. In a return to the Democratic Party fold, Bibb voted nearly 3 to 1 for the Democrats. The vote in Beats 10 and 11 was 355 to 118 for Stevenson-Sparkman. It was a Republican year in the nation. President-elect Eisenhower promised a trip to Korea to pursue a peaceful settlement of the war.

A January 1953 story about servicemen reported Rufus A. Parker of West Blocton, who had been in Korea four months, was recovering from wounds. Corporal Enos Franks was discharged with 18 months in Korea with the Korean and United Nations Service Medals, four Bronze Battle Stars, Korean President and President's Unit Citation. When the Korean armistice was signed on July 27, there was a community sigh of relief, but no local celebration of note.

In January, Governor Gordon Persons appointed local undertaker Chester Collins as Bibb's coroner. The *Press* reported this as a first, since the job was usually performed by the sheriff. Mayor Stanley announced plans for a fireproof jail and proposed paving of residential area streets. The high school's dance orchestra was in demand for proms at area schools. They were engaged by Vincent, Jemison, Shelby County, and for a Key Club dance at Centreville. A photograph of the ensemble appeared in the May 21 *Centreville Press* and identified the members as Margaret Robinson, Charles Glen West, Sylvia Clinner, Billy Liddle, Vera Beasley, Sara Frances Smith, La Rue Tibbs, James Ingram, Bobby Liddle, Charles Melsoni, Leo Horton, James Vernon, Ken Smith, Ronnie Hayes, Billy Vic McCool, Billy Allen, Yvonne Reach, and Ann

An unique perspective on the town was given to WBHS students of the 1950s when they took over town government for a day. Pictured are (left to right) William H. Wright, town clerk, Elizabeth Reach, Charles Beasley, and Mayor Stanley. (Courtesy of Alabama Historic Ironworks Commission.)

Oglesby. Later the group played for Sulligent High School's prom, and the band director let the group's bus be driven a few miles to the west before returning so the uninitiated could say they had visited Mississippi.

A stir of civic club organization occurred in 1953. The Elks Lodge established an Emblem Club, an organization for women who were wives, widows, sisters or daughters of Elks. Prominent citizens were recipients of a charter for Club #237 on April 18. Twenty members were immediately installed in the new club. The president was Mrs. Boyd Hartley.

On May 29, the West Blocton Civitans received their charter. Chapter organization planning had been initiated in late April. Jack Harmon, manager of the local Alabama Power Company office, was elected president. The Civitan charter banquet was attended by 200, and 38 enrolled members were identified in the *Press* item. By July, plans were revealed for the club to sponsor a Boy Scout Troop, and a chapter committee seeking industrial development information met with Alabama Power Officials on July 10 in Birmingham. A sizable group from the club was treated to a cruise on the Gulf States Paper Company boat on the Warrior River. In August a "Clean-Up" campaign co-sponsored by the city council was set with cash prizes to be awarded and was inaugurated by a high school band concert and a softball game. To make the campaign a community-wide affair, "bleachers were set up for colored citizens who were competing in a contest of their own." The campaign was declared a success. Prizes were awarded to Mrs. J. M. Turner, Mrs. Harry Adams, and Mr. and Mrs. C. P. Parks. The Boy Scout Troop was organized with Charles Haynes, scoutmaster.

In August, editor Oakley observed that the last steam engine in the Southern Railway System was taken out of service on June 17, 1953. The "fire was knocked out of the firebox" to end a 123 year use of coal on the country's railroads. The Southern Railroad had begun "dieselization" in 1939. A magazine article submitted by W. A. Stewart of West Blocton, a railroad company retiree, prompted this story. The following week's issue of the paper contained the editor's lament on the condition of mining in Bibb County, once the source of the county's largest payroll of about $3,000,000. He said there were only 16 families left in Belle Ellen, hundreds had moved from Marvel and that Piper was about gone. He wrote, "We are glad to see that the people of the county are wide awake in their efforts to attract new industries." Brent had just welcomed Cass Toy Factory from Athol, Massachusetts, and Centreville was expecting a garment factory to be constructed on the proposed southern bypass. Bonds had been sold and good things were in the works. Natural gas lines had been extended to Centreville and Brent. There was no such success in West Blocton. Black Diamond Mine was still at work but with a greatly diminished work force. The county paper did its best to give the town's mayor and council a boost in its search efforts. It cited West Blocton's up-to-date water system, new fire department building and new fire truck, a new two-cell jail, and two miles of new pavement with more to be added. Citing the clean-up campaign, West Blocton was proclaimed a clean town and the citizens very public-spirited.

In 1945, thirty-two West Blocton businesses were listed closing on Thursday afternoons during the summer. For Christmas two-day closings in 1953, there were only eighteen. The cut back in mining had taken a severe toll. Another prominent citizen announced his job resignation to move to greener pastures. D. C. Moore, postmaster for several years and active in church and civic affairs, took a position with Universal Atlas Cement Company in Birmingham

Coach Charles Sweeney, on the high school faculty for five years, elected to return to a commissioned officer's position in military service and resigned in mid-summer. The August list of new teachers at the high school introduced a new coach. Morris Higginbotham had been at Walnut Grove School the year before where he had a seven-win, one-loss and one-tie record. He had been a B-24 gunner during World War II, played professional baseball, and was a 1952 graduate of the University of Alabama. The new coach was to bring sports appreciation to new heights in West Blocton.

Black Diamond Evening Shift, 1954. (Courtesy of Billy and Nina Smith, Milton, Florida.)

The 1953 season was supposed to be an "off-year," reported Helen Raye Thrasher, school correspondent. The Tigers won all their games including a nail-biter 19-13 victory over previously unbeaten Bibb County to close the season. The team was west Alabama's only undefeated team and was named by the *Birmingham News* as one of the state's ten best. A town on the industrial decline could celebrate its newfound football fortunes.

In January 1954, William J. Murphy was appointed temporary postmaster, and was confirmed in that post by President Eisenhower the following August. In February, ex-Governor Folsom, looking to run again, spoke to a crowd at West Blocton's Strand Theater. In April, yet another young doctor, Carl Robinson of Bessemer, set up a two-hour per day practice in town. The State Highway Department announced in May that the contract to straighten Highway 5 was expected by August. Legal notices appeared in the *Press* concerned with the condemnation process for right-of-way acquisition. The major project on the projected route was the railroad bridge west of the old crossing at Woodstock.

In May, a representative of the Alabama Gas Corporation visited West Blocton to advise the town leadership of the company's policy on extending lines. At that time, it was restated that the town was not to be served by natural gas because extension of lines from Brent was too costly. In June, it was announced that West Blocton's application for natural gas service was officially turned down. In May 1954, advertisements for the J. F. Wood Plymouth and Dodge dealership no longer included West Blocton, the business had been moved to Centreville. The Frederick Chevrolet business had ended in the 1940s. Editor Oakley proclaimed 1954 the "most progressive year in the history of Centreville."

More mine closings were announced in April. The weekly paper said, "We understand that Roden Coal Company of Marvel plans to close down their mine permanently. The company has been in business for fifty years. We also understand that Black Diamond of West Blocton plans to close around the first of the month because of a contract loss and the competition of coal shipped in from other places." Black Diamond Number 9 had its last day of operation in May 1954. There had been a series of cutbacks causing many to move away seeking employment. Some miners had gone to Arizona, others to New Jersey and Illinois. Again, a mine company's departure meant the departure of workers and management personnel, the closing of the company store, and job loss.

M. C. Davie. (Courtesy of J. W. "Bill" Davie.)

The May 13, 1954, issue of the *Press* encouraged the Industrial Development Committee of West Blocton by offering the assistance of the people of Brent and Centreville. The lack of natural gas was mentioned as a problem. "It is needed as a drawing card. Industries don't just happen, you must have something to offer them," said the editor. Implication was that the West Blocton leadership had not worked diligently for installation or that it had shunned natural gas believing that would hurt the sale of coal by the few places still operating.

In May 1954, death took long-time civic leader and merchant M. C. Davie. A resident since he was ten years old, Davie had served several terms as mayor and many years on the school board as well as in other leadership roles. The Davie family was prominent since the early days of the town. The Davie Store was a part of the business community since its founding by J. B. Davie in 1889. After her husband's death, Mary H. Davie continued the store's management until failing health made it necessary for her to enter a nursing home. The store continued under local management directed by the oldest son, Marshall C. Davie, Jr. of California. It continued until May 7, 1988. It lasted ninety-nine years, the oldest retail business in one-family ownership in the county.

Mary H. Davie taught music in the local schools for more than fifty years. She also offered private lessons in piano, voice, and violin to children for an equally long time. She served as pianist and music director in the First Baptist Church and was a frequent soloist for numerous community programs. She was the recipient during her professional and retired years of many awards and honors for her contribution to the performing arts. Her confinement to the Bibb County Nursing Home left a conspicuous void in the town. The 1988 closing of the Ben Franklin store, always remembered as "Davie's," was the end of the family's presence in the community. The town has a richer heritage because of the special capabilities by this prominent, talented family.

The Civitan Club was under the leadership of businessman Frank Ferrire. The achievements of the year included the assistance in organizing a club at Woodstock and the continuing support of the boy scout troop. In October, civic leader Ferrire made a plea to the community: "The West Blocton Club has started a drive for new members. There is a considerable number of citizens in and around Blocton who owe it to themselves and their town to assume some of the civic duties and unite as a unit, put your shoulder to the wheel and all pull together instead of separate ways."

If things during 1954 were not bad enough, a pretty good WBHS football team lost to Bibb by a touchdown in the

November classic on the Centreville field. The Choctaws were led by a pair of college-bound, highly recruited backs, Stallings and Faucett. Faucett was to make future sports headlines for Georgia Tech. Bibb would not win again until 1959, but there were two ties along the way.

A January 1955 issue of the county paper included a commercial-civic round-up that highlighted the status of county municipalities. West Blocton's attainments were catalogued as: three and one-half miles of black-topped streets, a Main Street fire station and jail, a new fire truck with 1000 feet of hose and a 500 gallon booster pump, and water system extension and improvement. A long time boast was restated. The water supply was one of the best and purest in the state and had continued to receive State Health Department compliments. Special attention had been given the holding pond site including clearing, ditching, and pump maintenance. In the town limits special effort had been made to replace damaged drainage pipes and sanitary conditions were in better shape with attention given to keeping trash, weeds, and grass cleared from public areas and providing insect spraying in the summer months. A full-time policeman, Herman Clinner, was on duty who had other tasks assigned including maintaining stop signs, painting curbs, and lining off parking places.

Mary Huddleston Davie. (Courtesy of J. W. "Bill" Davie.)

These accomplishments plus the town paying off $2,000 in water works bonds were deemed highly commendable, especially since they were done in the face of a declining economy with the coal mines being first on slack work and finally completely shut down. Town officials at the time were Mayor Ben O. Stanley and Councilmen Rupert M. Bearden, W. D. Deerman, Charles B. Haynes, John M. Langston, and Raymond W. McCulley. Indeed, these were difficult times and the town's leadership was clearly saying, "we may be down but not out." Improvement of the community's infrastructure and support services was evidence that attempts to stay alive were in effect and continuing. The following November, as proof that the town government was still functioning, Mayor Stanley announced that the town was completely out of debt with the last payment made November 1, on water works bonds issued in 1935.

In the summer, the town was host to 300 delegates from the area's sixty-seven American Legion Posts. A 1955 news item contained information about re-routing Alabama Highway 5 and the construction of a new bridge at the Cahaba River crossing near Piper and, unfortunately, the further erosion of local businesses with Chester Collins' June 1955 announcement that he was building a funeral establishment in Centreville. The decision was not Collins' but the Brown Service Company's with which he was affiliated. The population decline in north Bibb County was the major reason for the move. Collins remained in Centreville until September when he sold the business and moved to Jasper. There had been two undertaking establishments in West Blocton for decades and now there were none. Residents of West Blocton still say they believe the twenty-six mile round trip for funerals to be the longest distance in the county.

An item in editor Oakley's September 1 column noted that counties with chapters of the NAACP were being petitioned for school integration. He noted that the organization did not have a chapter in Bibb and even if there were one "we believe a good group of Negro leaders would discourage any such move that would cause dissension between the races." He further noted the need for better school facilities for both races in the county.

While it was not mentioned in the editor's column, the condition of the public African-American school on the Heights in West Blocton had been deplorable for years, and to a lesser degree, so was the condition of the white school.

Chief Clinner. Herman Clinner has been the longest serving police chief in West Blocton history holding the office for more than twenty years. (Courtesy of Patricia Clinner Deerman.)

The mid-1950s was a late date to consider decent schools for the county's minority race. Interesting developments, however, occurred soon.

The best and biggest news was the 1955 Tiger football team. Coached by Morris Higinbotham, the team was untied and undefeated and a winner in November over Bibb County by the largest margin ever, 32-6, their 28th win in the last 30 games. The team was declared District 3 AA champs at a December 15 trophy presentation banquet in the high school auditorium by the *Birmingham News*. The offense amassed 268 points and the defense allowed 63 in nine games. Team members mentioned in the news articles were David Hicks, Roy Holsombeck, Walter Sansing, Tommy White, William Hubbard, and Travis Hall. Seniors on the team were Jack Wade, James Whatley, Cecil McClellan, Frederick Henderson, Charles Shattuck, and William Hubbard. Prospects for the 1956 team were proclaimed outstanding. In appreciation, local businessmen, Huey Frederick and Gene Hicks, staged special dinners for the players and their dates.

According to editor Oakley's column, by January of 1956 there were only 75 coal mining jobs in Bibb. In a continued effort to make the community better appearing, a June public hearing was scheduled to deal with unsafe or unsanitary housing in West Blocton's town limits. The purpose was to determine if a housing authority was needed in the town. No action was taken to form an authority.

The high school entered a self-evaluation program in 1956 under the leadership of its principal, Oscar Zeanah. The objective was accreditation by the Southern Association of Schools and Colleges. Zeanah took the occasion to report to the public through the county paper that the school failed to meet that organization's standards because it lacked a gymnasium for a comprehensive physical education program and a shop for manual arts training and possessed a library that was not adequate. He said there was no department in the school that was not in need of new and improved facilities. He stressed the deplorable lack of an appropriate school building. It was a plea principals had placed before the county and community's leaders since 1920. Dr. Marjorie Swift, Emerita Professor at the University of Alabama who co-directed the school study, and Mrs. Oscar Zeanah confirmed in 1998 that the plusses outweighed the minuses in the final evaluation report, and the school was awarded accreditation by the Southern Association. The official notice came in December 1958. Swift remembered the special musical talents of a student who sang for the final presentation event. The panel was very impressed with senior Jerry Carter.

A new school plant sought for decades by the school's principals, however, was several years away for West Blocton High School. Zeanah had managed to install indoor rest rooms earlier, and in 1957, he arranged to have the steam heating system in the abandoned Piper school dismantled and installed in the West Blocton school. The student reporter complimented the system in the school news column in April.

In spite of the handicaps that substandard facilities imposed on students and teachers, capable and talented students from the school continued to display their ability to overcome. One music student of the era made his mark on the community and state even before leaving high school. Jerry Carter was a member of a high school quartet, the Crownsmen. Other members were Charles G. West, Billy Roy Finnen, and Charles Carroll, Jr. He sang solo for local clubs and had an impressive solo appearance at the state band and choral festival. He pursued a music career after graduating from the University of Alabama in 1964 and served as a minister of music in Georgia, Alabama, Mississippi, and Arizona. Jerry also had a flair for writing and served as high school sports reporter. He had plenty to write about during his junior year when the Tigers ruled the gridiron. It was another undefeated, untied year with wins over powerful Greenville High School and a season ending win over Bibb County High, the first win on the Centreville field in eight years. Four of the team's seniors were Crimson Tide signees. Walter Sansing, Tommy White, Roy Holsombeck, and David Hicks all experienced playing time on Coach Bryant's teams. Coach Higginbotham's record at WBHS was 43-6-1, (1952-1958) according to a report in the October 25, 2000, *Centreville Press*.

In September 1956, the town reelected Mayor Ben Stanley, who ran on his record of roads, waterworks, debt payment, fire department improvement, and a new sanitary program. In that same year, the county announced the construction of a "new school for the Negroes" in West Blocton.

In the business community in March, Mr. and Mrs. Frank Ferrire moved their flower shop into the building next door to Davie's, a space formerly occupied by a unit of the departed Black Diamond Coal Company. The shop added fashionable clothing for the younger set for both men and women. For a period it was the place to shop in the county for the latest in clothing. Its show windows were equal to those found in the metropolitan districts. It was one of the few bright spots in a shrinking downtown shopping district. As the decline continued in West Blocton, good things were happening in south Bibb. A proposal for a Hill-Burton hospital had been taken to the point of voting for the necessary sales tax to support its construction. The tax passed by a large margin county-wide but in Beat 10 by only one vote. It was voted down in Beat 11 by 80 to 34. Intra-county sectionalism had raised its head again.

Another stab at seeking industry occurred late in 1956 with the formation of a North Bibb Industrial Development Committee. The Brent leadership that had been so successful in attracting the toy factory, joined with some forty business and local government leaders in West Blocton, Woodstock, and Marvel to work on the project. In an early December meeting, some twenty-five members of the committee met in city hall. In an appraisal of the situation, it was noted that regardless of all the mine closures, there were no vacant houses in West Blocton. It was observed that most of

WBHS Teachers. (left to right) Ruth Bevis, Eunice Keyes, Oscar Zeanah, Effie Ray, Lucille Yeager, and Morris Higginbotham. (Courtesy of John Crowder.)

The Crownsmen. (left to right) Jerry Carter, Billy Roy Finnen, Charles Glen West, and Charles Carroll, Jr. (Courtesy of Jerry Carter, Jackson, Mississippi.)

Coach Higginbotham and Walter Sansing and Tommy White. Victory over Holt, 34-0. (Courtesy of Billie Higginbotham, Birmingham.)

Newspaper Illustration, WBHS 1956 Football Team. (Courtesy of Jonette McCool Ross, West Blocton, Alabama.)

WBHS 1956 Football Team. (left to right) Front row—Thomas Lee Medders, Milton Lucas, Roy Jones, Bob Allen, Floyd Lippeatt, Travis Hall, Kenneth Broadhead, Jacob Marchant, and Reed Martin. Middle row—Kenneth Reach, Bobby Oglesby, Travis Dailey, Tommy White, Morris Moody, Paul Stewart, Alden Liddle, Ronald Burns, Steve Davis. Back row—Bill Cobern, Donald Holland, Franklin Melsoni, Bobby Hicks, Walter Marsh, Roy Holsombeck, Wayne Cobern, Jack Hicks, and David Hicks. Not pictured—Walter Sansing. (Courtesy of Jonette McCool Ross.)

those living in the town were working in Tuscaloosa, Centreville, Bessemer, or Birmingham. For example, there were 34 West Blocton men working at the Pullman Plant in Bessemer. The only visible help to the workers was the improvement of Highway 5 south from Woodstock that shortened driving time of commuting workers.

The *Press* reported in February 1957 that it had been told that a KKK organization had been formed in the northern section of the county. The editor observed that the Supreme Court was to blame for the race problems and it would be years before peaceful relations could be achieved because of the politicians and the court. Eventually, a sign appeared on West Blocton's Main Street that read "The Klan People of Bibb County Welcome You." By March bids were let on the new school for "Negroes" in West Blocton. By late summer it was almost finished and was ready for occupancy by September 20.

In July, Oscar Zeanah announced his resignation to take a position as principal of the high school in Enterprise. He had been in West Blocton for 13 1/2 years, principal for nine. He was considered one of the best principals in the state and so recognized by Dr. Ben Hubbard, the county school superintendent. By summer's end the state's principals endorsed that opinion when Zeanah was named president of the state principals' association. Following his stay at Enterprise, he was made superintendent of Andalusia schools and remained there until retirement. He was serving in an advisory role at the University of Alabama in Birmingham when illness forced his second retirement. The county paper editor wrote of Zeanah, "He has not only been a good school man, but is a leader in community and church affairs, and can always be depended upon to take an active lead in anything for the betterment of the county." Fred Justice, an Arkansas native was named principal and would serve two years. The community held a goodbye program at the town park for the Zeanah family. The student body president, Tommy White, made a short address of appreciation and received a bear hug from Zeanah. After saying farewell to the principal, the locals arranged an appreciation event in August for Coach Morris Higginbotham. He would coach one more year at WBHS. In August 2000, Tiger Stadium's field was named Morris Higginbotham Field in honor of the successful and admired coach.

Dr. Percy Nolen returned in 1957, purchased the old Murphree Drug Store, and set up his practice upstairs. The business was renamed the West Blocton Drug Store.

The *Press* editor reminded readers early in 1958 that Bibb County was "drying up." He quoted population figures noting that the rural areas and mining towns had lost hundreds of residents. In the 1920 census, Bibb had a population of 23,144 and projections for 1960 indicated only 15,700. He observed that farming seemed to be a thing of the past,

West Blocton Heights Junior High School. (Courtesy of Alabama Historic Ironworks Commission.)

citing the sale of large tracts of land to the paper and lumber companies. He also lamented the disappearance of Belle Ellen as well as the mine closing at Marvel and other mining towns.

Mayor Stanley suffered a stroke in June and resigned his post in October 1958. He had worked diligently for the town during hard economic times and had managed to make some progress. In July, word came that West Blocton had a chance for a garment plant to provide a possible 80 jobs. The Wes-Bloc Manufacturing Company that would produce a line of men and boy's sport clothes opened operations with fifty workers in mid-September. The old Wood Motor Company building, the ground floor of the Masonic Temple, and some of the Black Diamond property on Main Street housed the operation. The managers were brothers George and E. M. Roten, Jr. of Birmingham. The town issued $35,000 in 15 year industrial bonds to attract the firm. A welcoming ceremony featured former Birmingham mayor and then a vice president of the Alabama Power Company, Cooper Green, who had assisted the local committee in the industry hunt. An ill Mayor Stanley remained in his car for official handshakes and remarks. Stanley a sixty year resident of the town, died the following February. Dr. Percy Nolen was chairman of the industrial development committee.

R. G. Moore, mine operator and engineer, was appointed to the post of mayor to complete the Stanley term. Other events in 1958 included the disappearance of the old TCI ballpark. The American Legion divided it into twenty-one lots for sale. The local post also celebrated the dedication of an American Legion Home. The Church of Christ began construction of a meeting place on old Highway 5 in McGrawtown. Morris Higginbotham resigned to take a position at B. B. Comer High School, and the ten-year stay of Winston White as high school band director ended. The W. T. Morrisons and the senior Ferrire families moved to Birmingham. Both families had resided in the town for over forty years. During the school year, the students heard concerts by the University of Alabama String Quartet and the Montevallo Choir. A local student, Judy Harmon, was a member of the Montevallo group. Another WBHS student, Bob Allen, was awarded a scholarship to Auburn. Son of Mrs. Margaret Gentry, Allen had served as student body president, was a star football player, and an Eagle Scout. He was to appear frequently on Auburn's honor roll during his college years. Replacing Coach Higginbotham was Jack Smalley who had played football at the University of Alabama. In August, Postmaster William J. Murphy announced that a new post office would be built across the street from the Elks Lodge. The town took great pride in the continued success of Sansing, White, and Holsambeck who were playing for Alabama's Bear Bryant.

The post office was completed and dedicated on May 30, 1959. Congressman Armistead Selden was the main speaker. The building was financed by banker Josiah Kennedy of Centreville and leased to the government on a ten-year basis. A new post office gave the community reason to believe the old town was still going to be around a while longer regardless of the commercial decline.

Though it resulted in no road, a committee from Shelby County and West Blocton along with an Alabama College representative and officials from Alabama Power met with state highway chief, Sam Englehardt, to explore the possibility of a highway from West Blocton to Maylene. The objective was to establish a direct route to Alabama Power mine operations so Bibb County coal could be easily shipped to Maylene for processing. The locals were still trying to resurrect the coal mining industry, an exercise akin to whipping a dead horse.

By April 1959, the Wes-Bloc Manufacturing Company had so many orders it was necessary to move their operations into larger facilities. The cost was determined to be $80,000. The company had started with 50 people and now was working 175. With adequate space, they believed they could employ 325. For the previous six months of operation, the firm had paid salaries of over $100,000. It was not coal mining level pay, but it was something. By September land had been purchased for the new plant building and arrangements made for issuing the industrial development bonds. It was up to the citizens of the town and county to raise the funds. At the end of the year, construction was under way.

In July 1959, Principal Justice resigned and Clint M. Bevis, principal of the elementary school, was assigned to the post. In October, the Bibb County twenty-five bed hospital was dedicated in Centreville. In their first win since 1954, the Choctaws bested the WBHS Tigers, 24-7, (there had been two ties during those years). At the University of Alabama, local scholar Kenneth Smith, in his junior year, was elected president of the University Chapter of the American Chemical

West Blocton Post Office. (Courtesy of Cahaba Trace Commission.)

Society. In May, a legal notice appeared in the paper extending the town's corporate limits. In December, an old West Blocton lodge was revived. The Navaho Tribe of Red Men was slated for meetings on Sunday afternoons in the Knights of Pythias Hall. Officers were B. J. W. Mize, Sam B. Shuttlesworth, R. G. Carroll, W. H. Harper, G. E. Dailey, W. D. Brown, and T. Salvadori. To join the Red Men, a prospective member had to be white, believe in the Great Spirit (Christ and Christianity), speak English, be an American citizen, and not be a communist. Red Men had not been active for several years. No further news was found about this lodge so it probably was short lived. Its brief resurrection may have been related to the contemporary race problem.

By May the official census report was published. Bibb County showed the sharpest loss of population in the state during the decade. The official 1960 numbers for the county were 14,450, a decade's loss of 3,527. West Blocton was reduced to third in size among the county's municipalities after having been the largest town in the county since the 1890 count. It was stark evidence of the end of coal mining, the county's largest industry for over a half-century.

Dr. Ben Hubbard left the Superintendent of Education office to accept a professorship at Mississippi College in Jackson. Hubbard was said to have brought fairness and cooperation in the operation of the county school system. Under his administration both county high schools had gained Southern Association accreditation. Make-do arrangements had been used to squeeze West Blocton by the accreditation committee in the form of some band-aid work on the facility by adding indoor rest rooms and a central heating system. The school still lacked a gymnasium. These qualifications were of no concern at BCHS where central heat, rest rooms, and a gymnasium had been a part of that school's plant for years. A shock came to West Blocton's new school hopes in April when it was announced that funds from the state would be used for a new high school building in Centreville. West Blocton was to get two brick classrooms and a gymnasium. Other schools were to get small amounts for upgrading.

A new administration was elected in West Blocton in September. Floyd I. Morrison, a local contractor, won the office of mayor. The council was John M. Langston, Oliver K. Yeager, W. D. Deerman, W. E. Finnen, and Ellington P. Jones. The mayor and town council made immediate but modest plans to upgrade the town's streets and sidewalks. By December, more concrete and ambitious plans were adopted. A hoped for highway to Montevallo was in the works, a modern street with gutters and sidewalks from downtown to old Highway 5 was being surveyed, and to add to this street improvement, 3,000 feet of six inch cast iron pipe was to be installed to replace fifty year old four inch corroded water mains. The boldest proposal involved the matter of the need for a new high school. The mayor described the old 1935 wooden building as "built with second-hand lumber and materials and now a fire trap and home for termites." Appreciation was expressed for the funds for a gymnasium and two classrooms, but many felt those additions on the congested site would be an unwise investment. Instead, a plan was proposed to place a portion of the funds in escrow until a new school could be built on adequate acreage. This plan for the school had the endorsement of the City Council, Elks Lodge, Civitan Clubs of north Bibb County, County Commissioners, Probate Judge G. H. Stacy, and the mayors of Centreville and Brent. In a future plea for a new school in West Blocton, the Mayor of Brent, Fred H. Davis, later elected to the state legislature, would be an advocate for the project's completion.

Photographs were made at the high school dramatically showing the condition of the building and inadequate grounds. The committee's letter of recommendation and the photographs were presented to Bibb County's State Representative Virgis Ashworth (former West Blocton resident and WBHS graduate then living in Centreville) who made an inspection of the facilities. He added his endorsement to the plan. These facts and plans were presented to the governor by a delegation from West Blocton and the county on December 8. A photograph appeared in the mid-December *Press* showing the committee standing with Governor Patterson. Bids were let on November 24, 1959, for a new high school building in Centreville.

Early in 1960, first payments on the Wes-Bloc Company bonds were paid. Irene Lawerence decided to give up writing the local news for the weekly paper, and Mrs. Boyd Hartley took on the task. In addition to former Mayor Stanley other long-time citizens passed away during the year. The town bade farewell to W. A. Jones, Coca Cola plant operator; J. A. Johnson, depot agent; Sallie Thrasher, boardinghouse operator; Mrs. Joe Arih, an Austrian immigrant citizen since 1912 and active member of the local Catholic Church for twenty-five years; and Mrs. R. H. Wood, civic leader and Methodist Church steward.

The February 4, 1960, *Centreville Press* printed a story about Commander Fred E. Stewart, MSC, USN, administrator of the U. S. Naval Aviation Medical Center in Pensacola who had completed thirty-three years of service. Commander Stewart had begun his service career in 1927, had served aboard the battleship *Pennsylvania* and as an assistant to President Roosevelt's personal physician in the White House. Stewart was a native son of West Blocton, a member of the WBHS Class of 1925. His niece, Elaine Stewart Henderson of Montgomery, provided other details about this distinguished military officer's career. After high school he attended Columbia Military Academy in Tennessee for a year, returned home and worked in Pope's Drug store as a pharmacist for several months before his 1927 navy enlistment. He had assignments in Virginia, Washington, D. C., New Jersey, New York, Cuba, California, Okinawa, and aboard several ships. A major achievement was his authorship of the *Naval Hospital Corpsman's Manual.* Upon retirement from the Navy, he was appointed head of the State of Hawaii's sanitation department and after retirement there returned to Pensacola for a second retirement. Commander Stewart died in 1985.

West Blocton was experiencing local infrastructure problems in 1961. A particularly harsh late winter and spring ushered in the year. February storms dominated the news. West Blocton's small waterworks sustained considerable damage as a result of uncommonly low winter temperatures and excessive early spring rains. The mayor and council dealt with the problem as funds would permit. A March 2 *Press* news item was headlined, "West Blocton Water Supply and Main Reservoir in Dangerous Condition." A Birmingham engineering firm was employed to make a survey, and the mayor asked for help from Civil Defense, the State Board of Health, and the Geological Survey. A call to Fort McClellan produced immediate relief. A truckload of pumps, emergency storage tanks, and filters arrived promptly. Then more rain washed all temporary dams away. Landslides at the spring occurred when retaining walls collapsed muddying the water supply. When borrowed Army equipment slipped into the mud and water, Huey Frederick's wrecker pulled it out. The town's water supply was at only one-third capacity. In late April, the water supply was still in an emergency condition. The engineer's report was awaited to make repairs to the reservoir and pipe lines. In early June, Mayor Morrison and the council reported a complete overhauling of the waterworks system was nearing its end. The emergency was declared over.

The water supply was perceived to be adequate "unless there was a very long or a very severe drought period." Problems persisted, and by November, there was another water emergency.

Mayor Morrison had promised street projects, and in late July, the work of widening, surfacing, putting in curbing, gutters and sidewalks was well under way. The project was scheduled to be completed in three months, and was called the best street improvement in the history of West Blocton. Support came from the State Highway Department and by County Commissioner Emery Lippeatt. By August water main installation was in progress, and by December three miles of road paving east toward the Cahaba River was under construction. It was the first segment of a planned "highway to Shelby County." It extended to Ardella, an old railroad junction near the Cahaba River.

West Blocton has always taken pride in its citizens' achievements and in March 1961 applauded its first appointment on record to a United States service academy. Billy Hartley, WBHS senior, National Honor Society and Letter Club Member, son of Boyd and Margaret Lane (Edmonds) Hartley, was notified by Congressman Armistead Selden that he had been given a principal appointment to the United States Naval Academy at Annapolis, Maryland. As it turned out, Hartley did not attend the Academy but opted for enlistment in the Air Force, winning his pilot's wings and remaining in the military service for a commissioned officer's career.

Another "local-boy-makes-good" story was the elevation of Bibb County representative Virgis Marion Ashworth to the position of Speaker of the House in 1961, during the term of Governor John Patterson. The story of his success was printed in the May 4 *Birmingham Post-Herald.* Ashworth's family had been West Blocton residents for years, and he had finished WBHS in the early 1930s. He entered the University of Alabama but was a victim of the Depression's hard times forcing him to drop out to take a job in the coal mines. He worked in the mine until 3:00 in the afternoon and then drove to Birmingham to study law from 7:30 to 9:30 at the YMCA, the site of Birmingham Law School. He studied at the Birmingham Public Library and said he was there so frequently that he was given a key. With the approval of the librarian, he often opened and closed the building's doors. He passed the bar examination in 1937, and after brief employment by the FBI in the Washington, D. C. area, he returned to Bibb County to practice law. He became county deputy circuit solicitor in the early 1940s. He was elected to the State House of Representatives in 1958. While practicing law and when elected, he was a resident of Centreville but carried all the boxes in the northern beats by large margins.

A new high school was probably a hot topic for gubernatorial candidate Ryan DeGraffenreid when he visited West Blocton. With the candidate in front of the Vanity Beauty Shop are (left to right) Joe Arih, Floyd Morrison, Abston, Hop Little, John Kelley, Harvey Kizziah, Ryan DeGraffenreid, Mack Carr, Joe Lucas and Emmett McCool. (Courtesy of Martha Fancher Kelley.)

Racial Conflict and Other Civic Problems

With so much racial tension at the time, it is not surprising that conflict, though minor, would occur on West Blocton's streets. The specific circumstance is not fully explained in the county paper but Mayor Morrison felt compelled in late August 1961 to write the following to the community through the newspaper's columns:

> Friends, White and Colored:
>
> The tension is high on both sides due to so much publicity in our papers from outside our town and state. We have and can live happily together both in West Blocton and elsewhere, and keep our life, if we will just be calm and try hard to use good judgment. There is room for all of us to live in peace and happiness. We should be ashamed of the recent incident which took place in our town between some of our children, both white and colored. Careful investigation of these cases showed a positive lack of parental training and correction.

The implication in the remainder of the item indicated remarks had been made about girls' dress styles leading to their parents seeking law officers' intervention. For the times, it was but a blip on a large screen of racial clashes. Fortunately, West Blocton, through it all, was free of the major kinds of disturbances so much in the news in Birmingham, Tuscaloosa, Selma, and Montgomery.

The town council and other prominent citizens led by Mayor Morrison, and with the backing of county officials, rejected the county Board of Education's award of two brick classrooms and a gymnasium to be added to the wood frame, 1935 high school building. Funds were put in escrow for use on a new building on a larger site. In late 1961, the search for acreage for a school was in full throttle. The *Birmingham News* entered the scene with an expose' of the situation citing in detail the less than standard school plant in West Blocton. The story was prompted by the difficulty in locating an appropriate site for a new school. Six acres comprised the old school's campus. In that small, compressed area was the main building that housed ten classrooms, principal's office, library, and the auditorium. Adjacent separate units were a home economics building, a converted wood-frame former residence used for girls' physical education, a Quonset hut for special classes, and a steel frame field house and band room. The town leadership was seeking forty acres for a new school, but it would be another four years' wait for all needs to be met for adequate site acquisition and funding.

As the Patterson administration was ending, the mayor wrote him again making a plea for high school building funds. The Governor replied politely with the promise he would do all he could but that currently money was not available. He did say he would see the street project in West Blocton to completion. The *Centreville Press* editor also went to bat for the high school building and reported in his column of the urgent need. He had visited the governor and listed county needs including the new school project for West Blocton. In the fall of 1962 George Wallace was elected governor. His Bibb County vote was three to one. He won by comfortable margins in Beats 10 and 11.

A New High School Building at Last?

In late summer 1962, Bibb County legislator Fred Davis introduced a bill that would provide West Blocton with funds for a new high school. With funds in escrow from an earlier allotment, the additional funding would provide over $400,000 for the facility. Davis was successful in moving the bill through the House, and Senator H. P. James worked the bill through in the Senate. In the meantime, a forty acre site near the 1937 elementary school had been acquired. Governor Wallace had made a personal inspection of the 1935 WPA high school structure during the 1962 campaign. The real credit for a new school, said an area resident who was a high school student in the mid-1960s, goes to Fred Davis who presented pictures and grim reports on the old building's condition. A rather dramatic presentation and plea brought the House chamber to a positive vote on the project. State Education Department officials are reported to have said the wooden West Blocton High School building and surrounding support structures were among the worst in the state.

Italian Visitors. (Courtesy of Lillian Cinelli Harris, West Blocton, Alabama)

As though the poor facility were not enough, the town's water supply, while passing minimal State Health Department requirements, did not pass State Education Department standards. A *Tuscaloosa News* story carried a photo in early September of students drinking water from U. S. Army Lister bags that had purification tablets dissolved in the containers' drinking water supply.

The mayor's report in August 1962 continued to reflect the water supply problem but indicated much improvement had been made. Over $40,000 had been spent on new lines and new fire plugs. However, the water problem only got worse. By early November 1963 it was against the law to wash your car with the municipal water supply. A system in difficulty was further hampered by a drought year. Things were so bad 10,000 gallons of water were transported daily from Centreville. The problem was to be only partially solved by the summer of 1964.

From time to time the community experienced reminders of its European roots. Interesting visitors from Torino, Italy showed up in town in August 1961. Mr. and Mrs. A. Viarigge, with relatives from California, sought the help of Floyd Ambrose and Mrs. Frank Cinelli to locate his birthplace in "Little Italy" or the "Hollow." Mr. Viarigge's family had returned to Italy when he was five months old. This was a story with a different twist, a visit to one's American birthplace by an "Old Country" citizen. Mrs. Cinelli led him to the precise spot where Viarigge's parents' home had stood, and he filled a plastic bag with soil to take to Italy.

Black Diamond Number 11 Tipple. (Courtesy of Billy and Nina Smith.)

Henry Clay Randall. (Courtesy of Diane Kerr.)

West Blocton citizens were informed in January 1963 that the town's fire siren had frozen and had not sounded when the Frank Hardemon residence on Main Street caught fire. When the local volunteer firemen failed to show because there was no alarm, the home owners had placed an emergency call to the Centreville Fire Department who responded as quickly as possible. The Town Council was embarrassed and assured the public that repairs to the siren were being made. A notice in the *Centreville Press* signed "Town Council" asked the community to contact E. P. Jones or J. M. Langston before calling for out-of-town help.

Industrial Ups and Downs

The early days of 1963 found the town with modest local surface mining operations that kept the Black Diamond Company washer operating, and railroad switch engines could be heard working in late evening hours moving the coal to the rail yards in Woodstock. The Black Diamond Company also operated a slope identified as "Number 11" to the east of Smith Hill, generally in the area of old Number 3 TCI mine. Relatively few miners were employed in this operation.

The modest Wes-Bloc Company garment plant that had begun with so much hoopla and hope failed and left creditors holding the bag and bad checks. Bankruptcy was declared, and in December advertisements appeared offering machinery and other supplies for sale to cover the defaulted mortgage. Work was initiated through Alabama Power Company channels to reopen the plant that had been built with town industrial bonds. Cooper Green was again asked to help find a responsible, viable garment plant company. As late as June, no plant prospect had materialized, and editor Oakley joined the search effort with a plea in his column for everyone to work for a new industry in West Blocton. In October, efforts paid off when Santana Sportswear of St. Louis, Missouri announced plans to open in West Blocton. To prove its good faith, a check for $8,366.50 was presented to the town to pay principal and interest on the town's bonds on the building erected for the failed enterprise. The bold *Press* headlines announced: WEST BLOCTON GARMENT PLANT TO RESUME OPERATION SOON.

Two hometown scholars made the news in June and July 1963. Dr. Helen Raye Thrasher, daughter of Mr. and Mrs. George Thrasher, was awarded her University of Alabama medical degree in June at Denny Stadium graduation exercises. Her achievements and awards filled two newspaper columns. Dr. Thrasher specialized in ophthalmology and remains the Birmingham eye doctor of choice by many of her friends and relatives in and near West Blocton. In July, another hometown scholar, Henry Clay Randall, assistant professor of history at Birmingham Southern College, holder of a degree from the University of Alabama, two degrees from Cambridge University in England, and also M. A. and Ph.D. degrees from the University of North Carolina, was headed back to England for further research. His study was to be on the life of Benjamin Disraeli, a British statesman. Randall predicted his research might take five years, and he would continue to make summer journeys to England to see the project through. In retirement in the 1990s, Professor Randall was still making annual pilgrimages to England to continue his study of English history.

By mid-year interest in the Civitan Club had waned and their charter was surrendered. Another sign of community business decline was the loss of Dr. Percy Nolen and his family who moved to Helena. With Nolen went the town's drug store.

An incident reported in the West Blocton personals column of the *Press* of October 16, 1963, eventually was recorded in scientific journals and other publications that consider such matters. Clarence Franks and his son, Robert, took active parts in the Bibb County Fair that year. Clarence had on exhibit a live coral snake that he had caught near Belle Ellen. Tharon Camp, the assistant county agent, said the snake was the first of its kind known north of Jackson, Alabama. Robert won first place with his 4-H White Leghorn pullets.

Clarence Franks' nephew, Kennedy Smith, a student at the University of Alabama, took the snake to the University's biology department which verified the county agent's contention that the coral snake had been found farther north in the Cahaba basin than any of that species had been found before. The biology department reported the find to scientists at the Smithsonian Institution, which requested that the snake be dispatched to them. Franks returned to the site and captured two more snakes so the University could have a specimen. In the interest of sharing the wealth of biologically diverse Bibb County, the third snake went to Auburn University's reptile collection by way of the University of Alabama. Today, books on reptile facts still give Franks credit for his discovery.

West Blocton Reacts to President Kennedy's Assassination

Regardless of political affiliation, the citizens of West Blocton, like those all over America, were profoundly shocked at the assassination of President John F. Kennedy.

The national tragedy in Dallas on November 22, 1963, was met with mixed feelings in Bibb County. Darrell Finnen now an antiques dealer in Nashville, Tennessee, remembered his reaction that day. Finnen was a high school senior and student body president at WBHS. As do many others who recall that event, exactly where he was and what he was doing became etched in his memory. At the author's request, he wrote the following:

> It was a beautiful, warm November day. The sky was a bright blue without a cloud. I had been playing tag football during lunch. The end-of-lunch bell rang and I had just walked into the library when someone announced Kennedy had been shot. Sad to say, many of us shouted and clapped. Some of the girls were sad; most of the boys were happy. In retrospect those times seemed to mirror the War Between the States, the Federal government against Alabama and Mississippi. It was not just an issue of segregation but once again our whole way of life was being swept away. At that time it was difficult to say who was our greatest enemy, the Russians or the Kennedy's. So when he was shot the reaction was understandable. As the days and weeks wore on after his death, I felt very bad about my personal reaction. However, the bad feeling was a result of the fact that a murder had taken place. But he was still the enemy. I hate violence and degradation to the nth degree and that is what makes me sad about how I reacted on November 22nd 1963. The one single thing that is as vivid today as it was in 1963 was how blue the sky was that day.

Finnen added that he also deplored that Federal troops had fired on students at the University of Mississippi resulting in the death of a woman student whose life, he thought, was as precious as Kennedy's.

Twelve year old Joy Parker, who was a sixth grader in West Blocton Elementary School, remembered that her class was deeply shocked and sad over the assassination of President Kennedy. Alberta Quinn, her teacher, arranged for the class to have access to television so they could watch the events as they unfolded in Dallas and Washington. Joy remembers that the classroom was as quiet as a tomb and that some of the students cried while watching the Kennedy funeral broadcast.

Other students remembered that the episode was frightening because some thought the country might be under attack from foreign elements in Russia or Cuba. Those were the times when school children in West Blocton and across the nation had practiced atomic bomb drills by taking positions under desks, placing their arms over their heads with faces close to the floor. In Tuscaloosa several noon diners at the Stafford Hotel were said to have stood and cheered at the news about President Kennedy. There were reports that TV soap opera viewers had complained to local stations for interrupting their programs with the bulletins out of Dallas. It had been only seven months since the integration of the University of Alabama orchestrated by the Kennedys that featured Governor George Wallace's "stand in the schoolhouse door." Yet, others in Tuscaloosa attended later planned special prayer services in their local churches. If there were memorial services in West Blocton churches, it was not recorded in the newspaper nor was there mention of how the

Elks Little League. Down through its history, the Blocton Elks Lodge has supported a wide range of civic and charitable causes. (Courtesy of Martha Fancher Kelley.)

matter was handled in the following Sunday services. The troubled 1960s decade also witnessed the assassinations of Robert F. Kennedy in California and Dr. Martin Luther King in Memphis.

In late November, a $45,000 health center was announced for West Blocton. While Dr. Ethel Montgomery had served the community from her home office and enjoyed a fine reputation as a diagnostician, there had not been a full service doctor since Dr. Percy Nolen had departed. The proposed health center was to have a doctor associated with the Centreville hospital who would have special hours in West Blocton.

The low bid on the West Blocton High School announced in mid-January 1964 was from Dallas Construction Company of Selma and the architects were Pearson, Humphries and Jones of Montgomery. An architect's drawing appeared in the Centreville paper in April. Funds were still uncertain so construction was not begun.

Looking for any glimmer of hope on the industrial scene, attention was given to the opening of a rock quarry near Big Springs, five miles north of West Blocton. The operation was to employ forty to fifty workers. Over 305,000 tons of crushed stone for road building was supplied in an initial order. Greenville Quarries of Kentucky invested a half-million dollars for machinery. In the immediate vicinity Santana Sportswear had opened in January to produce boys' trousers. Thirty-five were employed to start up with the possibility of eventually hiring 200, and by late June, the plant was working 150.

The spring of 1964 produced more encouraging news. The legal notice of "Incorporation of a Bank" appeared in the *Centreville Press* on March 26. West Blocton was to have a state bank after being without banking facilities since 1932. Capitol stock was to be $50,000 and directors were Eddie G. Downs, Frank Ferrire, W. W. Findley, E. P. Jones, J. M. Langston, Howard C. Pate, and O. K. Yeager. The corporation was called Bank of West Blocton. The facility would not go into operation until mid-March 1965 following approval by the State Banking Department and the Federal Reserve System.

It was learned in April that Principal Clint Bevis of WBHS, in that position for ten years, would take a new post in Dothan. Herman Crider, elementary school head was made the new principal. Crider would have the task of planning for and opening the new high school.

An interesting visitor in June, back home from an overseas assignment in far-away Kabul, Afghanistan, was James Arthur Sumner, proof positive that West Blocton folks could be found all over the world. Sumner was employed with an engineering firm that had contracts for road building projects in the Middle East. Incidentally, Sumner reported in 1999 that his maternal grandfather, James Newton Erasmus Hayes, was once prominent in the local timber business and helped negotiate a land donation by TCI to the community to extend Mt. Carmel Cemetery. Hayes helped clear the several new acres for the right to harvest the timber. It is believed this is the area identified by local citizens as the "new section" of the cemetery. The acquisition probably occurred in the late 1920s.

The municipal election of 1964 had the incumbent, Floyd Morrison, opposed by local merchant, Frank Ferrire. Morrison ran on his record which some thought very good. His achievements were the addition of paved and widened streets, gutters, curbs, and sidewalks from downtown to old Highway 5 in West End; paving and grading of connecting county roads with Highway 25 to the east; the promise of a new high school; successfully attracting the Santana sewing plant; obtaining the Hill-Burton Health Clinic; resolution of emergency water system problems; good relationships with other towns in the county; and an easy relationship with politicians in Montgomery. One citizen who recalled the Morrison administration called him a "man with a vision." Morrison had family connections in West Blocton and after many years in the construction business in New Jersey, he came home. His mother had been a life-long town resident. Morrison knew how to get things done but in the process tended to be ahead of the small town, conservative-thinking, merchant leadership. The old "wait for better times" thinking predominated. Another asset was his wife, the former Clara Logan Desmond, a well-liked and respected businesswoman. Many times, a former associate said, the Yankee-accented mayor acted and then sought approval. He was nicknamed "Boom-Boom" because of his tendency to speak forcefully and move fast. When the votes were counted, merchant Frank Ferrire, son of an Italian immigrant family, was elected by a vote of 216 to 95. The new mayor announced his immediate objectives. He would obtain a doctor for the town, improve the water system, and attract an industry that would employ men. The major goal to obtain an industry was never achieved. However, through his perseverance other objectives were attained such as the development of a dependable water works system. Frank Ferrire served for twenty years, longer than any other individual in the position. The *Tuscaloosa News* said of him, "Frank Ferrire is a husky 44-year old man with a fountain of energy who enjoys doing good things for other folks and his town." Ferrire's councilmen were primarily businessmen, a continuation of the established tradition of drawing civic leadership from the merchant class and professional class. They were W. D. Deerman, drayman; A. N. Kitchen, insurance and general business; Ellington P. Jones, Coca-Cola bottling plant owner; Ballard Gilbert, mining; and Oliver K. Yeager, service station operator.

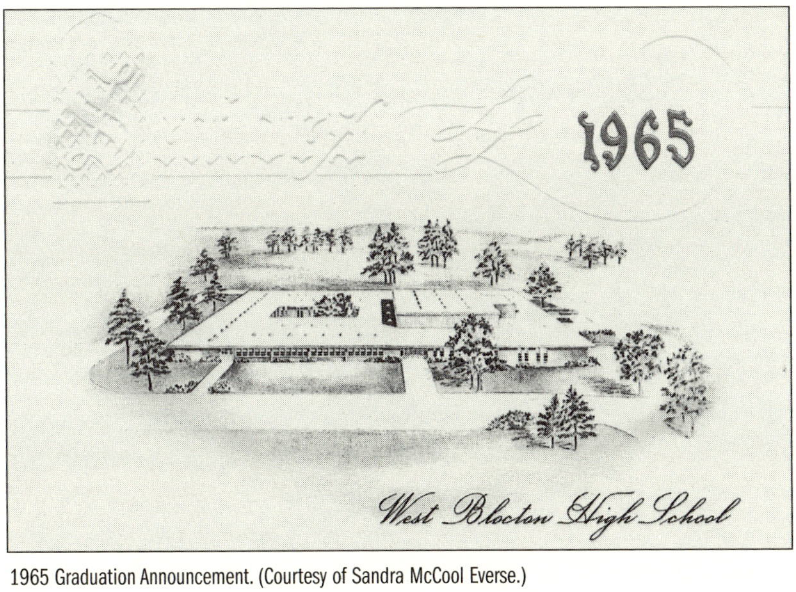

1965 Graduation Announcement. (Courtesy of Sandra McCool Everse.)

By 1964 the number of businesses in the community, most on Main Street, had further diminished. Mayor Ferrire was one of those still holding

Ribbon Cutting at the New Bank. (Courtesy of Ferrire Collection.)

on. A merchants' business promotional in December offered an automobile by raffle. Businesses participating were listed giving an indication of which concerns were still open. They were: O. K. Yeager Service Station, Mildred's Flowers and Clothing, Kitchen Cleaners, I&N Quality Store, Pierson Hardware Store, McIntosh Sisters Clothing, Burns Clothing, Vanity Beauty Shop, Charlie's Shoe Shop, R & R Grocery, Lawley's Beauty Shop, Benny's Barber Shop, Jones Hardware, Davie's Variety (the Ben Franklin Store), Whaley Furniture Company, Langston's Big Star Store, Huey's Cafe, Putnum Thermo-Gas, Smith Super Saver Grocery, Finnen and Jones Grocery, West End Grocery, Floyd Ambrose Store, Reach's Gas Station, McKinney's Parts & Service Station, Coca-Cola Bottling Company, and Boots Carroll Service Station.

Mid-decade of the 1960s in West Blocton saw a new, long sought, long promised, first class high school building under construction. The Class of 1965 thought they would graduate there but the contractor's schedule did not permit it. Theirs was the last class in the old 1935 building although their graduation invitations carried the architect's rendering of the new building. The community and its new mayor had to believe things were on an upswing with the breaking of ground for the new Bank of West Blocton that occurred on a cold, bright mid-January day. The photograph of the ceremony shows the bank directors, including Mayor Ferrire, in winter coats posing for *Centreville Press* coverage. With the bank building nearing completion, bids on the 1600 square foot West Blocton Health Center and Clinic were advertised in late March.

The bank ribbon cutting and open house took place on Sunday, May 16. Boy Scouts were on hand to direct traffic and admit guests at the door. Director Walker Findley's grandchild made the first deposit, and little Dinah Crowder was the first child to bring in her piggy bank to deposit her savings. The local reporter, Mrs. Boyd Hartley, wrote at the end of her story, "We in West Blocton are proud of this badly needed addition to our town and are happy to welcome Mr. Eddie G. Downs, the bank president." A boldface advertisement said, "The Bank of West Blocton will open for business at 9:00 a.m. on Monday May 17, 1965". A similar announcement had appeared in the *Blocton Courier* in November 1892, seventy-three years earlier, heralding the opening of the Blocton Savings Bank, Bibb County's first bank.

Mary Clyde Moon (Owens) and Nell Quinn Moon celebrated the dedication of the new high school with Governor Wallace. (Courtesy of Mary Clyde Moon Owens, Centreville, Alabama.)

Intra-county opposition that had surfaced in 1949 to a third bank in Bibb County to be located in West Blocton was somehow overcome. There had not been a new bank in the county since the opening of the Brent Banking Company in 1912. Proof of the viability and need of banking facilities in the town was shown just one year later when President Downs announced that the bank's assets had reached one million dollars.

In mid-July, Ferrire noted that work had begun on the new health center on the property where the "Old Green School House" had stood. The mayor also announced that new mercury vapor street lights had been installed on Main Street and plans were made for their installation in other parts of town. Prior to the new high school dedication, the *Tuscaloosa News* ran an illustrated article featuring the new school, the Santana Sportswear Company, the Black Diamond Coal Company Washer, the new bank, other business shots, an aerial town photograph, and a short town history.

The biggest event in the county in 1965, local folks thought, was the dedication of the new West Blocton High School that took place on Sunday, August 22. The main attraction was Governor George C. Wallace. Over 1,000 citizens crowded into the new facility's auditorium-gymnasium to participate in the afternoon ceremony. Mayor Ferrire welcomed the group to West Blocton, and Judge Harper Stacy welcomed the visitors to Bibb County. The *Tuscaloosa News* covered the proceedings in detail. Building a new school in West Blocton was one of Governor Wallace's campaign promises, and he said, "I came here while campaigning and it seemed to me no other place in the state needed a new high school more than this town. The old school was not safe for students, in my judgment." The building was open for tours conducted by students and teachers. After the dedication program, Governor Wallace toured the town inspecting the Santana Sportswear Company, the new bank, and the new health center.

At mid-year unsettling news of the Viet Nam War had begun to sink in on the county's population. Editor Oakley's column of July 29 carried the following entry:

> THE WAR IN VIET NAM
>
> The war in Viet Nam is very bad. We don't understand how to fight these people, and many believe that we cannot win. We may be able to reach a place where we can negotiate with the Communists but in the meantime many thousands of U. S. Soldiers will lose their lives. We lost about 55,000 men in Korea, and couldn't win that war and we could fight better in Korea than we can in Viet Nam because of the terrain.

> We are committed to fight the Communists everywhere we can but the Viet Nam situation could end up in World War III. It will cost billions of dollars and we will lose many men.

Among American soldiers to lose their lives in Viet Nam were two West Blocton youths.

The year 1965 began with the hope of its being a turnaround year for the town. A modest library continued operating in the building next to the mid-town city hall, the former location of the drug store. A reading club was organized and summer story hours were scheduled for children. The regional library bookmobile made regular stops.

In September, another stab at organizing a civic club was made. A Lions Club charter was presented to twenty-one members in September. Among the charter members were three African-Americans, perhaps a central Alabama "first." Mayor Ferrire was hopeful that the new County Road 24 bypass would soon be back into local and state road plans. An effort was underway to obtain a doctor for the town, and city mail delivery was seen as a possibility. It was also announced that the town had a good chance to have natural gas. The county paper editor urged the town's population to go after this service from the Alabama Gas Corporation. Centreville and Brent acquired natural gas service in 1952.

Intense cold weather in early 1966 presented the town's administration with the return of a water supply problem. Mayor Ferrire published a notice of appreciation to all who had assisted in what he termed "a water crisis." The February notice said the supply was low and the system could not furnish water for a major fire. The National Guard unit in the county helped patrol streets at night to prevent fires, and the mayor expressed special thanks for their help. The notice said further, "we have had the most severe weather in years and our old system could not hold up." The following November, the town received a $414,630 Federal Housing Administration loan for a new water system. Senators Sparkman and Hill and Representative Armistead Selden were given credit for exerting the weight of their offices to obtain the loan. Basic water rates were raised from $2.50 to $4.00 to repay the loan. A high-flow spring on the road to Scottsville identified as the Stewart Spring (others call this source the Williams Spring) and adjacent land were purchased as well as

Order of the Eastern Star, West Blocton. (Cahaba Trace Commission.)

a site for a new tank. The project was to be under way by the spring of 1967. The condemned water system, at last, was on the road toward replacement. The small Gillespie spring, hailed for years by residents as the purest water in the state and ingredient for the "best Coca-Colas in the world," could no longer supply an extended water system. A more dependable, greater flowing spring would replace the old Frog Level spring used for decades as the town's water supply. Not only was a resolution of the water supply problem addressed, a legal notice appeared in the February 24 *Centreville Press* granting a franchise for natural gas to the Alabama Gas Corporation. In March 1967, the mayor was advised that construction would start soon and that Alagasco representatives would canvass the town by April 1, taking applications of those who wanted access to natural gas. In late October 1967, Mayor Ferrire and company officials were on hand at the main valve for a photo-op on the occasion of turning on the gas for West Blocton.

In the spring of 1966 a glint of job hope came with the expansion of the Jones Timber Company pulpwood business which added more loading and storage space. Associated with Paul Jones were George O. Desmond of River Bend and local resident Floyd Morrison. The office was in the old Blocton Depot, by then greatly reduced in size. Other pleasant news was that the West Blocton Library was in its second year of operation. Local donations had allowed remodeling of the space next door to City Hall to accommodate the facility. Pledges were sought to keep the library open. A dollar-a-month club was proposed for its support. Long time merchant and Masonic Lodge member, Samuel B. Israel, was honored in a special gathering in March which included guests from across the state and Mr. Sam's immediate family. Dr. Culp, President of Alabama College, was the occasion's speaker. In July, Chapter 185 of the Order of the Eastern Star was re-instituted. By September the town could take pride in its new health center, new street paving, mercury vapor street lights, a new Southern Bell cable, and there was a rumor a pharmacist would soon locate on Main Street. The list of improvements also cited the new by-pass linking Highway 5 with the Montevallo Highway. A newspaper report of improvements said, "we in West Blocton are looking forward to better days ahead."

The community added to its "firsts" in February 1967. A recent law allowed women to serve on juries. As astonishing as it may seem now, prior to this time, women were banned from this court activity. Clara Jean Marchant, who the *Centreville Press* said was from West Blocton, was the first woman in Bibb County's history to sit on a jury.

In May the contract was awarded for the construction of the new water works. The engineering firm of Polglaze and Basenberg of Birmingham produced the plans and specifications. Blocton-born Polglaze's name was still associated with the firm even though he had died in the mid-1940s. The $321,000 project was to be completed in 180 days. Also to be constructed was a 200,000 gallon steel tank on Smith Hill to serve that community, McCulley Hill, and Hebron.

School Integration

The *Centreville Press* echoed the daily papers' major topic in March 1967. Its editor's column headlined the news, "ALL ALABAMA SCHOOLS MUST INTEGRATE THIS YEAR." Dire circumstances were predicted. "If the court order is carried out it will mean the complete destruction of the school system in Alabama." Pursuant to the court order, on March 31, the Bibb County Board of Education ordered Superintendent Francis Pratt to prepare and submit a desegregation plan. Pratt turned the job over to the board attorney and on April 3 checked into the hospital. The board delayed final decision on the matter with the superintendent absent. The editor observed, "Bibb County is now being run by the federal courts and not by Congress. The federal government intends to take over the school systems of this country and dictate how they should be operated." By May 18, Bibb County was not in compliance and thus was the last Alabama county to follow the court order to submit desegregation plans. The May 25 *Press* issue said, "Bibb School Mix Is Ordered By Three Judge Panel." Integration was outlined in the August 24 county paper. One minority teacher was assigned to Brent Elementary and West Blocton Elementary. One white teacher was sent to Faucett Junior High in Brent. Two white teachers were assigned to Davidson High School and two minority teachers to West Blocton High and one to Bibb County High. West Blocton High was to have 46 minority students transfer, and the larger Bibb County High School was to have 8. The county high school for African-Americans, Davidson High, was in Centreville. West Blocton students had been bussed there for decades. They were now to be served locally, accounting for the larger number assigned to

WBHS. School was to open on September 1, and the newly elected superintendent, W. R. McKenzie, asked all parents to stay away from school for the first few days to give the principals and teachers a chance to comply with the court order. All transfer students were to register at the school of their choice on August 24.

Myrtle Cole Jones, teacher and counselor at WBHS, remembered she was nervous on the morning of registration. Integration opponents made no overt threats to the court order, but so much was said in state and local papers and on radio and television about the event, all were understandably apprehensive on that August morning. Jones recalled that she looked up from her table and saw in line parents and their well dressed and mannered children from an African-American family she knew. They greeted each other warmly and courteously. The tenseness of the moment simply fell away to the routine task of registering students as the teachers and office staff had done in previous years.

All did not go smoothly when school opened. Teachers and principals could not monitor every minute of the school day and conflict on the playground did occur, nor were all white teachers openly sympathetic with minority discomfort. Too, minority students felt threatened as they walked to and from school. One individual in her adulthood reported that the fear etched in her mind during those years three decades in the distant past is still with her and on occasion creates a momentary "flashback." Vividly remembered were "white men with threatening looks" in a pick-up truck slowly following her and her friends on their after-school walks home.

By the basketball season in 1968, long time African-American community names appeared on the Tigers' team roster. Cutts and James were among immediately identified high-scoring athletes and so recognized, and a Banks family scholar made the academic honor roll list. While the situation was far from perfect, the school disaster predicted by the county newspaper did not happen in West Blocton.

Prior to the fall opening of the public schools, a private school had been announced for West Blocton. Alpha Academy was to offer classes from grade one through six. The owners were Boyd and Mary Elizabeth Bailey of Woodstock. Classes were scheduled for the basement of the First Baptist Church. Fees were $350 per child, and if forty were enrolled, the second child in a family would pay $250. Mrs. T. G. Medders was assigned grades one through three and Mrs. Bailey, four through six. It was the first "private" school organized in the county during the era. The owners eventually acquired the former John Kren residence for classes. Alpha Academy operated for more than a decade. At one point in its history it offered Latin, French, German, and advanced mathematics as well as regular subjects.

Community Improvement Attempts

Elks Lodge. The original lodge building was remodeled in the early 1960s and the second story removed. (Courtesy of Cahaba Trace Commission.)

The West Blocton Lions Club under leadership of Harmon Frederick and Howard Pate, had a list of accomplishments to report. They had sponsored a youth center at the old high school building featuring roller skating. The Lions Club Community Center also provided Ping-Pong and billiards. The Lions supported Girl and Boy Scout activities, contributed to sending the WBHS football team to the Sugar Bowl to see Alabama play Nebraska, and sponsored a Christmas lighting contest.

The Elks Lodge sponsored the first homecoming banquet in six years in 1967. The Elks were boasting of a remodeled club, fully air-conditioned with a new dance floor. After a year without a director, the high school band had reorganized under Bill Ronilo. Previous band instruction in the school had produced a number of outstanding musicians and in 1967, University of Alabama sophomore Freddie Grady was making a name for himself. He was featured in a *Tuscaloosa News* article as a versatile member of the Million Dollar Band, accomplished on several instruments, in particular, trumpet, trombone, and clarinet. Grady was also a member of the University Symphony Orchestra.

In September, WBHS dedicated its new stadium and football field at the new school by defeating neighboring rival Brookwood 35 to 13. On Veterans' Day, a parade honoring all veterans and servicemen in Viet Nam marched through town. It featured the high school band, cars and horses, and anyone wishing to march. A local veteran was quoted in a story saying, "The boys over there know what they are fighting for, and we know if the line is not held there, America will be next."

Thomas DeWitt Poole. (Courtesy of Willie Mae Poole, West Blocton, Alabama.)

John William Snelson. (Courtesy of the *Westonian*, 1968.)

The Viet Nam War Takes West Blocton Lives

West Blocton had sent many of its young men to the Southeast Asian war and several wounded had been reported, but the war really hit home in February. Sue Pickett, long time local news reporter, wrote of the death of a local soldier:

> Sp. E. 4 Thomas DeWitt Poole Killed in Viet Nam
>
> The Vietnamese War was brought close to home when we heard the news that a local boy was killed in action in Viet Nam.
>
> Spec. E. 4 Thomas Dewitt Poole, son of Rev. and Mrs. Mose Poole of West Blocton, died on February 12th in Viet Nam. Specialist Poole was with the 1st Cavalry Division. He was a graduate of

> West Blocton Heights Jr. High School and Davidson High School in Centreville. He was 21 years old. Thomas entered service on March 22nd, 1967. He is survived by five brothers and five sisters. Thomas' father, Rev. Mose Poole preaches at Ebeneezer and at Star Bethlehem in Wilsonville.

Mrs. Pickett spoke for the community in her weekly *Press* column: "Words cannot express our sympathy to the family, nor our gratitude to the boys over there who give the 'last full measure of devotion' to their country. May God bless and comfort you."

Thomas Poole was not the last West Blocton soldier to lose his life in the Southeast Asian war. Sgt. John William Snelson was killed while leading his men on a mission when ambushed on April 23, 1970. Snelson was a 1968 graduate of WBHS. He had been on leave about two weeks before his death. He told his mother, Sarah Snelson, he would be home on Mother's Day. Sadly, it was on that day his funeral was held in the Union Baptist Church. He is buried at Pine Hill Cemetery in Bessemer. Mrs. Picketts' personals column in the *Press* said, "Our sympathy to Mr. and Mrs. Elisha Snelson. The Snelsons have borne their grief heroically, and remind one of the words of President Abraham Lincoln who said, "with charity for all and malice toward none." Mrs. Snelson responded to expressions of sympathy from the whole community by paraphrasing Lincoln. "I love everyone. What we need is more love in this world and less hate." Mrs. Pickett added, "The Snelson's deep faith in God is an inspiration to all."

Henry "Cakeman" Jordan. (Courtesy of Zelda Jordan Norris, West Blocton, Alabama.)

With the war continuing in Southeast Asia, it was business (or lack of it) as usual. Early in 1968, Mayor Ferrire thanked some twenty businesses for supporting the Christmas lights in downtown. In another case of West Blocton folks being all over the world, a Tokyo, Japan magazine featured a story on the children of Radioman Reed Martin, Jr. and his wife who were on overseas duty. The Martins, from West Blocton, lived next door to a new restaurant. Their children, Miranda and Mike, the magazine article said, "seem to be a special attraction here." When the children and their mother returned home in early 1970 after five years in Japan, the children, fluent in a second language, sought individuals with whom they could speak Japanese. It may be safely assumed they did not find them locally. The Martin children brought a distinct and interesting Far Eastern flavor to West Blocton.

The community was shocked in mid-February by former mayor Floyd Morrison's sudden illness. He was reported to be in Shelby Memorial Hospital, a stroke victim. The county newspaper reviewed his devotion and achievements for the town while he served as mayor.

On May 9, Mrs. Pickett wrote in her *Press* column, "This morning we heard the sad but expected news of our Governor's death. She [Lurleen Burns Wallace] was a gallant, brave and gracious lady. As I wept at the news, I said, I didn't realize that it could hurt me this much. I am sure that many, many others shared this feeling with me." Mayor Ferrire and a delegation from West Blocton attended the last rites. Lieutenant Governor Albert Brewer was sworn in and almost immediately had to consider a run for the office. When campaigning got under way, he made an April 1970 visit to West Blocton. A photograph of the event near city hall appearing in the April 30 *Centreville Press* shows several hundred listening to him speak. A sizable caravan accompanied the governor who was on tour en route from Birmingham to Tuscaloosa. In the May primary, Wallace won Beats 10 and 11 by two to one. West Blocton owed George Wallace too

West Blocton First Baptist Church. (Courtesy of Cahaba Trace Commission.)

1969 Fire Truck. Mayor Frank Ferrire gave the keys to the new fire truck to Chief E. P. Jones on January 11, 1969. Members of the Town Council and volunteer fire department pictured are A. N. Kitchen, L. B. Frederick, James R. Fitts, Ballard Gilbert, Emmett Carroll, Paul Stephens, Murphy Reach, Clarence "Nump" Franks, Herman Clinner, J. Leonard Carroll, Charles Evans, William Wright, Howard Arnold, L. P. Parker, O. K. Yeager, Milton McKinney, Kenneth Burns, and Boots Carroll. (Courtesy of Ferrire Collection.)

much to vote otherwise. The county went for Wallace 2,588 to 1,790. In June, the census count for the county was listed at 13,764. Ten thousand residents had left the county over the decades that marked coal production cut back and eventual cessation of underground mining.

The Black Diamond Coal Company, still operating its small Number 11 mine east of town, announced in April that it would construct a new preparation plant. Its cost was projected to be $375,000 dollars. It was reported that a night shift would be added giving local miners job opportunities. The reality of danger even in modern safety-conscious mines was revisited on the town in August when 56-year old Henry Jordan was killed in Number 11 when a tremendous "horse-back" rock fell on him. The community was accustomed to the danger in mining, but it was shocked and grieved to lose this good citizen. It was the only recorded fatal accident in this mine. This was one of the last mine accident deaths recorded in the West Blocton area.

The First Baptist Church congregation built a new, ground-level church at a cost of $82,000 and dedicated the facility on June 30, 1968. The pastor was P. F. McGuire. A new bank, new school, new post office, new churches, water system improvements, and a new fire truck were all evidence that the old town was still alive. However, from time to time in Mrs. Pickett's column, it was noted that all was not well in the business section of Main Street. She wrote in May 1970, "the empty stores along Main Street have a very unhappy look, so it is always good to see new stores opening." The reference was to Virgil Lawley's purchase of the old Smith Hardware building for several proposed businesses. To add to the town's business district woes, Whaley's Furniture store was robbed during the last week in May. Sunday thieves had entered the front door, broken open the store's safe, and made off with cash and checks.

Morris Moody. (Courtesy of Morris Moody, West Blocton, Alabama.)

Further school desegregation orders came from the courts for the 1968–1969 school year. The new building erected but a few years past housing the West Blocton Heights Elementary School for African-Americans was ordered closed. West Blocton Elementary School was assigned two minority teachers. The recently built school building eventually became a private club named, "The Continental."

In September, Mayor Ferrire was re-elected. Two months later, the Tigers beat the Choctaws for the first time since an extra game played in 1964 and the first regular season game since 1956. The one-sided score was 41 to 7. For WBHS grads and students who took great stock in the outcome of football games, it made 1968 a good, memorable year. Wins over the down county rivals were few, so it was something to celebrate.

Early in 1969, West Blocton welcomed the addition of a new, modern fire truck, an American-LaFrance center-mounted pumper with a 750-gallon tank capacity. `The pump was mounted on a Chevrolet chassis. The old truck was retained giving the town two fire fighting units for the first time in its history. The town would not be left unprotected when other communities asked for help in fighting fires. The purchase price was $16,820, much of which was raised by public subscription. Serving as fire chief at the time was E. P. Jones. Another attempt was made to join a countywide industrial development drive. West Blocton representatives on the committee were Frank Ferrire, Eddie Downs, Emery Lippeatt, Jack Cain, E. P. Jones, A. N. Kitchen, and John Langston. Visitor to his grandmother's home in July was Dr. Kennedy Smith who was serving as visiting assistant professor at Louisiana State University after having completed his Ph.D. in chemistry there. Smith, son of William and Nina Smith, was a successful scholar having graduated from WBHS and the University of Alabama. He was to join the staff of the University of California at Berkeley in the fall. At some point in his remarkable career, he also attained a law degree. Morris Moody, coach at WBHS for several seasons, was appointed principal in 1969, a post he filled until 1996, the longest known principalship tenure of any Bibb County educator. The new coach was Norman Sutton who reported that his 1969 team was small, and in November it was payback time for the south Bibb eleven, the locals were bested 43 to 0.

In September, the FBI came to West Blocton to investigate the case of a $10,000 extortion. The *Birmingham News* gave the details. Bank president Eddie Downs had received a telephone call threatening his life if he did not deliver $10,000 in a bag to a secluded spot in the "old cemetery." He hurriedly did so and returned to the bank where he had a second call saying the delivery was unsatisfactory. He was instructed to retrieve the bag of cash and take it to another location south of Woodstock and that order was obeyed. When the FBI agents searched the area the money was gone. The case was solved in early 1971. The *Press* of February 4 reported that FBI agents in Miami, Florida had arrested Thomas Wayne Hosmer of Greenpond in connection with the case. Another suspect, J. W. Perry of Centreville, was in the Bibb County jail under arrest for grand larceny. The FBI also arrested him for the bank extortion. A May trial date was set. No money was recovered.

A New Elementary School

In the late 1960s, the West Blocton Elementary School, built with Works Progress Administration (WPA) funding in 1937, had become over-crowded. There was no longer an auditorium because it had been divided to make two classrooms. The building leaked and was in general disrepair. Students who attended the school during its last years said when it rained the water "just poured in" and the floors had huge holes patched over with plywood sheets. Parent Clarence Franks headed a group of sixty other school patrons to present a statement of needs to the county board. They asked for four more classrooms, thirty-one desks, two more bathrooms, roof repair, and painting inside and out. The county board chairman, C. E. Hornsby, a Centreville banker, and W. R. McKenzie, the new superintendent, were sympathetic but stated there were no funds to accomplish the delegation's request. The meeting, however, ended on a hopeful note. County Representative Walter Owens announced he had an appointment with the governor for the next Wednesday to discuss the possibilities of a new school building for West Blocton. *Press* editor James Oakley voiced support for the group and promptly reported the details of the needs expressed by the citizen group.

In 1971 at a rally in Centreville promoting the new private school for south Bibb featuring Governor Wallace, the governor announced he had placed in the education appropriation bill $350,000 for a new elementary school for West Blocton. Senator Givhan and Governor Wallace had promised the project during their 1970 campaigns. Givhan had photographed the building inside and out for a presentation to support the new school. He expressed shock at the facility's condition. Franks and his parent group had made their points well. In April 1972, the Board of Education report reflected that proposed drawings of the new school had been rendered. Funding was a problem. A new school would cost more than had been allocated. Another glitch in December postponed the acceptance of bids. There were minimum wage problems, questions about soil percolation test results, and questionable assurance of a water supply. Construction finally began in January with the expectation it would be ready for occupancy the following school term, 1973. Cost of the school was $502,478, and land was purchased for $3,500. There were architectural fees and equipment costs of over $50,000. The A. D. Lovelady Construction Company of Plantersville won the contract. Open house was held in August 1974 with county and state officials present. State Board member Victor Poole represented Governor Wallace. Proud and smiling students who were to be in the new building when school opened gave tours of the new, air-conditioned building. Yet more work was needed since the site lacked a driveway or parking lots. Parents had difficulty in delivering their children to the school's doors. Eventually, with PTO efforts through various fund raising events, the situation was improved. The school building in current use has had classrooms added over the years.

Total, Not Token, School Integration

Consuming much of the County Board of Education's time and concern was the order for complete desegregation of the county system filed on October 23, 1969. The initial, partial integration of the system had been determined not

Mt. Carmel Church. (Courtesy of Cahaba Trace Commission.)

to be acceptable. The new federal judges' approved plan for disestablishment of the county's dual system was published on February 19, 1970. New names, though they proved to be temporary, were given county schools. WBHS became North Bibb County Secondary with a projected enrollment of 420 white and 160 minority students, and BCHS was named South Bibb Secondary with a projected enrollment of 398 white and 265 minority students. The immediate result in the county seat was the planning for a private school for south Bibb County. The private Alpha Academy began its third year in West Blocton.

Civic Concerns and Changes

Continuing its reputation as a patriotic community, West Blocton's American Legion Post and Lions Club sponsored an Avenue of Flags for Veterans' Day 1969. The dedication included ministers Jack Cain, Bobby Jones, and Mose Poole as well as Charles Stewart, Legion Commander, William Pratt, Lions Club President, Mayor Ferrire, Floyd Morrison, and Representative Walter Owens. Flags were flown from power poles from old Highway Five near Ambrose's store through downtown. Men from the area who had lost their lives in service were remembered in the ceremony. In December, plans were announced for a shrine to honor the area's soldiers and sailors who died while serving their country from World War I through the Viet Nam War. While the idea was not acted on immediately, the result of that initive was a memorial stone placed at City Hall as a part of the nation's Bicentennial celebration. During the fall, the town took pride in the news that Johnny West and Joe Farrar were awarded athletic scholarships at Troy State University.

In November 1970, local telephone subscribers received questionnaires relating to the town's phone service being added to the Birmingham exchange. A 938 exchange drive that lasted for months and years was promoted by Sue Pickett and other citizens. It was not an easy, quick-results task, but perseverance eventually paid off.

The Mt. Carmel Baptist congregation dedicated a new church on December 24, 1970. The community continued to show faith in their church membership and in the town's endurance. This was the third building for the congregation that had been founded in 1869, the oldest in the area. In November 1997, the West Blocton Improvement Committee dedicated a marker to commemorate Mt. Carmel Baptist Church, the oldest church in West Blocton.

A threat loomed in early 1971 to erase one of the few improvements that West Blocton had obtained from federal and state sources. Rumors were rampant that lack of funding threatened the closing of the new health clinic. Floyd

Morrison, recovering from an illness, went to bat for his town by making a plea through the county paper for the clinic's continuance. Assurances were given that while the county had found it necessary to cut back on the County Health Department, the north Bibb clinic would remain open.

In April, showing confidence in the local economy, the Bank of West Blocton announced it would build a larger, more modern bank building. Immediately Mayor Ferrire and the council negotiated with the bank directors to buy the 1965 bank building for use as a city hall. The long-time use of old storefronts for municipal headquarters was soon to come to an end.

Only speculation could account for town ordinance No. 267 adopted in April 1971. In part it read:

> Section 1. It shall be unlawful, in any primarily residential city block or other primarily residential area in the city for any person, firm or corporation to keep, harbor, maintain or confine any ducks, geese, chickens, guineas, turkeys or other fowls in any stable, building, structure, corral, pen or enclosure, from which or from the chicken or other fowl so confined or maintained therein, vile, foul, unwholesome, offensive unpleasant or disagreeable odors or smells arise and spread or are so wafted to the homes or occupied primacies of other people living on such city block or to such other primarily residential area or in the vicinity thereof.

There were other parts of the ordinance. Essentially, no chickens or other fowls could be kept in town. Only those households on the edge of the corporate limits could hear the bucolic crowing of roosters and cackling of hens harbored by over-the-line neighbors. Cows and hogs were long gone and now in 1971, so were chickens and their ilk.

As new buildings were erected, an old one came down. The ninety-year old wooden, tin-sided Coca-Cola plant building located next to the Elks Lodge, survivor of the town fire, was demolished in 1971. An April 8 legal notice stated that the board of directors of the West Blocton Coca Cola Bottling Company, had voted to dissolve. The W. A. Jones family had operated the plant since 1912 having acquired it from a Mr. Holifield. In its early days, it was named the Standard Bottling Works and had been opened by W. R. Young, Sr. and Dr. L. E. Peacock. When Peacock left the company in 1910, one of Birmingham's future business magnates in the soft drink business, Crawford Johnson, Jr., was associated with Young in its operation. When it closed it was recognized as the smallest Coca-Cola franchise in the world. After the closing of mining operations to the east, the bottling plant's market was greatly reduced. Its territory was always small, serving only those communities in the north and eastern parts of Bibb. In 1971 the plant was located in the old Wood Motor Company building, its third location. There had been a move from the original frame structure in the 1950s to a concrete block building facing Florida Street. Among knowledgeable collectors, the West Blocton Coca Cola bottle is a rare item, always bringing a premium price. Ellington P. Jones was operating the plant when it closed.

When the bottling works closing was announced, news came that a plastics plant would occupy the vacant building and that the next door empty dry cleaners building would also be acquired for the new company. The firm was called Artistic Plastics. Don DeCraft was owner of the plastics company that initially employed fifteen workers. Probate Judge Fred H. Davis was given credit for landing the plastics plant. West Blocton had the only building available and cooperation between town and county officials produced success.

A daring bank robbery by two Centreville men occurred at the West Blocton Bank on Monday June 1 just at closing time. The heist netted $8,100. Fortunately, there were many witnesses so the men were identified and apprehended in short order, and the money recovered. It had been less than two years since the bank president had been threatened and money extorted.

The operation of a small sewing plant and a few jobs at the plastic manufacturing enterprise, largely paying minimum wages, could only make a dent in the huge loss of better paying mining jobs. It was noted in a September paper item that the Segco mine at Maylene in Shelby County had closed placing yet other West Blocton commuting miners out of work. With this closing, all underground mining had ceased in Bibb and Shelby Counties. However, the Black Diamond Company's presence in West Blocton was still evident. The paper also reported that compliance with water pollution laws had been achieved with the installation of a large pump and pipe extensions to a settling pond under the management of superintendent Collis Crane.

After being without a local drug store since the departure of Dr. Nolen in 1963, negotiations were successful in attracting a Bessemer firm to bring this long needed service to town. Albano Drugs was announced in April 1972 as an achievement of Mayor Ferrire and the council. The location was to be next to downtown city hall in the old library location. When it finally opened in December, it was called the West Blocton Discount Drug Store.

A loan approval and a grant to expand West Blocton's rural water system came in August 1972. As a result, city water was piped to Smith Hill, McCulley Hill, and Hebron, and a new water tank on Smith Hill was to be added to the system. Another water extension project was funded in April 1973. The additional communities involved were Red Eagle, Lucile, Frog Level, Primitive Ridge, and Cedar Grove. Approximately 225 families would be served by what engineers said would be plenty of water. An abundant supply of water was promised for an area that had experienced water shortage during times of drought because of low flowing springs and dry wells.

Proving the town had been home to the infamous as well as the famous, a Southern Airways DC-9 jet was hijacked to Cuba in November 1972. One of the hijackers the FBI identified was 27-year-old Louis Moore alias Lewis Cale, an African-American, who was said to have been born and raised in West Blocton. The plane was hijacked in Birmingham. While it was seized, the flight took it over the United States, Canada, and made two trips to Cuba. A two million dollar ransom had been paid. It was recovered when the hijackers were apprehended in Cuba. It was the wildest and longest hijacking to date. The FBI turned up several people in West Blocton who knew Cale's family, no longer area residents.

Signaling another end to the coal mining era was the December 14, 1973, ruling of the Alabama Public Service Commission closing the Southern Railway depot and freight office in West Blocton. The closing was strongly protested by Mayor Ferrire, A. N. Kitchen, George O. Desmond, and Jack Harmon who made a trip to Montgomery in September. West Blocton was thereafter served from a base station in Bessemer. There had been a depot on the same site in old Blocton since the founding of the Cahaba Coal Mining Company in 1883. A ninety-year-old service disappeared.

In the early 1990s, the rails and trestles along the route from Blocton to Big Springs were salvaged. The line had been used for several years to park surplus "rolling stock." The musical sound of trains in the Caffee Creek Valley was at an ignominious end. There was no official protest made by county or local governments when the announcement was made declaring the line's demise. All but three miles of Truman H. Aldrich's major, seven-month engineering feat of 1883–1884 was gone in a matter of weeks. Today, the roadbed is overgrown and is being slowly eroded, and only concrete piers remain at former creek crossings to give evidence of the historic and bustling rail line.

The town took pride in congratulating native Lyda M. Keyes who was recognized for publishing a volume of poems entitled, "Thoughts of Love." Mrs. Keyes had served as West Blocton's *Centreville Press* correspondent for several years. She taught school, wrote a history of the West Blocton First Baptist Church, and led a drive for a monument to the memory of West Blocton and area war veterans. Mrs. Keyes was a member of the United Daughters of the Confederacy and a charter member of the Alabama State Poetry Society. She held a certificate of merit from the editors of *Who's Who in Poetry* based in London, England. She was also designated a member of the International Platform Association, an organization of speakers and reporters. In 1975-1976, she served as local Bicentennial Year coordinator.

Another industrial disappointment for West Blocton occurred in 1972. A glowing announcement in bold *Press* headlines indicated that a Carlisle, Pennsylvania company was moving to town and a factory would be built that manufactured snow plows and other grading blades near the old coke ovens. The announcement caused considerable excitement, but for reasons undocumented, the project never developed. During the year Wayne Callahan and Randall Goggins opened a taxidermy business on Main Street and word came from Miami, Florida that Sam Silver, a West Blocton native, had been made circuit judge of Dade County.

The *Montgomery Independent* in January 1973 reported a success story of a former West Blocton area citizen. Roy R. Mason, a 1951 graduate of WBHS, had expressed in a high school social studies project that his goal was to be a millionaire by the time he was forty. Twenty-two years later that had been accomplished. The newspaper singled out

Mason as a top success story of 1972. While he was involved over a period of two decades with a number of real estate and financial institution enterprises, his major business success was with petroleum distribution. In 1973, he was recognized as a top petroleum distributor in the Southeast. Mason's children attended the University. An admirer of Coach Bryant, Mason provided significant funding for several projects. One involved the athletic program and another the Bryant Continuing Education Center. The latter facility contains a Roy R. Mason conference room on its second floor. This generous businessman contributed the cost of the room's construction and its furnishings. As noted earlier, there is an Elmo I. Ellis Conference Room in the Communications School, Reese Phifer Hall, honoring another altruistic West Blocton native.

Another business success story is the career of Bill Rice. Born in Piper to parents Johnny and Nannie Rice, he grew up in West Blocton and like other sons of coal mining families, experienced the ups and downs of the coal industry in the area and also faced the fright of having a father injured in a mine accident. At age thirteen, he shined shoes at Parks' Barber Shop on Saturdays and later worked in Main Street grocery stores. He lettered in WBHS sports, played in the band, served as cheerleader, and belonged to the school's scholastic honor societies. His post-secondary school training included study at the University of Alabama and stints at vocational and business schools. After completing his education, he spent twenty-six years in the lumber industry as a wholesale salesman. In 1986, he launched Transload Limited, a reload business that by 1990 became a national business operation. He also served as a founding director of the international association, Transload Distribution Association. He gives credit for his achievements to his father's wise guidance, high school principal Oscar Zeanah, and a strong faith.

The town can point with pride to the accomplishments of these native sons. They have left their marks.

The welcome news that the Viet Nam War had ended created an emotional release for the community that was expressed by the *Press* staff and reporter Sue Pickett in the February 1, 1973, paper. Editor Oakley compared the quiet reaction to the war's end to the celebration atmosphere that characterized local reaction to the end of World War II. Mrs. Pickett wrote, "Peace, peace has come and we are numb, muted, quelled almost. No bands marching, no flags waving, no sirens and shouts of joyous people as in times past when wars ended . . . peace has come for America, the killing has ceased. Our men will return home, prisoners will be released and we should rejoice."

Believers Quartet. (left to right) Billy Finnen, Billy Roy Finnen, Dewey Williams, and Brian Finnen. (Courtesy of Billy Roy Finnen, Sr., West Blocton, Alabama.)

The family of Corporal Thomas DeWitt Poole who had lost their son in February 1968, stated they were glad the war was over. A family member regretted that their loved one would not return but expressed happiness that, "we don't have to worry about more brothers or sons dying in that war." The war also took the life of resident Sergeant John William Snelson in April 1970. West Blocton soldiers Poole and Snelson are among the thousands of names on the Viet Nam Memorial Wall in Washington, D. C., and they also are listed on an Alabama Memorial to those who gave their lives in the war.

In April 1973, a new municipal garage to house the town's fire trucks was erected back of the bank that was to be the city hall when the new bank

Taking part in the ground breaking ceremony for the new Bank of West Blocton were O. K. Yeager, Eddie Downs, bank president, Howard Pate, and Mayor Frank Ferrire. Ferrire, Pate, and Yeager were members of the Board of Directors. (Courtesy of Ferrire Collection.)

was finished. Bibb County Vocational School was to be built half way between Centreville and West Blocton. A modest victory was claimed privately by Mayor Ferrire in the choice of the school's location. It was initially announced for Centreville, but the argument was advanced by Ferrire and others that it should be equally convenient to the students at WBHS and BCHS. While the facility was under construction, the first vocational classes were held in the old West Blocton Elementary School.

Neighboring city Brent was hit by a F-5 tornado on early Sunday evening, May 27, 1973. Concern for friends and relatives was immediate. Lives had been lost, and there were numerous injuries. Much of the town of Brent had been blown from the map, and major damage extended into Centreville. On Monday morning, the West Blocton fire alarm blew "loud and long" to alert its citizens to gather clothing, food, and money for the many victims. Four truck-loads of donated material were dispatched immediately. The mayor and council sponsored additional relief projects. The town's water works supplied three army tank-trucks with fresh, treated water to be transported to Brent, every thirty or forty minutes for as long as needed.

A musical group was organized in 1973 called the Believers Quartet consisting of Billy Roy Finnen, Sr., his sons Billy Roy, Jr. and Brian and Dewey Williams of Anniston. The quartet had concerts in Anniston, and in several locations in Georgia and Florida. In the fall the Alpha Academy opened for its seventh year. The Bank of West Blocton announced that its holdings had reached five million, a remarkable achievement for the seven-year old institution. Ground was broken in October for the new bank building. In November, a West Blocton branch

Faith Baptist Church. (Courtesy of Cahaba Trace Commission.)

bank was announced for Woodstock. In October water came to Smith Hill. Fire plugs were turned on to flush out the system before residential water could be used and residents gathered on the community store's porch to witness the event. Mary Lee Martin celebrated the occasion by removing her shoes letting the hydrant water flow over her feet. To promote the drive for phone connections with the Birmingham exchange, petitions were placed in West Blocton businesses in September. Local calls to Birmingham and much of Shelby County would be available to north Bibb subscribers when the tie-in was accomplished.

A new congregation, Faith Baptist Church, was constituted in February 1974. Meetings were held initially in a Main Street store front, and eventually the church occupied the St. Thomas Catholic Church that had closed in 1972. In March, a new Alabama Power Company office and appliance sales store was announced with projected completion by June. It was built next to the post office. Among town improvements evident in April was the addition of a second full-time policeman. The chief was James Jackson and assistant, Roy George. A new, automatic garbage truck had been ordered and was in operation in late March. The concerned populace that complained about garbage disposal along roadsides expressed hope that a regular garbage truck pickup would help eradicate this problem.

Local student-athlete at Troy State University, Donnie Jones, son of Ruth Carr Jones, was given the school's "Attacker Award" in the fall of 1973, and in May 1974, another West Blocton son at Troy, Joe Henry Farrar, was honored with the James Burns Clements award for having the highest scholastic average for a graduating athlete. He was a chemistry and biology double major and had attained a 2.7 grade point on the school's 3.0 system. Farrar was on the science honor roll, had been awarded a National Science Foundation Scholarship, and had been a two-year starting linebacker and a member of the Trojans' 1971 and 1973 Gulf South Conference championship teams. He was also a member of Omicron Delta Kappa, a national leadership honorary fraternity. His father, Morgan Farrar, had died in February. His mother was Jean Israel Farrar, a local schoolteacher. Another sports item involving a local scholar-athlete was on the local paper's pages in April. Wayland Blake had been appointed to the head coach's job at Gordo High School. He had played football under Coach Morris Higginbotham at Livingston State and had attended Florence State and the University of Alabama's Gadsden Center. He had completed requirements for the master's degree at Florence State, the degree awarded in July.

On May 26, 1974, the Bank of West Blocton held a ribbon cutting to celebrate the opening of the bank's new headquarters. The newspaper account of the event said there was an overflow crowd in spite of the spring rain. President Eddie

Downs, Mayor Frank Ferrire, and Vice President E. P. Jones greeted the guests. All the bank officials and employees were dressed in formal attire for the gala occasion. All members of the town council were in attendance. Local caterer, Carolyn Griffin, was responsible for the refreshments. The day gave the community a lift and hope for further business revival. By July, the new city hall was in use. Minor remodeling and redecoration of the 1965 bank building had been accomplished. It proved to be an ideal facility for the town government headquarters. By August, the old 1935 high school building, in disrepair and too unmanageable to maintain as a youth center, was on the auction block, and the 1937 elementary school was victim of the same fate. The mid-1970s were proving to be a time of change.

It was also a time of change in the nation. The "Watergate Scandal" had proved to be the downfall of President Richard Nixon. He had resigned in disgrace on August 9. Some of West Blocton's citizens took issue with Nixon's departure. Mrs. Pickett, local news correspondent, and her husband David, jointly signed a letter to President Nixon with friends Barney and Lorraine Champion of Montevallo. It read:

> Dear Mr. President,
>
> We have grieved for you as you have been crucified. Of course you made a grievous mistake, but we believe you were thinking of America's good, and that many who have worked to destroy you are guilty of far worse than you did.
>
> We believe you have been one of America's greatest presidents and will remember with gratitude all you have done to bring peace to the world.
>
> Our prayers are with you and your family. May you go with God.

Perhaps this was the thinking of a segment of local residents, but the record does not reflect the opinion of others. It was a time of changing party allegiances so feelings may have been along party lines as was the case across the nation.

An incident causing community disunity and anger occurred in early 1975 when four Bank of West Blocton directors, including Mayor Ferrire, O. K. Yeager, E. P. Jones, and Howard Pate were ousted. Pate believed it came about because of a difference with the bank president, Eddie Downs, over an employee retirement plan. "The rift caused wounds and would leave scars," said the local reporter. The belief that the bank was no longer "home owned and operated" was indicated as a reason for distrust of the move. Two new directors were businessmen from Vestavia and Centreville. The disagreement may have initially damaged the bank, but it survived the removal of town funds from its holdings and the withdrawal of accounts by a number of patrons.

To counter-balance the disturbing news of the bank controversy, a grant announced in late 1974 became reality in February 1975 with the opening of a primary care facility in the West Blocton Health Center. Medicaid screening and treatment was available to children of Bibb County residents. Adults were to be added in a month. The Appalachian Regional Commission, the University of Alabama, the West Alabama Planning and Development Council as well as support from the town of West Blocton and the county jointly financed the program. Dr. F. Douglas Scutchfield and a nurse practitioner started the service and were assisted by Dr. Mark Alexander from the University and Dr. William Owings of Brent.

Noteworthy items appearing in the local news columns in early 1974 found the high school basketball team having success. In January, the team record was 9 and 0. The first winning season in years was under way. Jim Lees, a 1974 WBHS graduate, was appointed to the U. S. Air Force Academy. In March, the North Bibb County Voters League was meeting at the Liberty Baptist Church. In April, the community learned that former resident Marshall C. Davie, Jr. was working with a California "think-tank" at the Rand Corporation. The WBHS Football Banquet attendees in May 1975 were struck with "surprise and disbelief" when Joe Willie Namath, former University of Alabama quarterback and then of the New York Jets, was introduced as the event's speaker. Namath was a schoolmate of Jack Hicks at Alabama where they had become close friends. The visit engineered by Hicks was kept secret until the evening of the event.

The *Centreville Press* editor, James Oakley, Jr., (Oakley, Sr. died in October 1972) in May voiced the opinion of many when he proposed a tax on strip mining. Open strip pits showing on the road side near West Blocton were identified as a mess, and he cited the damage to county roads. There was a series of complaints and suits involving the major culprit, Burgess Company, stripping coal in the county. Editor Oakley proposed a severance tax like that imposed on timber and pulpwood.

African-American citizen Ed Brown organized a baseball team known as the West Blocton Barons and their write-ups appeared in the summer of 1975. The team was a part of the first annual North Bibb County Chapter of the NAACP Fourth of July outing. Announced in July as a new member of the Blue and Gray Softball Hall of Fame in Montgomery was Sam Govie, a late 1940s graduate of WBHS and brother of Mrs. Edith Hicks. As a football player, he was a member of Coach Oscar Zeanah's mid-1940s undefeated team. In Govie's senior year the Tigers football team lost only to Tuscaloosa County High.

Much of the news column space in the daily and county papers was dedicated to preparations for the up-coming year's celebration of the nation's Bicentennial. The mayor received the appropriate forms from Governor Wallace in August 1975, and they were completed with committee appointments and plans for West Blocton's participation. Lyda Keyes was made chairperson for the local events. There was to be an emphasis on local history and general improvement in the community. As a result of proper registration, Representative Walter Flowers announced the American Revolution Bicentennial Administration officially recognized West Blocton as a Bicentennial community. Without delay projects were announced. An avenue of flags would be flown on all patriotic holidays to honor the town's deceased servicemen, and a veterans memorial monument was to be placed near city hall.

While it was not developed during the Bicentennial year, the idea of a trail to the old coke ovens was listed as a potential project. The developments in the mid-1990s at the ovens site had their beginnings in the mid-1970s. There was a listing of a number of "firsts" associated with the town relating to coal mining, schools, and churches. Mrs. Pickett revived an essay for the *Press* titled "Wave Flags" that concentrated on the town's veterans and their records. It had appeared in an earlier *Birmingham News* magazine section. The townspeople were encouraged to celebrate Veterans' Day on November 11. Mrs. Keyes published a booklet titled, "Historic Notes and Patriotic Poems." One of the verses had World War II hero Kent West as its subject. The official town song adopted for the observance was, "God Bless America."

The official Bicentennial kickoff at city hall occurred in the week before Christmas 1975 with the presentation of the Bicentennial Flag and plaque by Gus Baldwin of the state committee. A planned parade could not be held because of a steady downpour. During the ceremony, Mrs. Keyes encouraged students to make a special study of the town's history. Morris Moody, high school principal, was master of ceremonies. Plans were announced for the compilation of names of citizens to be honored and recorded in a ledger kept at city hall.

Considerable effort was exerted in 1975 to provide improved medical service to the town. In September, a mobile clinic program was proposed. In October, the Bank of West Blocton backed the establishment of an office for Dr. Stephen Barton of Tuscaloosa who was temporarily assigned to the West Blocton Clinic to get it running effectively. Also assisting with the doctor project was the local druggist, Morris Crawford. The University of Alabama Community Health Sciences and the West Alabama District Health Department joined the active recruitment of a resident physician.

Bicentennial preparations and the doctor search were important news stories in December. A first for the community was an arrangement by the West Blocton Bank management to have Santa Claus arrive in town by helicopter. The town's children, young and old, were thrilled.

Another Bank Felony

There was other news, not good, that filled newspaper pages shortly before Christmas also involving the bank. It was a bizarre robbery of the Bank of West Blocton, and was the third crime episode there since its opening in 1965. The crime started with the attack and detention of a man accosted by masked men in his home on Highway 25 near Centreville. His car and two trucks were taken. A fire burned the nearly completed pastorium at Mt. Carmel Baptist Church, apparently set by the culprits to create a diversion.

It seemed the bank was a small town target on the underworld's hit list. Three men in ski masks armed with shotguns and rifles took the bank employees by surprise as it opened on Monday December 22. They cleaned out three cash drawers with cashiers lying on the floor where the bandits had instructed them to position themselves. As the thieves left the bank, they were fired on by a local man with a deer rifle who had been alerted a robbery was in progress. A county deputy, Billy Sims, was also alerted that the getaway car matched one that had been taken from a Centreville residence. He spotted the car on the River Bend road and gave chase. Speeds were reported to have reached 100 miles per hour during the chase. The three left the car on a logging road and escaped on foot. No money was found in the abandoned vehicle. Eddie McMillan, from whom the car was taken, had gone through the frightening experience of being gagged and tied up in his own home on Sunday night. He chewed his way free from rope restraints and called the county police. His captors had taken all his vehicles from the property but left the stolen pickup and dump truck in the woods in nearby River Bend.

A massive manhunt ensued in the area aided by local airplanes and two helicopters, one from Tuscaloosa and the other from the State Highway Patrol. Over fifty law enforcement personnel were on the trail of the robbers by noon Monday. Bloodhounds were ordered from Atmore Prison to assist in the search. The robbers eluded capture and by January 1, 1976, law enforcement could not claim success.

Automatic cameras in the bank had recorded the crime. Masks covered faces, but the camera had revealed other clues. The local press believed there were suspects and that apprehension was near. The FBI, Bibb County Sheriff Smitherman, and three local officers were exerting major effort to solve the case. On March 18, the paper announced that George E. Young and M. Plotmisk were in custody, and J. Wallace and R. Simpson were being sought. Young was from Brent, and Simpson and Wallace were from Centreville. Adding to the crime record of Young was his identification as a kidnapper in another case from the previous December involving University football player Paul Harris and his girl friend. By March 25, all suspects were in custody. Plotmisk was originally from Massachusetts and was absent without leave from the army. The last two were taken in London, Kentucky charged there with a local kidnapping. The sheriff speculated that $20,000 taken from the West Blocton bank would not be recovered since most of the money had been spent. In June, the four in U. S. District Court in Birmingham entered guilty pleas.

Mining's Last Gasp in West Blocton

What appeared to be the last mining operation locally came to an end in January 1976. A contract cancellation by U. S. Steel with the Black Diamond Company and Burgess Surface Mining Company caused the announced closing of both operations. In West Blocton it meant the closing of the coal washer processing plant. Some of the local miners were transferred to other Black Diamond operations in Jefferson County. Some mining continued through the summer as indicated by a strike confrontation in September. Burgess, constantly in labor and ecological litigation, was involved in surface mining near West Blocton and at Boothton. An August "wildcat" strike caused an open confrontation between company and union officers at both locations. The resolution of the conflict was not reported. Mining was at a virtual end in the area.

A few good things in the news helped take the edge off economic bad news. Winner of the Bibb County Junior Miss contest in March was Denise Shadwick, a senior at the University of Montevallo, daughter of Mr. and Mrs. Leslie W. Shadwick of West Blocton. Denise had been valedictorian of her class and had served as a cheerleader. She represented the county in the state contest.

July Fourth 1976 in West Blocton was a special day in the town's celebration of the Bicentennial. At 1:00 p.m., three boy scouts began ringing the town's version of the "Liberty Bell." The bell was from old St. Francis Catholic Church and more recently from St. Thomas Aquinas. It had been in Mayor Ferrire's front yard since the church closing in 1972.

As a part of the day's celebration, the official flag was raised on a new flag pole, that was put in place on Flag Day in June. The assembled group sang the town's Bicentennial theme song, "God Bless America," led by councilman A. N. Kitchen. Lyda Keyes was introduced and praised for her work with the year's celebration. Reverend Althoff, the Methodist preacher, addressed the group on a patriotic theme, and the Bicentennial flag was raised for the first time.

Bicentennial Flag Raising. A local reporter wrote, "Mrs. Keyes and A. N. Kitchen pulled the rope to raise the flags. The wind was kind, it blew the flags gently causing them to unfurl and wave proudly over the gathering." One of the last remaining company houses is visible behind the group, though it is abandoned and dilapidated today. (Courtesy of Ferrire Collection.)

Town Council Oath of Office. Sworn into office in front of a picture of the *USS Ross Gray* in city hall were Randel Goggins, Alex Williams, Frank Ferrire, E. P. Jones, Roy Jones, and Felix Parker. (Courtesy of Ferrire Collection.)

Also a part of the day's events, a photograph-mural of the naval ship named for Ross Gray was placed on the lobby wall of city hall. In later years this image was moved to the Cahaba Lily Activity Center. Earlier in the year, special programs were held in the schools. The winner of an essay contest on the history of West Blocton was Shianne Jones, daughter of Mr. and Mrs. Roy Jones. Another event in October, a special sale of Bicentennial souvenirs was announced for Trade Day on Main Street. A special program featured Country Boy Eddie of television Channel 6 in Birmingham and recording artist, Robert Cunningham, a West Blocton native. The West Blocton High School Band also played for the celebration.

In June, the local plastic plant changed hands. Unique creations there included a transparent elephant with a tiger in its stomach for 'Bama fans and a tiger with an elephant in its innards for Auburn backers. The firm had a standing order from Ohio State University that featured a buckeye in a clear plastic cube. Other business news during the year included the opening of a branch of the West Blocton Bank in Centreville and the building of McKinny's Bumper to Bumper Auto Parts Company on old Highway 5 near the high school, the first significant new business addition in years. A double-wide trailer was placed near the health clinic to house medical services. It contained five fully equipped examining rooms and one room to handle minor surgery and more complicated diagnostic examinations. Open house was held in late June.

Another "first" in Bibb County occurred in the August municipal elections. Felix Parker, candidate for the city council, won enough votes to make the September runoff and in that vote, he won by a two to one margin, the first African-American to hold an elective office in the county. When the town's streets were renamed in the 1980s, Parker and other African-American leaders were instrumental in having town streets named for Dr. Martin Luther King and Rosa Parks, prominent Civil Rights Movement leaders. Mayor Ferrire won an unprecedented fourth term in the municipal election. In the November general election, Beats 10 and 11 as well as all of Bibb County voted for James Earl Carter of Plains, Georgia for president. The county remained in the Democratic fold by 606 votes.

BEACON (Beginning Educational and Creative Opportunity Now) School opened in West Blocton in September to serve twenty-six handicapped children unable to attend regular public school. Governor George Wallace was scheduled to be present at open house for an inspection of the facility. Mary Ann Moody was project director as a part of her duties as Special Education Director for the county.

In October, the town received a grant of $80,000 to partially fund payment for extension of city water to Stacks Hill, Number 7 Hill, River Bend, and Hebron. Mayor Ferrire and county officials received the grant check directly from Governor Wallace's hands in Montgomery. In December, the West Blocton Health Clinic received $103,795 from the Appalachian Regional Commission to continue its operation for another year. In January 1977, the Health Center welcomed Dr. P. A. Del Vecchio of Birmingham to its staff for one day a week. This made it possible to have access to a doctor at the clinic from Monday through Friday.

Making the pages of the *Tuscaloosa News* sports section was WBHS athlete Marvin Wesley, West Alabama's leading high school basketball point-maker. WBHS in mid-February had a decent 16-5 record, one of the best in years. Wesley had a high scoring night of 53 points to help add to his average of 24 per game. He was a member of the National Honor Society chapter having been selected by his teachers and principal with a solid grade average of 85. He was also a split end on the football team, the baseball team's short stop, and he ran the half-mile in track. Ordinarily a quiet leader on the court, he said he had attained toughness by playing with competitive older brothers on a make-shift back yard court. His coach Alton Johnson said he was a "lets-go" player, shunning loud, hot-dog kind of playing. The coach believed Wesley was college material. He made a Birmingham paper's 2-A first team his senior year. Marvin Wesley played his college ball at Livingston. Now he is in the computer business in St. Louis. His brothers and sisters were equally successful. Older brother Thomas, in the 1990s, is a coach at Alabama A&M in Huntsville. He was the first African-American recruited by the University of Alabama though his playing time there was limited. Thomas transferred to Sneed Junior College in Boaz where he was a star.

A String of Drug Store Robberies

City and small town drug stores were targets of those selling and using controlled substances in the 1970s. West Blocton's new drug store was plagued with a series of burglaries. The town's only policeman was criticized sharply by the storeowner, Jerry Albano of Bessemer. The mayor was apologetic but had no answer to the problem with inadequate manpower in the revenue strapped town to patrol the business section day and night. The council backed the mayor saying all were deeply concerned with the problem. The drug store was located in a storefront belonging to the city, next door to the old mid-town city hall. An announcement was made in May that the business no longer stocked narcotics, and a sign was placed in its windows to inform possible intruders. An on-going conflict with the city government was made more evident by the unusual notice that read, "Due to the lack of enough policemen in West Blocton we cannot stock any controlled drugs or narcotics." The store owners had been robbed six times starting in 1974. Eventually, the West Blocton Discount Drugs moved into the old West Blocton Savings Bank building where it had access to the still operable Mosler bank vault as a secure keeping place for prescription drugs. That move was made in December. Even after the use of a bank vault for storing controlled substances was accomplished, the break in problem persisted. In March 1978, thieves entered the front door by breaking two dead bolt locks. They did not get the prescription narcotics in the giant safe but cleaned out the store's shelves. The management said a sophisticated alarm would be installed to discourage future break ins.

West Blocton Talent Makes the National Scene

The town needed something to boost its morale in 1977. The disturbing drug store robberies, a diminished economic base because of the disappearance of mining jobs, and the loss of downtown shops made for an element of despair. During the year, the old Murphree Drug Store, a virtual 1920s museum, owned by an absentee landlord, was stripped of its rare and unique 1928 furnishings, some of which were later spotted in use in a Pelham antique shop. The town retained its

Owanna Burt. (Courtesy of Owanna Burt Victory, Charlotte, North Carolina.)

City Hall Monument. (Courtesy of Cahaba Trace Commission.)

pride in spite of setbacks. Good things did occur that uplifted the community's spirit. Two attractive and intelligent young ladies had made it to the state finals of the Miss USA Pageant in Auburn in mid-May. Regina Shadwick, no stranger to pageant successes, won the Leadership Award and was one of the top twelve finalists among thirty contestants. The most startling and exciting news was that Owanna Burt, daughter of Mr. and Mrs. Mallory Burt, had won the position of first alternate and was to be one of two young women to represent Alabama in the national Miss Teen USA pageant scheduled for Shreveport, Louisiana on August 25-28. A proud community and fellow students at WBHS followed the events closely. Word was telephoned daily from Shreveport that Owanna was making great progress and had passed through the judging to the finals. The remarkable news came on the evening of August 28 that little West Blocton's own Owanna Burt had been crowned Miss USA Teen, a national honor.

The report of the event with photographs of the winner was printed in a number of publications and a syndicated story appeared in many newspapers across the nation. A reception was held in her honor in West Blocton, and in the fall, she entered the University of Alabama. Friends of the honoree, Annie Mae and Carl Jones, headed a move to erect a "Welcome to West Blocton" sign at the Highway 5 south entrance that carried the prominent wording, "Home of Miss U. S. Teen 1977-78, Owanna Burt." She was featured in the University's homecoming parade in October riding in Joe Namath's red Cadillac convertible. Both Owanna and Regina have enjoyed successful academic and professional careers. Regina made the state finals in 1978 as first alternate, and she was a finalist in the essay competition.

Other good news during the year involved more water expansion funds. A $171,000 grant, announced in April, allowed service to 100 more area homes. In September, over $144,000 in grants was awarded to provide primary health care at the West Blocton Clinic. The town extended congratulations in the spring to one of its favorite sons, Joe Henry Farrar, who graduated from the University of Alabama-Birmingham School of Dentistry. He later attained an advanced degree in orthodontics. Dr. Farrar has successfully practiced his profession in North Carolina.

Jones IGA Opening. Myrtle and Tom Jones proudly watched while daughter Monette cut the ribbon held by Mayor Frank Ferrire officially opening the new IGA on November 2, 1978. (Courtesy of Ferrire Collection.)

The monument to memorialize the town's veterans and coal miners was put in place in November 1977 near City Hall. Mrs. Keyes, who chaired the past year's Bicentennial celebration, gave credit to Leslie Shadwick and his daughter, Regina, for collecting the final contributions from patriotic citizens and merchants to pay for the engraved stone. There was more local good news in November when the Tigers outscored the Choctaws, 22-6.

The newly elected County Commissioner was Carl Jones who won the place vacated by Emery Lippeatt. Jones brought off a distinctive improvement for the district and the town by arranging for a county shop on property where the 1937 elementary school had stood. He obtained a portion of the building materials through a donation from Hammermill Paper Company, secured a land donation from the county Board of Education, and received other funding from the County Commission. The shop was identified as a meeting place for various service groups, and it provided a secure place for county vehicles. It was seen as a permanent voting place for Beat 11. The building was in use by early June 1978 and was said to be the first shop built on county property. Jones was off to a good start. Unfortunately, he fell seriously ill in the fall and did not recover. Many local citizens advocated Annie Mae (Tibbs) Jones, his widow, for the appointment to fill out Carl's term. The governor, following county political leaders recommendations, awarded the post to a Brent resident, the runner-up in the past election. Mrs. Pickett voiced the opinion of many in her news column of December 21, that the appointment was an injustice showing again the disregard for the wishes of voters from north Bibb. The intra-county division was again made evident. Mrs. Pickett wrote that some have spoken of "secession" from south Bibb and "Why, many ask, is north Bibb ignored in so many things?"

While the town was welcoming yet another short-term physician, Dr. Roshdy Habib, the friends of Dr. Daniel W. Ray took pride in learning that he had advanced academically and in military rank. In the summer, he received a diploma for specialization in prosthodontics from the University of Texas and was advanced to Lieutenant Colonel in the Air Force. Dan was the son of Effie and Harry Ray and a 1961 WBHS graduate. Dr. Ray now serves as a military medical professional in Denver, Colorado.

In November 1977, the Jones IGA Store opened in new quarters on the Cahaba Drive bypass. The grocery had been housed on the ground floor of the Masonic Temple for several years. The town government, challenged on the use of a mosquito-spraying program, won the right in court to continue the project. Plans were laid for a city park at the site of the old high school, once the location of West Blocton's first baseball facility, Roycroft Park. Detailed designs were considered for the park in March 1979, and $78,095 in federal and other funds were awarded for the project. The park was completed, put into operation but did not last. Funds were not made available locally to maintain and further develop the facility.

Local Telephones Connected to the Birmingham Exchange

One of the best things to come to West Blocton and north Bibb County was the Alabama Public Service Commission's approval of the proposal to add the 938 exchange to the Birmingham and Bessemer exchange. It had been a long, difficult project. Sue Pickett first promoted the idea in 1970 and periodically reported the proposal's progress and urged citizens to sign petitions. When the project lagged, she nudged it back into action. By 1978, the movement had shifted into high gear in the Woodstock-Greenpond area. Over 500 jammed into the Strickland Memorial Civitan Center in August to make plans for the final push to gain the Commission's approval. That move came in late October. During 1979 equipment reworking in the local telephone exchange was made and by November, it was announced final apparatus orders were in the works by South Central Bell. Installation was to take several months. At last, on Sunday, July 20, 1980, the lines were open. Long distance charges were no longer necessary for calls to Birmingham, Bessemer, and other communities in Jefferson and Shelby Counties.

Police Chief George Fleming received the keys to a new patrol car from Mayor Frank Ferrire with Alex Williams, Jr. looking on. (Courtesy of Ferrire Collection.)

It was a continuing and not-so-successful effort to obtain and retain a doctor in West Blocton. The municipality had also employed a number of short-term police chiefs. A search for a police chief in the spring of 1979 found retired serviceman George Fleming, and he was to remain in the chief's post until winning the county sheriff's office in 1994. He was reelected in 1998. Fleming's wife is the former Della Price, a West Blocton native. The chief had spent Army service time as a military policeman for 18 1/2 years. He served in Germany, France, and Viet Nam. In the latter assignment he was involved in a battle for the U. S. Embassy when it was occupied briefly by the Viet Cong. The police chief served as a de facto assistant to Mayor Ferrire during his final term in addition to being a full-time chief. During his tenure, the town accomplished another local and county "first" in hiring Rheinda Whitsett as a patrol officer. With her hiring, a politically correct title was introduced to the community with a "patrolperson" on duty. Whitsett stayed on the force for only a short time. She moved to Louisiana where she is still in the law enforcement profession. Fleming was a good match for the community and he served long and well.

Word came from an August meeting of the County Commission that there were three partially funded projects available to the county: a farmer's market in Brent, a library for Centreville-Brent, and a municipal building for West Blocton. The West Blocton project was to serve as a nutrition site, voting place, and small library. The projects required local support. In November, it was announced that the West Blocton facility was to be a reality. A business change on Main Street occurred in early 1979 when grocery store owner John Langston retired after 33 years in business selling his store to J. D. Mosley.

West Blocton's football team topped Bibb in 1977 and 1978, the latter was the locals' last win over the down-county rivals. The 1979 contest, with Bibb heavily favored under the expert tutelage of new coach Stan Moss, a Crimson Tide veteran, was a tie on a rain swept field in West Blocton. Bibb won one-sided victories during the remainder of the

WBHS Baseball Champions, 1980. (Courtesy of Billy Lippeatt, West Blocton, Alabama.)

series that ended in 1983. Based on enrollment, BCHS advanced to Three-A ranking and WBHS remained Two-A. School size advantage for Bibb and the need to add schools of equal size to both schools' schedules were given as reasons for the series termination. The 56-year football rivalry begun in 1925 (there were no games in war years 1943-1944) ended on two sour notes, the Tigers were badly beaten and Bibb later had to forfeit the game because of having played an ineligible star athlete. West Blocton school officials blew the whistle on the infraction. BCHS had to pay a modest fine to the Alabama High School Athletic Association and forfeit all their year's wins. Bitterness over the penalty and the way it was imposed was expressed in the *Centreville Press* but soon faded because there was no longer an annual game to bring up the memory.

Tiger Baseball Rules

The WBHS Tiger's 1979 baseball team had advanced farther in state play in their division than any other in the school's history, and in the 1980 playoffs, they won it all. The team had several members that played together from little league through high school. The little league coach, Roy Jones, was with the team for several summers. Jones noted the talent of many of the young players and predicted they could be champions. As the spring schedule unfolded, the West Blocton Tiger baseball team inched closer and closer to a perfect season. They won the way to the finals with victories over Cordova, Susan Moore, and Stevenson High Schools. A loss in the second of three games with Stevenson was the team's first defeat of the season. In May, they were headed to the state finals with Headland High School near Dothan. A long bus ride exhausted the over-eager boys, and the first game of the three-game series was lost. Rain prevented the Saturday double-header making it necessary to return home. The playoff was continued the next weekend. One game down, the team again boarded their yellow school bus to southeast Alabama where they won the next two games and the

Santana Ceremony. Taking part in the ceremony commemorating the final industrial bond payment were O. K. Yeager, Major Frank Ferrire, Howard Pate, Marc Novoson, Michael Novoson, Glen Jones, R. G. Moore, J. M. Langston, E. P. Jones, James Atcheson, and Charles Haynes. (Courtesy of Ferrire Collection.)

state Class 2-A title. A late-night gathering of excited and proud parents, elated schoolmates, and friends awaited the team's arrival at the high school gymnasium. An escort convoy made up of a town fire truck, police car, and a score of other vehicles met the bus at the county line and made its way home with sirens and horns blowing full blast. The team was met by a large, enthusiastic turnout and was fed refreshments including a huge decorated cake that proclaimed their achievement. It had been almost twenty-five years since a WBHS team had won state athletic honors. The school activities and welcome sign carried the notation, "1980—Two-A State Baseball Champions" until it faded after years of weathering. Members of the 1980 championship team were: Harry J. "Jimmy" Adams, Jr., Barry Sloan, Mike Morris, Rodney Crimes, Arthur Cottingham, Robbie Carroll, Kenny Preston, Mike Hubbard, Luther Cottingham, Paul Thompson, Johnny McKenzie, Don Cook, Kenny Walker, Calvin Brown, Kenny Thrasher, Mike Seagle, Calvin Washington, David Smith, Stewart Martin, and Norman Richardson. The head coach was Billy Lippeatt, assisted by Kenny Brown. The team's unofficial theme song was "We are Family." Five of the starters won college athletic scholarships.

In July 1980, town voters gave Frank Ferrire his fifth term. On the occasion friends dropped by his residence to offer congratulations. Martha Raye Kelley brought a cake with "Congratulations Mayor Ferrire for Twenty Years Service" embossed on it. Mrs. Pickett wrote about the occasion saying, "The office of mayor especially is certainly a public service requiring much hard work, long hours and very little money. One must really love his town to offer his service."

In August, the mayor and council joined with owners of Santana Sportswear Company to observe the final industrial bond issue payment in special ceremonies at the company's quarters in West Blocton. The company survived textile industry ups and downs until its departure for a cheap labor location south of the border in the 1990s. The

Town Officials. (left to right) Terry Acker, Martha Fancher Kelley, William H. Wright, and Dave Israel. Mayor Frank Ferrire is seated. (Courtesy of Martha Fancher Kelley.)

small Artistic Plastics plant closed in November 1980 after eight years of operation. It had failed to develop into other than a small business offering a handful of low-paying jobs. Encouraging predictions appeared in the August 28, 1980, *Centreville Press* concerning growth of business and population in north Bibb County. It was anticipated to be another Shelby County, Alabaster-Pelham-Helena area development. The belief that the increase in population and that industrial development for the immediate future was sure did not prove to be immediately correct. Real, perceptible growth in the sector was another decade and a half away.

While addressing a September high school assembly, Mayor Ferrire collapsed. The sudden illness was a stroke that left him seriously paralyzed. He spent several months in a Birmingham hospital and in October was officially sworn into office in his hospital room by his stepson, attorney J. Fred Wood, Jr. Council member E. P. Jones was selected to be Mayor pro tem while Ferrire was ill. In January 1981, Ferrire attended a mid-month council meeting for the first time since his stroke. He remained in Birmingham for much of the time until April to be close to rehabilitation services. He returned for more sessions at Lake Shore Rehabilitation in August and returned home in September.

One of the first "trade days" was held in April 1981, to bring the public to downtown local businesses. Also in April, William J. Murphy announced his retirement from the postmaster position after serving for twenty-seven years. Faye Pope Hickman was assigned the position as acting postmaster and was confirmed permanently in July. Early in 1981, the official Bibb County census count was listed as 13,200. That was a new low for the county since 1880. The 1990 numbers showed a modest increase to 16,576.

In the 1980s, among other problems caused by a lessening population, was loss of the tax base. The immediate local problem was the lack of revenue to run the county and especially finding matching money for funded programs. The threatened victim of early 1981 funding problems was the West Blocton Health Center. A detailed story appeared in the local paper of February 5 suggesting the center would close. This had been a continuing threat since the facility opened. An obviously poor community in a seemingly poor county lived hand-to-mouth. The situation was compounded since the country was in recession as a result of national economic policy. The Reagan administration had taken measures to "wring the inflation out of the economy." By 1982, Bibb's unemployment rate topped the state statistics at 24.3%. A smaller staff and other cut backs were instituted and the clinic remained open.

West Blocton Native Witnesses History-Making Event

Wayne Lucas, a West Blocton native, served as a Federal mediator for the Department of Labor and was stationed in Chesapeake, Maryland. On Sunday, March 29, 1981, Lucas departed the Birmingham airport for Washington, D. C. to attend a National AFL-CIO Buildings and Trades Conference. While on that trip he was a near-witness to a tragic page in American history. The after lunch speaker on Monday was President Ronald Reagan. After the speech, Lucas and friends were visiting in a hotel corridor when, but a short distance from where they were standing, an attempt on the President's life took place. In an interview later he said, "we heard the shooting outside the building just twenty feet from where we were standing." He and associates got to the front entrance just in time to see the President's limousine speed away. Lucas related, "We could plainly see the three men lying on the ground and the group of men wrestling with a man near the wall." Lucas said that had he not stopped to talk, he would have been in the doorway near the President when the shooting took place. "When I learned President Reagan had been shot, it was sickening to me," he said. The rest of the conference was canceled and he returned home the next day. Lucas said it was an experience he would never forget as long as he lived. His eyewitness story appeared in detail in the *Birmingham News* and was reprinted in the April 9,1981, *Press*. West Blocton friends and neighbors sought out Lucas for days following to hear first-hand his experience in Washington D. C. the day President Reagan was shot.

Considerable local government energy was expended in 1982 to purchase a new fire truck. Fire Chief E. P. Jones reported it was time for a replacement. A drive to reach former residents with a nation-wide plea to make contributions was explored. The request was printed in the county weekly newspaper. The drive paid off two years later when the town purchased a $61,000 fire truck in November 1984. The acquisition gave the town two workable trucks.

From the business sector, the local newspaper reported that two machines worth 3 million dollars each were in north Bibb County excavating coal, one near McCulley Hill. In March 1982, the Bank of West Blocton changed its name to the First State Bank of Bibb County. In April, the county again led the state unemployment statistics list at 24.3%. West Blocton merchants sponsored trade days on the river near old Piper with Colleen Weaver Parker as chairperson of the committee. Parker was also active in the formation of the Bibb County Chamber of Commerce. In June 1984, she was recognized for her leadership in recreational and other civic affairs by being appointed by Governor Wallace to the State Advisory Council for Recreational Resource Planning. Parker operates one of the few businesses left on West Blocton's Main Street, Colleen's Flowers.

The best news came in June 1982 when the promised $100,000 was reconfirmed for the construction of the West Blocton Municipal Building. The site for the structure was to be next to the old Baptist Church. Also announced later in June was the remodeling and expansion of the town's medical clinic.

West Blocton Mourns Loss of the Bear

Several West Blocton athletes had played for Coach Paul Bryant. They and other community members joined many other Alabamians in mourning the "Bear's" passing in late January 1983. Those former Tiders from West Blocton were among the congregation in the First Methodist Church of Tuscaloosa for the coach's last rites on January 28. Another old hometown boy, Dr. W. R. White, Jr., was also among the mourners. He had served as the team's official dentist during most of Bryant's tenure. Others from the community stood among crowds of other Bryant admirers on Interstate overpasses near Caffee Junction and Vance to pay respects as the long funeral procession passed beneath making its way to Birmingham's Elmwood Cemetery. It was particularly touching to see hand-lettered banners hanging from the overpasses bidding the "Bear" good-bye. In a column appearing in the February 3 *Centreville Press*, West Blocton resident Mary Ann Lees wrote poignant recollections of her visit to a practice session on the University campus and her experience in asking the coach to permit her to photograph him. He approved her request saying, "Won't cost me anything, will it?" Mrs. Lees' exact words are worth remembering for West Blocton fans of Bryant:

> Last Friday, Coach Paul "Bear" Bryant was laid to rest. We won't have to hurry to get to the stadium anymore to see the familiar figure standing by the goal post watching his Crimson Tide team "work out" before the game, or planning his game strategy or what ever it was he did while standing there. Whatever it was, it worked—just to see him there boosted team spirit and made you convinced that ALABAMA could beat anybody that day. We may be that way again, but it won't be without our friend Coach Paul W. "Bear" Bryant. His influence had imbedded itself in Alabama football and in everyone even slightly interested in the sport. He gave so much to so many and left us with the feeling that he would be satisfied to know that through his influence we would ALL try to be better sports and better people.
>
> Words cannot express the sorrow I feel at this great loss, but I know I am not alone. There will always be a warm feeling and perhaps even a smile when I remember the afternoon I spent with the Crimson Tide practicing on the field and being watched by Coach Bryant. To me, he'll always be there.

Coach Bryant was the subject of an article in November 1982 written by West Blocton native son Elmo I. Ellis who had made a trip back home with his wife Ruth to stay overnight with cousin Jean Israel Farrar. During the visit he was taken on a tour of the town. He wrote about that occasion:

> We needed my cousin's help, because even small towns change considerably with the passing of years. Trails and alleys that I walked as a boy are now paved with asphalt, and there are some roads that didn't exist in my youth.
>
> Many of the homes are new, too, and the old ones are lived in by families whose names I no longer recognize.
>
> So as a bedroom community, West Blocton appears to be alive and well. But there is virtually no business life. Nearly every store in town is closed, and many of the show windows are broken and boarded up.
>
> The Strand Theater showed its last film some ten years ago [October 21, 1962]. They say the operator, Earl Philpot, would still open the cavernous showplace in the evenings to sell popcorn and old movie posters bearing faces of such stars as Clark Gable, Myrna Loy and Tyrone Power. But a couple of weeks ago Earl died, so the popcorn machine doesn't pop on Main Street, and all is quiet.
>
> That's the dark side of this dusty mining town where the coal mines no longer operate. The brighter side greeted me in the morning when I arrived at the high school. It's a modern facility, comparable to the best in Marietta or Atlanta. They even have computers and word processors—newly arrived—which they'll be using to prepare students to function in the electronic environment of the future. It is understandable that pride and determination to excel are in short supply. These kids have been growing up in a town that reeks of depression and hard times."

Ellis used the story of the life of Coach Bryant to illustrate the possibilities one can have regardless of humble beginnings.

Elmo Israel Ellis had two speaking assignments during his visit, one at his high school alma mater and one in the College of Communications at his university. While in Tuscaloosa, he had a remarkable interview with Coach Bryant.

Municipal Building. (Courtesy of Cahaba Trace Commission.)

Alabama and its coach were dealing with three losses in a season that Ellis characterized as "not a recession but a disaster." The day before Ellis had spoken to WBHS students, his theme was along lines that urged them to take pride in themselves and to have determination to excel.

Ellis ended his column by musing: "How long would my hometown remain asleep if a leader like the Bear came to stay? I bet he'd wake it up and make it a winner! Regardless of his present problems, Coach Bryant knows how to do that as well as anyone anywhere." The former resident's appraisal of the condition of West Blocton in the 1980s needs no elaboration. Ellis, an accomplished and professional wordsmith, captured the moment extraordinarily well.

Unemployment figures for Bibb County in 1983 were high on the list of the most depressed. The figure released in April was at an all-time high, 26.5%. Those empty businesses observed by Ellis on Main Street in their various states of disrepair bore stark witness to the continuing decline of the town once characterized as "Bibb's Magic City." A new phenomenon in those years, Citizens Band radio users, "CBers," with a language all their own, had dubbed the town "Snuff City."

An even more telling appraisal of the town's condition in the 1980s was the account reported in the *Birmingham News* of the experience of the newest doctor. Thirty-three year old Dr. Thomas W. Ozbirn had chosen to practice in a small town, he told a reporter. The opportunity to be guaranteed a salary of $55,000 a year by the Health Development Corporation placed him in West Blocton. He described the community and his patients with a reality that stung. He portrayed a town with boarded up stores, and ex-miners with black lung cards. Thirty percent of his patients were unable to pay, some without running water in their homes. He described a town with many of the characteristics found in the depressed coal fields of Kentucky and West Virginia. "I always look in their ears, and you can see the coal dust settled there and if it's there you can imagine what it's like in their lungs," he said. He sensed a different kind of stress in some of his patients. In some places stress came when one had a fender-bender accident. Here, he related, stress came from simply trying to survive. Ozbirn, like the other doctors for two decades or more, did not stay long at the town's clinic. His newspaper interview, painfully true, met with mixed reviews in the community.

In March 1983, it was announced that the County Road 24 by-pass in West Blocton to the Cahaba River was designated as a Memorial Drive to honor veterans of the Viet Nam War. In late May several veterans associations held a formal dedication assisted by Holt High School's ROTC unit and the WBHS Band. The high point of the ceremony was the erection of a sign that read: DEDICATED TO VIETNAM VETERANS WHO FOUGHT TO MAINTAIN AMERICA'S FREEDOM.

In April, the West Blocton Municipal Building was dedicated. The facility was built as a meeting place, nutrition site for senior citizens, and a limited library facility. The city council, ministers from local churches, and the high school band

Jerry Claude Deerman was a talented driver with a host of local fans. (Courtesy of Jo Deerman, West Blocton, Alabama.)

Sue Pickett. West Blocton's reporter to the *Centreville Press* for more than three decades, her columns were a major source for this book spanning the years she wrote "the news." (Courtesy of Davene Pickett Smyth, West Blocton, Alabama.)

were participants in the ceremony. Councilman Felix Parker and Representative Walter Owens were designated as ribbon cutters. Owens was given major credit for the success in obtaining funding for the building.

Long before the University of Alabama opened its Bryant Museum, three sisters from West Blocton pooled resources and private collections of Bryant memorabilia to open a museum to memorialize the famous coach. Mildred Bureleson led the planning and was joined by her sisters, Margie Bush and Marie Nelson. The former American Legion Hut was freshly painted appropriately in crimson and white with a large outdoor sign picturing the coach and his wife. The sign read, "A tribute to a legend. He won 323 games, lost 83, tied 17. He won six national championships and 12 SEC titles. Thanks for the memories, Bear." The sisters named the spot, Bear Hall #1. The small Main Street building caught the attention of newspapers across the state and was featured in numerous articles. The mini-museum had an "Astroturf" floor with football field chalk lines, and walls lined with various photographs and painted portraits of the Bear. It also displayed photographs of each of Alabama's national championship teams and one of the coach's hound's tooth pattern hats. Visitors were serenaded with recorded Alabama football music and a ballad about the "Bear." The facility was an interesting attraction for several years.

Another tribute to Bryant appeared on the race-car roof of local driver Jerry Deerman who was getting notice in auto racing circles in the summer of 1983. Deerman was a frequent participant at the Birmingham International Raceway and enjoyed several significant wins. He had a number of cowbell ringing fans from West Blocton and sponsorship from a number of town businesses.

To further expand the town's water resources, a grant for $106,000 was obtained from the Appalachian Regional Commission for a previously announced reservoir tank on Smith Hill. Working on this project was Mayor Pro Tem E. P. Jones standing in for Mayor Ferrire who was again undergoing rehabilitation treatments in Birmingham. Local citizens expended a lot of energy in 1983 promoting projects under an organization called the Cahaba River Authority. The group acquired the old Strand Theater, a gift from Rev. Don Jernigan of Centreville. Unfortunately, funds were not found for its planned renovation as a concert and meeting facility. Later, the old Murphree Drug Store was also given to the Authority by Brady Belcher of Centreville. Another Authority project involved an attempt to acquire lands along the west bank of the Cahaba River near Piper for a park. Land swap arrangements between the University and a paper company were actively pursued without success. An event promoted as Old Timers Trade Day was scheduled on November 5 and dedicated to long-time citizens Sue and David Pickett. The community presented the Picketts with an appreciation plaque. Their attractive, talented granddaughter, Vickie Hallman, a television country music star from the popular "Hee Haw" series, sang at the gathering.

Two events making the news in late 1983 were the Wet-Dry vote (Bibb remained in the prohibition column by 75 votes), and Centreville, Brent, and West Blocton sued the county commission for funds that had been received by the county for distribution to the municipalities but kept by the commission.

CHAPTER 10

Town Centennial and Beyond

Of most significance in 1984 to West Blocton citizens were the results produced by a committee organized in late 1983 to plan a celebration of the town's centennial in 1984. Information provided by Dr. Rhoda Ellison's county history reflected that Blocton had been named in February 1884 when the first coal was shipped by the Cahaba Coal Mining Company. Actually, the town was born in the summer of 1883 when the incorporation of the company was formalized in Woodstock. Technically then, the celebration was a year late. The anticipated event was officially inaugurated by a town council resolution in early December 1983. The text follows:

Information provided by Dr. Rhoda Ellison's county history reflected that Blocton had been named in February 1884 when the first coal was shipped by the Cahaba Coal Mining Company.

> WHEREAS, Nineteen Hundred and Eighty Four marks the hundredth anniversary of the opening of the Cahaba Coal Company in this area; and
>
> WHEREAS Mr. Truman H. Aldrich, [the original document misnamed the founder as W. F. Aldrich, a brother] that company's president named the town Blockton after a one-ton block of coal was removed during early mining operations; and
>
> WHEREAS, since the 1884 recorded event represented the beginning of the community that was incorporated sixteen years later as West Blocton;
>
> NOW, THEREFORE, BE IT RESOLVED, that 1984 be designated as the OFFICIAL CENTENNIAL YEAR for West Blocton, Alabama.
>
> BE IT FURTHER RESOLVED that this action will serve to promote an appreciation of this town's history and the accomplishments of its citizens past and present.
>
> BE IT FURTHER RESOLVED that growth in civic pride will result from the attention given this milestone in West Blocton's unique and rich history.
>
> DONE this Fifth Day of December 1983, by the Town Council and Mayor Pro Tem [E. P. Jones] of the Town of West Blocton, Alabama.

A general invitation was issued in the January 22 *Centreville Press* asking all interested in helping with the centennial to attend. It was signed by a symbolic, fictional coal miner, "Wes Bloctoner." The meeting was attended by some two-dozen enthusiastic individuals on the afternoon of January 29. In that and subsequent meetings, plans were made for a year of activities and celebrations.

In an early meeting, Faye Pope Hickman was selected as chairperson of the Centennial; co-chair was Myrtle Cole Jones; and finance chairman was William J. Murphy. The beautification committee was led by Jackie Mosley, and Sue Pickett volunteered to direct publicity. One of the first projects came through the efforts of Carey Scurlock with the reprinting of the complete *Blocton Enterprise* of July 14, 1927, that reported the July 12 town fire. Copies were sold as part of a fund raising program to support the Centennial. Scurlock was also the artist that created the official Centennial

Centennial Logo. (Courtesy of Carey Scurlock.)

Willie Jo Parker played a vital role during the Centennial year by conducting research into African-American landmarks and heritage. (Courtesy of Felix Parker, West Blocton, Alabama.)

logo which featured a pick, shovel, mining car, and safety lamp in a circular emblem carrying the lettering, "West Blocton 1884-1984." The logo appeared on decals, car tags, a limited-edition coffee mug, and on the official printed program.

The first Centennial announcements listed several events including a poster and essay contest for school children, a flag raising ceremony at city hall, a special program on the Cahaba Lily, and an antiques identification and appraisal program. A call was made for pictures and newspaper clippings that would contribute to an understanding and recognition of the town's history. One of the earliest responses to the celebration was a poem submitted by E. H. Gentry. Its ending verses were:

Happy Birthday, West Blocton
Many Years have come and gone,
Where ever we may live or may roam
We still love that thrill of coming home.

Happy Birthday, West Blocton
You will always be home to me,
May we all remember you in our own special way,
Come home and celebrate your 100th Birthday.

The West Blocton Centennial kick-off event was the flag raising ceremony on Saturday, March 24. Congressman Richard Shelby was the star performer. He presented a flag that had been flown over the U. S. Capitol Building to Mayor Ferrire. The flag was raised to the sprightly music of the "Star Spangled Banner" rendered by the WBHS Brass Ensemble. Congressman Shelby made brief, appropriate remarks congratulating the town on the celebration of its rich past. The author of this history, standing in the bed of a pickup truck and lacking a podium to hold papers and provide support promised by one of the organizers, offered a short address on the town's history and hopes for a brighter future. A virus had him in its grip, and the offering was a shaky-hands, trembling-voiced attempt to say something significant. Fortunately, the text was printed in the local weekly. It read better than it sounded.

Contributing to the Centennial observance, a number of local citizens offered historical facts about the extinct churches of Blocton and West Blocton permitting a series of short histories in the county weekly. Fortunately, photographs surfaced to add to the stories' interest.

In May, a letter from Carla Boland appeared in the paper extolling the virtues of the town witnessed during an extended visit with her grandmother, Willie Oglesby Canterbury. Mrs. Boland wrote that she had once lived on the old Woodstock Road, and subsequently had lived in Charleston, South Carolina, Phoenix, Arizona, visited in California, and lived near Tampa, Florida. She paid high respect to West Blocton when she wrote:

> There is not another town like West Blocton anywhere in the world. It may be an out of the way place and the people may not have a lot of money, but the little spot on the map is the richest place I've ever been to. It has something more precious than money. This is a town whose people really care about each other and would give their last dime to help them make it through a difficult time. You are something special West Blocton and I love you.

Kathleen Boyd took the occasion in May to write the story of her mother's life in West Blocton. At age 95 and in the Bibb County Nursing Home, Martha Gilbert had related to Kathleen the story of her parents arriving in Blocton in the 1880s and the saga of their family living through good and hard times. She recalled the time of strikes and being removed from the company housing at the time gardens were beginning to produce food. Under the cover of darkness, she said, they slipped back to their garden plot to gather vegetables. She spoke of black lung and the toll it had taken on the lives of the men in her family. Her husband and sons-in-law died young of the incurable disease. She told that she and her family were proud to be a part of West Blocton. Everybody knows each other and are always willing to help in times of trouble. She ended the statement saying, "West Blocton is a good place to live."

During the year, the mayor and council continued to seek funds for a new fire truck. By May 10, almost $20,000 had been donated. Several large checks came from the state by way of discretionary funds through the offices of Senator Ryan deGraffenried and Representative Walter Owens. The town's merchants and citizens and county financial institutions also pitched in contributing significant amounts. By November, a $61,000 truck was delivered, an unofficial Centennial achievement worthy of note. Another project, an additional water storage tank, was placed on Smith Hill. This made three tanks to serve a wide spread area reaching from Hebron to Primitive Ridge to Lowerytown and other communities. The new stand-pipe tank carried the Centennial year dates 1884-1984 and the bold words, WEST BLOCTON.

The special Centennial program organized and presented by the Alabama Wildflower Society was held on Saturday, May 26. Featured was a special film on the Cahaba River and a lecture by George W. Wood, Jr., forester for the University of Alabama and widely-recognized wildflower expert. Wood, whose father had once worked in West Blocton as the Southern Bell Telephone office manager, spoke about the famous Cahaba Lily and listed other rare and unusual plants that grow in the Cahaba Basin. This session was an early idea-producing experience for a number of local wildflower lovers. In 1990, an annual Cahaba Lily Festival was initiated. The late May event in West Blocton is co-sponsored by the George Wood Chapter of the Alabama Wildflower Society, the Cahaba River Society, and other environmentally concerned groups.

Frank Ferrire had been the longest serving mayor in the town's history having served twenty years. He had become severely handicapped as a result of a stroke at the beginning of his fourth term. It was his decision that he could no longer devote time and energy needed to be a candidate to serve for another term. West Blocton, celebrating its 100th year, was to have a new mayor. Candidates announced in May were Vic Deerman, William J. "Bill" Murphy, Mildred Burleson, and Felix Parker. In the July election, Murphy ran first and Parker second. There was no clear majority. Murphy won in the July 31 run-off by a vote of 360 to 213. It seemed appropriate that a municipal election would fall in the Centennial year. It was a show of the town's diversity.

The new mayor had deep roots in the community. His maternal grandfather, J. H. Blake, had been a pioneer resident of Blocton as a blacksmith with the Cahaba Coal Mining Company, and he continued in that position with TCI. Blake was so impressed with the company owner that he named a son for Truman Aldrich. Blake subsequently moved to the country west of town and operated a farm and dairy there for several decades. Murphy served for three years until he resigned because of health reasons. The council appointed councilman Robert Franks interim mayor.

Meanwhile, the celebration of the Centennial was undertaken in the schools. Senior high students examined the town's history, family genealogy, and some made oral history recordings of grandparents and other older citizens. Classes invited Macon and Harvey Pope to learn first hand about the town fire of 1927, and Mayor Ferrire told students as much about the town history he and his parents remembered. The school library displayed old newspapers and photographs of places and events in the town's history. An essay contest produced several outstanding papers. Some of the winners were John Sloan, Gail Johnson, Benje Grady, Ethel Jones, Scott Rhodes, Tim Crocker, Ray Rogers, Renee Hoggle, Delana Hutchins, and Barbara Holt. Considered the best papers were those by Grady, Sloan, and Rogers. A poster contest produced interesting subjects. Poster winners were Paul Deerman, Keith Majors, Ashantis Cutts, Virginia Williams, Suzette Stone, and Sandra Stone.

Robie West. (Courtesy of Robie West Burt.)

George R. Burton, Sr. was the featured specialist in the Centennial antiques program held in the municipal building on June 24. Burton, a West Blocton native, was trained as an antique appraiser in some of the premier certification programs in the country. In 1984, he was associated with a Birmingham appraisal firm, Burton, Stone, and Horne. As a favor to the planning committee and his home town, he contributed his time to educating and entertaining a very responsive audience. Participants were invited to bring small items for appraisal and date identification. Some fifty odd young and old antique fanciers and holders of personal heirlooms attended the successful session. Burton also described what he believed a turn-of-the century Blocton or West Blocton household might have contained. He said the most frequently seen items might have been heavy, golden oak pieces in living rooms, and iron bedsteads were likely the most common bedroom furnishing. A popular product of the times, he related, was "O-Cedar" furniture polish that had a darkening, greasy effect on wood furniture after years of application by housewives. Most early kitchens were dominated by a coal range, a plain, utilitarian kitchen table, a Hoosier cabinet, and straight ladder-back chairs. Situated close to the kitchen stove was a tin tub for bathing. When stored, it might have been hung outside near the back door. He also reminisced about the depression days in West Blocton when even the possession of a common garden hose was considered an indicator of upscale economic status.

In hindsight the committee raised the question: "Why didn't we have a Centennial Queen contest?" It might have added more interest and some committee income support. As it happened, another West Blocton beauty won a place in the year's Alabama Miss Hemisphere Contest held in Birmingham on June 12-16, 1984. Robie Ann West placed third in talent, second in the state costume portion by wearing a coal miner's outfit which typified her home town, and second in evening gown competition. She was also the photogenic and overall beauty winner. She represented Alabama in the finals that were held in Miami Beach Florida. The expense-paid trip was a part of her prize for winning in Birmingham. Her parents were Douglas and Dorothy West. From hindsight, Robie should have been called the town's reigning Centennial Queen. Now she resides in West Blocton with her husband, Kyle Burt.

Another feature of the Centennial celebration was Church History Sunday. Many congregations in and near West Blocton took the opportunity to recall and record their histories. Several churches had not previously compiled histories and the occasion added to an understanding and appreciation of the town's religious history. Special stories featuring extinct churches were printed in the *Centreville Press* including sketches of the Episcopal and Catholic churches and the "League of Brothers" synagogue.

There were unexpected Centennial developments. A local talented artist, Herman Rhodes, painted a giant sign-board sized mural on the side of a downtown store depicting phases of the town's history and an image of mythical miner, Wes Bloctoner. There were special volunteer fix-up and clean-up projects by individuals and groups and many lent family

Centennial Chairperson Faye Pope Hickman and Senator Howell Heflin. (Courtesy of Faye Pope Hickman, West Blocton, Alabama.)

photographs and other items for display. The mayor made a downtown storefront available as a temporary museum that was visited by hundreds. As a University retirement recognition, the author's colleagues and friends contributed funds for a West Blocton Centennial $1,000 scholarship that was awarded to WBHS senior, Ryan Propst.

While most Centennial events were carefully planned and executed, there were other surprises. Mayor Ferrire initiated congratulatory letters from President Ronald Reagan, Senators Heflin and Denton, Congressman Shelby, and Governor Wallace. The committee was not aware the letters had been solicited. They remain on display in city hall, prized evidence of the celebration. The amount of coverage in state dailies was another surprise but the biggest was the degree of involvement by the community. The credit for the Centennial's success belongs to its energetic and creative committee. The event was planned with the expectation of attracting hundreds for its final big week. However, thousands attended.

There was an early major surprise. Initially, the Centennial Committee had decided that one weekend scheduled in conjunction with the high school homecoming game would end the year's celebration. Mayor Ferrire attended only part of the planning session, went to his office, and called Senator Heflin. He immediately reached the Senator's office and extended an invitation for him to be the keynote speaker for the weekend prior to the homecoming date. The Senator accepted. Plans had to be altered to include a week of activities rather than just three days. In addition to the Sunday, September 23, opening ceremonies, Tuesday Open House was set at First State Bank of Bibb County, and on Wednesday, the centennial chairman, Postmaster Faye Hickman, held a Patrons' Appreciation Day at the Post Office. A clever model of Blocton's first post office was constructed in the lobby, and school children took the occasion to make field trips to the replica that displayed historical postal material and a special stamp collection. So, rather than opening on Friday and ending on Sunday, the celebration started on Sunday and ended on the following Sunday. By July, the program was complete.

The committee evolved as a relatively small working group. It accomplished far more than was thought possible. Faye Pope Hickman, Myrtle Cole Jones, Carey Scurlock, Mayor Ferrire, and Mayor-Elect William J. Murphy have been identified. Others made significant contributions. Frances Kornegay's organization of the homes tour was one of the most successful events, and the opening day's historical fashion show and promenade planned and directed by Sandra McCool Everse

Many people got caught into the Centennial spirit and donned period costumes. Pictured (left to right) Margrette McCool, Robbie Findley, Laura Russell, and Billie Deason. (Courtesy of Faye Pope Hickman.)

was a spectacle. Billy Mack Roland had a major role in organizing an "Old Timers Baseball Game" and has continued planning the game as an annual event since 1984. Rather than honoring a particular class, the 1984 high school homecoming was dedicated to all former graduates. The half-time show featured the Centennial celebration in addition to the traditional homecoming themes. The high school homecoming court was a part of the Centennial parade. The Centennial parade coordinator was A. N. Kitchens. The mayor again had the old Catholic Church bell displayed and rung. It was placed for this occasion on one of the town's newest fire trucks. With the mayor in the right front seat and the bell ringing constantly behind him, the town and its visitors gave the twenty-year civic leader round after round of applause as he made his last official trip down Main Street.

In the September 27 *Press*, Sue Pickett, the consummate local news reporter, summed up the opening day of the Centennial week events:

> Centennial Week! It began Sunday, September 23rd. The Town of West Blocton brought together in love, rejoicing in our history, and determined to do our part in keeping our loved town alive and well, for our children's children, so that they might celebrate [its] Bicentennial. The big auditorium seemed filled to capacity. Many stood in doorways; and happiness and pride was a part of the air. A wonderful, interesting, nostalgic program for the hundreds who came from here and many states.

And on October 4, she wrote:

> Centennial Week has ended! A once in a hundred years week. What joy there has been, what love shown, what memories shared by the hundreds who came back to our and their loved town, West Blocton, to celebrate this wonderful birthday. Crowds all week long. Saturday the streets were filled with visitors. Old friends seeing each other after many years. Even the most ardent of us have been surprised at the tremendous response. Every one who had a part deserves our appreciation.

> There is a new spirit in West Blocton, or perhaps the old one just renewed. Let us go on working together.

The *Tuscaloosa News* editorial page for September 26 may have phrased it best:

> CELEBRATING A CENTURY
>
> Like the big seam of coal which undergirded the town, dreams ran strong through West Blocton.
>
> Shortly after its founding in 1884 during a coal mining boom, West Blocton was the fastest-growing town in Bibb County. By the turn of the century, the company town had been transformed into the hub of the area's social and cultural life.
>
> Many of the families of this area have deep roots in West Blocton. It was a place which valued hard work and self-sufficiency, a town where some family fortunes were established for decades to follow and where others were launched in the era of entrepreneurship and growth of the New South.
>
> West Blocton was devastated by a fire in 1927. It never fully recovered from its effects and from the economic hardships of the Great Depression. But its spirit remained and endures to this day.
>
> It was a spirit exemplified in the town's 100th birthday observance which began Sunday and continues throughout this week. The West Blocton Centennial, with its strong emphasis on history, is a reminder of flush times and growth, of adversity and testing, of family ties and relationships.
>
> Most of all, it is a celebration of the hardy spirit which kept the town and its people alive through good times and bad over the past century, which fired its hopes and fueled its dreams.
>
> It is a spirit which makes West Blocton's hopes as strong as its memories; a thing worth celebrating.

In the years following the Centennial, a small group of dedicated citizens formed the West Blocton Improvement Committee with its main objectives being the furtherance of community pride and development. Heading this group has been Myrtle Jones, a member of the town council. Among the committee's projects have been an annual Christmas parade, a summer Wild West Blocton Days event that now serves as a trade day, an occasion for street entertainment, and as a reunion weekend for the high school's alumni. This early summer event is also the occasion for the "Old Timers Baseball Game" now attracting hundreds of former residents and baseball players. On the Friday evening before the game, the organizers headed by Billy Mack Roland, sponsor a large fish fry. It is not unusual to see several hundred visitors and home folks taking part in this festive gathering.

Victor Deerman, mayor from 1988 to 1996, gave enthusiastic and encouraging support to the Improvement Committee and the Old Timers game. A first class area rescue unit has been supported and continued through several municipal administrations. Its "Jaws-of-Life" machine has been a great asset to the community. The organization is made up of devoted and well-trained citizens. Mayor Deerman supported the transformation of the old I&N Department Store into a community center. He also laid the groundwork for the sewage system that is now partially funded through the efforts of former Mayor Jerry Pope. Under the leadership of the current mayor, Jabo Reese, and the Town Council, this system will soon be a reality.

As an outgrowth of the Centennial program on local wildflowers, the Improvement Committee sponsors an annual Cahaba Lily Festival that has received regional and national attention. In 1999, that activity celebrated its tenth year. The committee claims its annual celebration has contributed significantly to the Cahaba Lily now being the poster wildflower of state conservation groups. Encouraging news broke in May 2000 indicating that Congressman Spencer Bachus had sponsored legislation that would make the area near the Piper bridge a National Wildlife Refuge. A ten-million dollar expenditure for the project is budgeted.

In the late 1980s, the committee worked with the town council and mayor in transforming the old I&N Quality Store into a meeting space that can accommodate an audience of up to 300. It is the site of the annual Cahaba Lily Festival and is appropriately named the Cahaba Lily Activity Center. The building houses many photographs and other items that

Cahaba Lily Activity Center. (Courtesy of Cahaba Trace Commission.)

reflect the town's interesting history. The committee added a public library to its list of accomplishments in the mid-1990s. Emma Hicks, one of the Improvement Committee members, headed this project and serves as the facility's librarian.

The Italian Catholic Cemetery, established in 1901, has had clean-up support from the Committee, and a wrought iron gate was erected at its entrance in 1996. A historical marker was erected in 1996 identifying the cemetery. Following a program in Faith Baptist Church, the former St. Thomas Aquinas Catholic Church, a re-dedication of the cemetery was conducted by a priest from Montevallo and nuns from Cullman. The re-reading of an ages-old ritual was a moving spiritual experience. Following a nomination by the Cahaba Trace Commission, the cemetery was placed on the National Register of Historic Places on April 22, 1999.

While the early 1980s Cahaba Trace project was slow to develop, it gained new life under the state endorsed Cahaba Trace Commission. West Blocton is an official part of the designated trace route. A major reason for the town's inclusion has been the development of Bee-Hive Coke Ovens Park. A remarkable transformation

Secretary of State Jim Bennett served as keynote speaker during the Blocton marker unveiling. Pictured (left to right) Mayor Jerry Pope, Secretary Bennett, Joyce Franks, Mistress of Ceremonies, and Councilwoman Myrtle Jones. (Courtesy of Alabama Historic Ironworks Commission.)

West Blocton Marker Dedication. (left to right) Charles Allen, Howard Franks, Jerry Pow, Mayor Jerry Pope, Rev. Bob Praytor, Myrtle Jones, and Arthur Finnen. (Courtesy of Alabama Historic Ironworks Commission.)

Town flag

has taken place at the 1880s coke ovens. Assisted by University of Alabama-Birmingham industrial archaeology students and their professor, Dr. Jack Bergstresser, a number of local citizens have uncovered some dramatic and significant nineteenth century industrial remains. The area was overgrown to the extent that no evidence of the massive brick and sandstone construction was visible until this ambitious and creative group applied themselves to this Herculean clean-up task. An impressive historical marker has been placed at the park entrance on County Road 24 giving a short history of the ovens and the company town, Blocton. The marker was a project of the Alabama Historic Ironworks Commission and the West Blocton Improvement Committee. A companion marker reflecting historical facts about West Blocton has been placed near the town hall, sponsored by the Cahaba Trace Commission and the town.

Through the research efforts of Reverend Bob Praytor, First Baptist Church pastor, historical data about the coke ovens has been found, compiled and published. The material is available to the whole world on the Internet. The fact sheets his investigation produced were pivotal in obtaining two substantial grants to further develop the park. The town's official flag designed in 1996 by renowned Birmingham graphic artist, Guy Arello, symbolically depicts the coke ovens, the Cahaba River and the Cahaba Lily.

Uncovering, celebrating, recording and marking local history are worthy endeavors. As this work ends, it may be appropriate to remember the words of the Roman historian Marcus Tullius Cicero, "History is the witness that testifies to the passing of time; it illumines reality, vitalizes memory, provides guidance in daily life, and brings us tidings of antiquity."

Though mining has ceased around Blocton, area men have continued to mine coal. Steve Battle, Jim Walter's Brookwood Mine Number 7, 1990. (Courtesy of Nelle Battle, Lowerytown, Alabama.)

EPILOGUE

The town of West Blocton has made little progress in the past decade. The Main Street business district has become like a ghost town movie set with many of the 1927–1928 buildings, erected after the 1927 fire, now in ruins. Roofs have given way, and windows are broken out and boarded up.

The nearness of the giant Mercedes-Benz plant and its plans to expand have not affected the town's old Main Street business district. Yet, West Blocton has a state-of-the-art sewage system, its water works could support a town four times its size, and updated Alabama Power Company high voltage connections are available. Too, Alabama Gas Company serves the area. According to the 2010 census, the town lost a small number of residents, but all other incorporated places in the county have gained population. While there has been no movement to expand the town limits, there are enough people within a four-mile radius to more than double the current population count.

There are some other noteworthy developments:

The Cahaba Lily Festival, described previously in Chapter 10, celebrated its twenty-second year in 2011. Since its founding, the Cahaba Lily has become Alabama's "poster" wildflower—a symbol of the state's wild, undisturbed,

Town Hall. (Courtesy of Cahaba Trace Commission.)

George "Mule" Suttles. (Courtesy of the National Baseball Hall of Fame Library, Cooperstown, NY.)

and scenic places. Nationwide, botanists and environmentalists have become aware of the need for the lily's preservation through journal articles and lectures by scientists and botanists such as Samford University's Dr. Larry Davenport. Concern for the lily's preservation led directly to the establishment of the Cahaba River National Wildlife Refuge in 2002. Future plans for the refuge include its expansion to protect more lily habitat and the building of an interpretative center in or near West Blocton. Credit for these exciting developments may be given to the committee that gave birth to the festival idea. The refuge and Cahaba Lily habitat are four miles east of West Blocton.

The community also takes pride in the development of the Blocton Coke Ovens Park, now established at the site of the Cahaba Coal Mining Company's coke producing operation, which began in the late 1880s and continued until the early 1900s. State tourism grants have served to improve the park—the latest grant provided for construction of a walk-over for easy viewing of the massive brick construction. Picnic and camping grounds are also being developed. Only recently, a grant provided playground equipment.

The Annual McKinney Wrecker Service's Truck and Auto Show on Truman H. Aldrich Parkway has become a part of Wild West Blocton Days, held the weekend before the Fourth of July. The show features old and customized vehicles and attracts hundreds of visitors.

In 2006, local and national sports columns and magazines revealed that Blocton was the March 1900 birthplace of George "Mule" Suttles. Recognized for his professional baseball career by posthumous election to the National Baseball Hall of Fame in Cooperstwon, New York, Suttles played for a number of teams throughout the leagues for twenty-three years, compiling a .327 lifetime batting average. A first baseman and outfielder known for his power, Suttles once played for the Birmingham Black Barons. As a youngster in Blocton he would likely have attended the company's Cadle School where the sport was enjoyed on the playground. Suttles died in 1966. A photograph of the museum's plaque honoring Suttles is framed and displayed in the town's Cahaba Lily Center where other photos and artifacts of the town's past line the building's walls.

Mention must also be made of the success of Sammie Hill, currently a professional football player with the Detroit Lions. Prior to spending his college-playing years at Stillman College in Tuscaloosa, Sammie was a standout on West Blocton High School's football, basketball, and baseball teams. Hill was featured in an April 27, 2009, story in the *Tuscaloosa News* (sports section citation).

The Coach Nick Saban statue at Bryant-Denny Stadium was created by a West Blocton artist. Jeremy Davis, a

27-year-old art student at The University of Alabama was recommended by his professors to sculpt the Saban statue now standing in front of the stadium. Davis is quoted as saying, "For me, it's an honor just to be a part of something like this. Being from such a small town, it's something you never expect." Jeremy has also been commissioned to do the Nick Saban bust which will be placed in the Mal M. Moore Athletic Facility on campus. *The Centreville Press* reported that Jeremy said, "Where I am from, we don't even have a red-light—but I am rubbing shoulders with Nick Saban and designed something that is going to be enjoyed by Alabama fans from now on."

We now know that Mr. T. C. Tuggle's story on pages 111–113 concerning the notorious swindler Charles A. Ponzi of Boston is true. Ponzi did indeed spend time in Blocton, confirmed by both his biography (*Ponzi's Scheme,* by Mitchell Zuckoff, published in 2005 by Random House, Inc., N.Y. pp. 50–54; *The Rise of Mr. Ponzi,* Charles Ponzi, originally self-published in 1937, republished by Inkwell Publishers, Naples, Florida, 2001, pp. 35–38).

Apparently, after a stay in Birmingham, which involved an illegal injury-insurance scheme, Ponzi needed to move on. He traveled in 1912 to the "Little Italy" settlement in Blocton, some forty miles southwest of Birmingham, where he was among Italian immigrants. There, as recounted by Zuckoff, he: "for several months scraped together a living as a translator, a part-time bookkeeper, and, occasionally, a nurse to injured miners. The Italian camp in Blocton reminded him of small-town life back home, always celebrating a christening, a marriage, or a holiday, and he felt 'embraced by a brotherhood of common interests and endeavors and brotherly love.' But conditions were only a step above primitive. The camp's ramshackle wooden houses had no electricity or running water. Still dreaming of riches, Ponzi began laying plans to make himself a local czar of light and water." However, this Ponzi scheme was not adopted by the Blocton community.

Still there in October 1912, Ponzi learned of an accident in the small company hospital in which a nurse was burned. As he had been befriended by the facility's director, Dr. M. C. Thomas, he asked about the accident victim's condition. The doctor admitted it was not good. A skin graft was needed but no skin donors had been found. Ponzi offered the backs of his thighs. Two surgeries were required. The operations were successful but put the donor in the hospital for several weeks. Local citizens talked of his nomination for a Carnegie Medal, but that effort faded. So, while most stories of Ponzi's exploits were of the criminal sort, the stay in Blocton offers at least one indication that he had a generous and caring bent.

Another story from T. C. Tuggle involves his hiring a thirteen-year-old boy from Mississippi to work in his tailor/photography business. West Blocton residents later visiting Meridian were attracted by the Jimmy Rodgers museum. They were surprised by a display of a set of postcards postmarked Blocton, from the famed Father of Country Music indicating that he'd been employed in his boyhood by Tuggle in the summer of 1911. Biographer Nolan Porterfield's *Jimmy Rodgers: The Life and Times of America's Blue Yodeler* (Nolan Porterfield, University of Illinois Press 1992, p. 23) tells of Rodgers's stay in West Blocton:

> Undismayed at being 100 miles from home and entirely on his own, the plucky thirteen-year-old soon found a job in West Blocton, a small hill town southwest of Birmingham (Author's note: At the time, West Blocton and Blocton had a total population of approximately 4,000). 'I'm working at Mr. Tuggle's,' he wrote to Dora (his aunt). The Tuggle Brothers, T. C. and J. M. were tailors—specialists in made-to-order garments—and proprietors of Tuggle's Yellow Front Photograph Gallery. The combination of needle and camera was a clever one, ingeniously appropriate to an age fascinated with the gadgets of fledgling technology and the inventions of modern science. The gallery not only did a brisk business in 'the mug trade'—portraiture—but advertised 'a measuring system' which was, to say the least, highly unusual. 'We take a photograph of each customer which enables us to insure an exact fit.' 'Mr. Tuggle is paying me 50c a day and Bord (*sic*) and I like my work.'

Panoramic view of the December 27, 1896, train wreck over the Cahaba River near Blocton. (Courtesy of Louisville and Nashville Railroad Records, University Archives and Records Center, University of Louisville.)

One can only speculate about Rodgers's business efficiency, but socially he was a fast worker: "I have many friends here and a sweet little girl to (*sic*)." Just how long he stayed in West Blocton is not known, but it is unlikely he was there for more than a few months.

Photos of the December 1896 train wreck on the Cahaba River railroad trestle taken the day of the wreck (see pages 11–12) have been found in the University of Louisville's library collection of the Louisville and Nashville Railroad's archives. Permission has been granted to reproduce the panoramic photo of the wreck site on the day of the wreck, December 27, 1896.

West Blocton is proud of all its military veterans. In particular, the story of Thomas Dewitt Pool, who was killed in Vietnam (recounted on pages 275–276), is made more complete with his being awarded the Distinguished Service Cross in 2008. Known as the DSC, this medal is the second highest award for valor bestowed on a soldier. The award was based on his heroism in a 1968 Vietnam firefight, which cost him his life. In the battle, he eliminated three of the enemy and then, not thinking of his own safety, went to the aid of a wounded comrade lying in an open area. While kneeling to help his platoon member, he was suddenly fatally wounded.

A correction: The local *Blocton Enterprise* announced in 1911 that Annie Oakley, the famous woman sharpshooter, was to appear with a circus in West Blocton (page 50). It has since been learned from a television program that a number of circuses had their own Annie Oakleys. An exchange of emails with the Annie Oakley Museum determined that the real Annie Oakley was nowhere near Alabama at any time in 1911. So, the 1911 circus in West Blocton had a bogus markswoman performance.

Well, what about the town's future potential? Here are a few considerations:

Underlying the ridges and valleys of the Cahaba Coal Field are millions of tons of high grade, clean burning, and low-sulfur coal. Current long-wall mining methods cannot be applied to extract the Cahaba Field's riches because of unfavorable geological formations. Who is to say, however, that someday methods will be developed to reach this valuable resource? Worldwide, coal is the most used energy source. Yet, deep coal seams still lie near West Blocton, awaiting the development of economical mining processes.

Unlimited industrial water stands in the vast caverns of old mining operations. A small operation that exploited old slate dumps needed water, and lots of it, to remove usable coal from the mine waste. By drilling into mines deep under the surface, potable water was pumped to the surface. After use it was returned to the underground source. Dr. Jim Parker, an engineer and part owner of the company, reported that within a short time the water was returned to its original purity. It appears that the problem of industrial water and its pollution problem may be solved by this proven experience.

Meanwhile, possibilities exist that favor the development of the area as a tourist attraction. The Cahaba Lily is one of the major reasons for the Cahaba River Refuge development. The river is a living laboratory for scientific study of fish, snails, and mussels. Near the refuge is the Bibb Glades, a remarkable, recently discovered site of rare flora. Some plants were so new to botanists that they had to be given names.

West Blocton is close to hundreds of acres of timber land (and mineral rights) owned by The University of Alabama. These lands are in a semi-circle to the south and east of the town. Such acreage so close to the Mercedes-Benz plant has the potential for development as an industrial site for automotive parts companies.

Perhaps, then, a new era of growth and prosperity may be in West Blocton's future. "Alabama" is an Indian word. Some sources say it means "weed gatherers" others say "to gather up." Let's adopt the latter and say: "Gather up" your resources, West Blocton, and move on.

Please permit me one heartfelt, special note. This reprint by The University of Alabama Press would never have happened nor would the first edition ever have been published without the help and encouragement of a now-departed dear friend. Sandra McCool Everse was my advocate from the first day that the possibility of a book being published emerged about the town for which we both had great affection. Sandra was a special person; may light perpetual shine on her.

Charles E. Adams
Tuscaloosa, Alabama

APPENDIX I

Mayors of West Blocton

1901-04 E. D. Reynolds
1904-06 Dr. L. E. Peacock
1906-08 W. H. Logan
1908-10 Yancey S. Smith
1910-12 Marshall C. Davie
1912-14 Thomas A. McCool
1914-16 William H. Wright, Sr.
1916-19 W. T. Griffin
1919-20 Marshall C. Davie
1920-21 E. M. Bishop
1921-24 William H. Wright, Sr.
1924-30 Marshall C. Davie
1930-32 Thomas A. McCool
1932-44 William R. Ray
1944-47 Dave S. Evans
1948-49 J. Wilcox Randle
1950-51 Victor Young Foster
1952-58 Ben Stanley
1959-60 R. G. Moore
1961-64 Floyd T. Morrison
1965-84 Frank T. Ferrire
1984-86 William J. Murphy
1986-88 Robert Franks
1988-96 Victor W. Deerman
1996-2000 Jerry D. Pope
2000- Jesse "Jabo" Reese

APPENDIX II

[1910]

YOUNG AND CO.'S
BUSINESS AND PROFESSIONAL
DIRECTORY OF THE CITIES AND TOWNS OF ALABAMA

Blocton Pop 2,000. Bibb Co. L&N Ry

Alexander, J. F., drugs
Ambrose, W L, groceries
Costello, S groceries
Friesner C A, pool and soft drinks
Green Bros. sawmill
Guisseppi, M. Genl mdse.
Harris T T. Dry goods
Hill Creek Coal Co, miners. (sic)
Holland W B, meats
Howell Mining Co., coal
Italian Cooperative Co, genl store.
Jones L C, physician
Lee, N A, genl mdse.
Lewis R, sawmill
Lowery C E, blacksmith
Miller J W & Co, coal miners. (sic)
Norrtis Lumber Co, mfrs.
Pratt Consolidated Coal Co.
Pritchett P P, soft drinks.
Steele R Y, genl mdse.
T, C, I & R R Co's Store S J Morgan mgr.

West Blocton
Pop. 4,000 Bibb Co. L & N Ry.

Allen C E, pub Blocton Enterprise
Allen Furniture Co, retail
Bishop E M, attorney at law
Blocton Enterprise, The, newspaper, job printing, C E Allen prop
Brown F, pool and soft drinks
Caldwell H G, barber shop
Dark T S, attorney at law
Davie J B , goods and cloting
Dorre Chas, restaurant and bakery
Douthit House, J W Stewart prop
Dunnabie Robt, genl mdse
Dyar W A, blacksmith
Finnen Bros, genl mdse
Finnen J M, meat market
Garrett Mrs L, millinery and notions
Gates Jewelry Co, Fred Gates prop
Gentry A H, hardware, N P and ex-off J P
Harvey W E druggist
Hays W M , confectionery
Hydrolic Power and Light Co, W A Harvey pres
Israel A, genl mdse
Israel W, genl mdse
Jones J F, drugs, genl mdse and livery
Keith R B, barber shop
Krentzman, J, General Merchandise
Lagrone R C, meat market
McCool T A & Son, genl mdse
McCray E C, groceries and restaurant
McCumber Thos, watchmaker, jeweler and photographer
Miles W, blacksmith
Norris A. shoemaker
Odess M M, dry goods and clothing
Ogletree H C, general merchandise
Parker Hotel, Mrs R C Parker prop
Peacock, L E, physician
Perry W W, millinery and notions
Pope M F, druggist
Price S W, livery and feed stable
Quinn O J, genl mdse, feed, sale and trade stable, also

dealer in hides and furs
Randall Geo W, dentist
Roycroft H genl mdse
Schauer M A, genl mdse
Sessions W M, groceries
Shannon Hardware Co, Jas and P Shannon
Singer Sewing Machine Co, S J faulk mgr
Sou Bell Tel and Tel Co, Geo Wood mgr
Sparks W D, genl mdse
Standard Bottling Works, W D Hollifield mgr
Thomas M C, physician
Trigg A W, physician
Tuggle Bros., tailors and photographers
Walters & Gamble, millinery
West Blocton Savings Bank, E D Reynolds pres
Wright W H, attorney at law
Young W R, fire insurance

Author's note: The directory lists seventy-six entries for Blocton and West Blocton. The population numbers are not official, both are exaggerated. From newspaper reports West Blocton business list is incomplete. By 1910 Ben Edmonds had opened a dry cleaning business and the Tuggle Theatorium was showing silent films and featuring vaudeville acts. Likely there were others.

BIBLIOGRAPHY

Books

Armes, Ethel. *The Story of Coal and Iron in Alabama.* 1910.

Brant and Fuller. *The Memorial Record of Alabama.* 1893.

Brown, Edwin L. and Davis, Colin J. *It is Union and Liberty: Alabama Coal Miners and the UMW.* 1999.

Cline, Wayne. *Alabama Railroads.* 1997.

Corolla. University of Alabama yearbook. 1896.

Elam, V. Donald. *Bloody Bibb.* 1985.

Ellison, Rhoda C. *Bibb County, Alabama: The First Hundred Years, 1818-1918.* 1984.

Ellison, Rhoda C. *Place Names of Bibb County.* 1993.

Faulk, Billie. "Socialism in Alabama." Thesis in progress. 1998.

Flynt, Wayne. *Poor But Proud.* 1989.

General Acts of Alabama. No. 976; No. 933. 1900-1901

Griffith, Lucille. *Alabama College, 1896-1969.*_

Letwin, Daniel. *The Challenge of Interracial-Unionism.* 1998.

Link, Authur S. *American Epoch.* 1955

Lovett, Rose Gibbons. *The Catholic Church in the Deep South: The Diocese of Birmingham in Alabama.*

Owen, William. *History of Alabama and Dictionary of Alabama Biography.* 1921.

Owen, Marie Bankhead. *Story of Alabama.* 1949.

Persinger, Jessie Cole. "Autobiography." 1988.

Rogers, Ward, Atkins, and Flynt. *Alabama: The History of a Deep South State.* 1994.

Squire, Joseph. *Report on the Cahaba Coal Field.* 1890.

Squire, Joseph. "Autobiography." Handwritten circa 1908.

State Inspector of Mines Biennial Report. 1892.

War of the Rebellion: Official Records of the Union and Confederate Armies. Various volumes.

Walker, James W. *The Struggle and the Joy: A History of Piper, Alabama.* 1993.

Ward and Rogers. *Labor Revolt in Alabama.* 1965.

Westonian, 1946 WBHS annual.

Periodicals

Alabama Heritage. University of Alabama. Fall 1986; Spring 1988.

Alabama Review, July 1952; July 1998.

Alabama University Monthly. June 1887.

Journal of the North Alabama Conference. Methodist Church.

Newspapers

Alabama Sentinel, Birmingham

Bibb Blade, Six Mile

Birmingham Age-Herald

Birmingham Ledger

Birmingham News
Birmingham News Magazine
Blade-Enterprise, Centreville
Blocton Courier
Blocton Enterprise
Blocton Journal
Blocton Messenger
Centreville Enterprise
Centreville Press
Chattanooga Tradesman
County Press, Centreville
Labor Advocate, Birmingham
New York Times Magazine
Montgomery Advertiser
Montgomery Independent
New Orleans Times Democrat
Shelby Sentinel, Columbiana
West Alabama Breeze, Northport

Letters and Pamphlets

Adams, Harry J. Letter to sister on war's end in Europe, 1945.

Academy of Motion Picture Arts and Sciences. Letters and newspaper clippings on Lois Wilson, actress.

Allen, W. B. Cahaba Coal Co. Records Extract, 1932. Birmingham Public Library, Department of Archives and History.

Alumni Records. Rensselaer Polytechnic Institute, Troy, NY.

American Institute of Mining Engineers. Brochure. May 1888.

Annual Report, 1900. TCI Co.

Blackout instructions, 1942. Mimeographed sheet in collection of author.

Cadle, Cornelius. "Sketches of War History."

Cather, James Pat. Notebook of Mrs. Wilson of Blocton.

Clements, Jerry. Autobiographical sketches.

Dailey, Jewel Jones. Letter about Dr. Clyde White Jones.

Davie, J. W. Various letters and family history.

Ellis, Elmo Israel. Various letters re: town fire, synagogue, I &N Quality Store.

Episcopal Diocese of Alabama. Minutes of the Executive Council.

Henderson, Elaine Stewart. Letters on Commander Fred Stewart.

Lloyd, Stewart L. A Brief History of the Cahaba Coal Field. 1957.

Moon, James D., Jr. War Time and Personal History, 1994.

Peacock, Dr. L. E. 1911 Annual Report to the State Department of Health.

Penhale, Ken. Document Collection.

Praytor, Robert. Number One Blocton Baptist Church; Research on Truman H. Aldrich, Blocton numbered mines, and Blocton beehive coke ovens.

Quinn, Alice. Letter to Willis Pope, 1943.

Red Men, Order of. Nebe Tribe Bylaws. West Blocton.

Riley, Henry. Letter to Walter Riley, 1945.

Theatorium handbill: "The Ten Commandments," silent movie.

Smith, William. Letter, 1996. Indian artifacts.

Swindall, J. B. Letters relating to president's plane, Air Force One, 1998.

The Coleman Company, Inc. 1987 Annual Report.

Wadsworth, J. B. Letter to his wife, 1899.

West Blocton Methodist Church Register, 1885 and other church records.

Young and Company's Directory, 1905.

Oral History

Adams, Harry J. Organized labor, early WBHS football.

Battle, Nelle Thrasher. Depression mine working conditions.

Battle, Stephen M. Mining and UMWA history.

Bodner, Steve. 1927 town fire.

Brand, Regina Parker. African-American local history.

Burton, George R. Church and family history.

Caffee, Dorothy Adams. 1927 town fire.

Carroll, Will H. sports, family history.

Collins, Bill. Lucille prison fire.

Crowe, J. Turner. 1927 town fire, railroad history.

Ellison, Rhoda C. Bibb County High School location.

Fancher, Herman Richard. POW World War II experience.

Desmond, Tulon E. Various local history items.

Deerman, Lewis. 1927 town fire, 1920s KKK.

Evans, Dave Jr. Coal mining.

Finnen, Arthur. Early waterworks, UMWA, Wesleyan Methodist Church.

Gregg, Goldie Thrasher. Town fire.

Harris, Joy Parker. President Kennedy's assassination.

Israel, Dave. 1927 town fire.

Israel, Oscar. Transcription of interview by son, Bill Israel, 1985.

Jennett, Ben W. Pacific War Zone World War II.

Jones, Myrtle Cole. School integration.

Lawrence, Jack. Town fire, mining history.

Lovejoy, Frances. County government.

Murphy, William J. Various local history items, J. H. Blake.

Neal, George M. TCI Schools.

Parker, L. P. Fire department history.

Parker, Willie Jo. West Blocton African-American community.

Patridge, E. L. Bibb County banking history.

Perry, Edward. Lucile prison fire.

Pilegge, Joseph. Municipal government.

Pope, Macon. 1927 town fire.

Stewart, Dorman. N. A. Stewart biography.

Sumner, James. Mt. Carmel Cemetery expansion.

West, Gene T. Lucile prison.

Wright, William R., Jr. Electric service start and 1927 town fire.

Young, George H. West Blocton Savings Bank.

Zeanah, Helen Lawley. 1940s, 1950s, WBHS administration.

INDEX

Abbott, Forney, 36
Abston, A., 58
Abston, D. A., 40, 57–58
Acker, Terry, 297
Adams, Dorothy, 178, 233
Adams, Elder, 57
Adams, Harry, 117, 139, 151, 161, 177
Adams, Harry James, 12, 86, 150, 162, 202, 232–234
Adams, Harry J. "Jimmy", 296
Adams, Jim, 117
Adams, John H., 32
Adams, Mrs. Harry, 251
Ada's Chapel Cemetery, 236
Adger, 66
Alabama–Blocton Coal Mining Company, 155
Alabama Entertainers, 198
Alabama Gas Corporation, 253, 273
Alabama Great Southern Railroad, 9–10, 12–13
Alabama Historic Ironworks Commission, 14, 312
Alabama Power Company, 132, 148, 159, 240, 251, 260, 266, 285
Alabama Wildflower Society, 305
Albano Drugs, 282
Albano, Jerry, 291
Aldrich, 33
Aldrich, Georgia, 31
Aldrich, Herbert, 31
Aldrich, Marie, 31
Aldrich, Martha, 84
Aldrich, Mrs. Truman H., 55
Aldrich, Truman H., 1–2, 5–11, 14, 20, 24, 26, 31–33, 54, 64, 282, 303
Aldrich, Truman, Parkway, 32, 134
Aldrich, William, 33
Alexander, Dr. J. F., 97, 106, 127, 133
Alexander, Dr. Mark, 286
Alexander Family, 59
Alexander, J. F., 122
Alexander, W. T., 60, 192
Allen, Adell, 207
Allen, Billy, 250
Allen, Bob, 258, 260
Allen, C. E., 98, 119, 187
Allen, Charles, 311
Allen Furniture Company, 145, 149, 158, 162, 176, 183, 184, 219
Allen, Mary Curd, 26, 158
Allen, Mel, 62, 201, 239
Allen, Mrs. C. E., 56
Allen, Mrs. Hunter, 84
Allen, Patrick D., 135
Allen, S. S., 135
Allen, W. B., 14, 19
Allgood, Dr. H. W., 133, 136
Almon, Miss, 89
Alpha Academy, 274, 280, 284
Althoff, Rev., 288
Ambrose, Floyd, 247, 265
Ambrose (Floyd) Filling Station, 181
Ambrose (Floyd) Store, 270
Ambrose (W. L.) Store, 149
American Legion Hut, 145, 302
American Legion Post, 147, 207, 225, 230, 234, 254, 260, 280
Ammons Barber Shop, 148–149
Ammons, Bill, 148
Ammons, James, 40
Ammons, James Thomas, 148
Andress, Eli, 136
Ann's Flower Shop, 239
Archer, George, 137
Ardella, 263
Arello, Guy, 312
Arih, Joe, 263
Arih, Mrs. Joe, 262
Armes, Ethel, 7, 13, 20, 31
Arnold, Howard, 277
Artistic Plastics, 281, 297
Ashworth, A. A., 213
Ashworth, Savanna, 51
Ashworth's Crossing, 6, 50, 165
Ashworth, Virgis, 262–263
Atcheson, James, 296
Auten, Leonard, 54
Auton, John, 33
Auton, Len, 36
Ayers, Ernest, 237
Bacon, D. H., 90–91
Baer, Samuel, 61
Bailey, Boyd, 274
Bailey, D., 58
Bailey, Mary Elizabeth, 274
Bainbridge, G. B., 118
Bainbridge, George, 36
Balasco, Janie M., 50
Balch, J. T., 166, 192
Baldwin Locomotive Works, 9
Ball, Edward, 134
Ball Family, 56
Bama Skippers, 210
Bank of West Blocton, 242, 268, 270, 281, 284–285, 287
Name Change, 298
Robbery, 281, 287–288
Banks, H. B., 153
Banks, J. R., 147
Banks, T., 49
Barclay, J., 33
Barker, Major, 9
Barringer, Alice, 58
Barringer, J. E., 58
Barrus, Lucy Ann, 8
Barton, Dr. Stephen, 287
Basenberg, Frank C., 45
Bass, Griff, 78
Bates, Mrs. Nora, 152
Battle, Nell Thrasher, 203
Battle, Steve, 312
BEACON School, 291
Bearden, J. T., 85
Bearden, Pete, 247
Bearden, R. M., 189
Bearden, Rufus, 84
Bearden, Rupert M., 254
Beard, Rev. Thomas, 56
Bear Hall, 302
Beasley, Charles, 251
Beasley, Vera, 250
Beavers, Independent Order of, 119
Beckwith, Rev. C. M., 56
Belcher, A. D., 40, 61, 72, 82
Belcher, A. J., 134
Belcher, Brady, 302
Belcher, Dan, 84
Belcher, Marilyn, 63
Belcher (W. E.) Lumber Company, 242
Believers Quartet, 283, 284
Belle Ellen, 12, 21, 27, 99, 103, 124–125, 127, 137, 157, 164–165, 192, 206, 222, 244, 247–248, 251, 260, 266
Belle Ellen Prison, 125
Bell Top Saloon, 36, 76
Bennett, Jim, 310
Benny's Barber Shop, 270
Bently, James W., 234
Bergstresser, Dr. Jack, 312
Berlin, Abe, 143
Berlin Brothers Sundries, 149
Berlin's Store, 145
Bessemer, 145, 159, 176
Bevis, Clint M., 260, 269
Bevis, Ruth, 256
Bibb Baptist Association, 52
Bibb County Banking and Trust Company, 92, 219
Bibb County High School, 107, 153, 202
Football Team, 254
Bibb County Vocational School, 284
Bible Methodist Church, 62
Bice, Verna Harper, 51
Bicycle Club, 37
Biddleman, Josephine L., 145
Biggi, Oedo, 115
Big Springs, 78, 134, 268, 282

Birmingham Furnace & Mfg. Company, 15
Birmingham Mineral Railroad, 10
Birmingham Southern Railroad, 65
Bishop, E. M., 90, 96, 98, 117, 138, 140, 152
Black Brothers Bus Line, 163, 194
Black, Bush, 42
Black, Chester, 159
Black Diamond Coal Company, 59, 212–213, 219, 222, 230, 242, 245, 247, 252–253, 266, 278, 281, 288
Black Diamond Coal Company Cafe, 159, 213
Black Diamond Coal Company Store, 213, 239
Black Diamond Women's Store, 248
Black, Dr. W. F., 42
Black, Hugo LaFayette, 187
Black, Percy, 50
Blake, George R., 145
Blake, J. H., 58, 144, 192, 305
Blake's Mill, 132
Blake, T. A., 147
Blake, Truman Aldrich, 144, 305
Blake, Wayland, 285
Blanton, Cooper, 187, 189–190, 200
Bleidt, Nell Finnen, 230
Blocton Baptist Church, 57–58
Blocton Baseball Team, 33–36, 121
Blocton Cahaba Coal Company, 179
Blocton Coke Ovens, 14–16, 20, 22, 27, 287
Blocton Company Store, 22–23, 40
Blocton Depot, 68
Blocton Flood, 133–134
Blocton Girl's Basketball Team, 154
Blocton High School, 42–44
Blocton Home Band, 116–117, 133, 136–137, 139–140, 142, 143, 146, 160
Blocton Hotel, 40
Blocton Ice Company, 149, 163
Blocton Junction, 10
Blocton Methodist Church, 52–54, 59, 85, 121, 162, 164, 187, 203
New Building, 207
Blocton Mining Company, 206, 213
Blocton Newspapers
Blocton Courier, 39, 42
Blocton Enterprise, 39, 181, 187, 189, 197, 200
Blocton Intelligencer, 39
Blocton Messenger, 39
Blocton Times, 39
Labor Advocate, 39
Blocton Number 2 TCI Band, 37
Blocton Post Office, 5, 110–111
Blocton Presbyterian Church, 59, 86, 152, 199
Blocton Red Ash Coal Company, 191–192, 195, 197–198
Blocton Savings Bank, 70–72, 84, 87, 92, 119–120, 219
Blocton Speedball Team, 110
Blocton Superintendent's House, 32, 34
Blocton Vigilance Committee, 78
Blocton Waterworks, 118
Bloomer, Dr., 120
Bloomer Family, 56
Bloomer's Theatre, 149
Blue Goose Cafe, 186
Bluff Trestle, 10, 136
Bodnar Residence, 178
Bodnar, Steve, 177
Bogus Row, 59
Boland, Carla, 304
Boley, Pat, 60
Book Club, 119
Boom, Ed, 147
Boothton, 145, 244, 288
Borum's Jewelry Store, 172
Bowen, Charles Hugh, Jr., 243
Bowen, Lavada Lawley, 243
Boyd, A., 50
Boyd, Kathleen, 305
Bracknell, Albert, 19
Bracknell Family, 60
Bracknell, W. B., 42, 72
Bradford, Felix, 206
Braehead, 103, 124
Braehead Coal Company, 155
Braehead Mines, 99
Branch, O. W., 147
Brandon, W. W., 103–104, 152, 157, 187
Brannon, Elizabeth Jennett, 144, 181, 223
Brennan Hotel, 152
Brent, 225, 284
Brent Banking Company, 271
Brewer, Gov. Albert, 276
Brewer, Roselle, 222
Brewer, Tucker, 239
Brierfield, Blocton & Birmingham Railroad, 10–11
Brierfield Furnaces, 2–3
Bright, Andrew, 127
Briles, Birt, 115
Broadhead, Tod, 22
Broadhead, W. O., 49, 118
Brookwood, 212
Brown, A. J., 153, 160
Brown, Albert, 139
Brown, Billie Ruth, 248
Brown, Calvin, 296
Brown, Cora, 47
Brown, Ed, 287
Brown, Kenny, 296
Brown, Mable, 207
Brown, Marion, 168
Brown Residence, 177
Brown, Rev. W. W., 207
Brown's Drug Store, 163
Brown, W. D., 261
Brown, Willie Mae, 207
Bryant, Coach Paul "Bear", 256, 260, 298–300, 302
Bryin, R., 33
Bryles, A., 33
Bryson, Andrew, 12, 69
Buck Town, 23, 58, 84
Bucktown Methodist Church, 201, 206
Bumley, George, 50
Burgess Mining, 287–288
Burleson, Mildred, 302, 305
Burmah, 51
Burmee, 51
Burnap Family, 56
Burns, Annie (Creel), 51
Burns Clothing, 270
Burns, Kenneth, 277
Burns, Ronald, 258
Burrows, J. G., 200, 204
Burrows, Louise Blake, 204
Burt, George, 230
Burt, Kyle, 306
Burt, Mr. & Mrs. Mallory, 292
Burt, Owanna, 292
Burt, Robie West, 306
Burton, Bob, 19
Burton, George Campbell, 212
Burton, George R., 212, 306
Burton, Harry, 211
Burton, Mrs. Percy, 60
Bush, Margie, 302
Caddell, Festus, 76–77
Caddell, Mamie, 76
Cadle, Cornelius, 1–2, 5–9, 13–14, 24, 31, 33, 66
Cadle School, 8, 46
Cadle's Silver Cornet Band, 8, 31, 33, 35–37, 104, 115–116
Caffee Cemetery, 1, 47
Caffee Creek, 9–10, 15, 26, 84, 134, 176, 218
Caffee, Dorothy Adams, 86
Caffee, Mrs. C. C., 136
Caffee, S. R., 72
Cahaba Coal Mining Company, 1–2, 4–5, 8, 12–14, 19, 23, 24–27, 30–31, 33, 39, 54, 67, 85–87, 127, 303
Cahaba Lilies, 249–250, 304–305
Cahaba Lily Activity Center, 4, 290, 309–310

Cahaba Lily Festival, 305, 309
Cahaba River, 7, 10–11, 161, 218, 249
Cahaba River Authority, 302
Cahaba River Drive, 94
Cahaba River Trestle, 11
Cahaba Trace, 14
Cahaba Trace Commission, 310, 312
Cahaba Train Wreck, 11–12, 316
Cain, Jack, 278, 280
Caldwell, Gaz, 22
Caldwell, Henry, 179
Caldwell, H. G., 58
Caldwell, Mary, 179
Caldwell, Norma, 248
Caldwell's Field, 194
Cale, Lewis, 282
Callahan, Elias, 7
Callahan, Wayne, 282
Callen, S. L., 43
Calloway, Athlete, 207
Campbell, Adewin, 85
Campbell, Charles, 84–85
Campbell, C. M., 113
Campbell, J. N., 84–85
Campbell, Rev. Thomas F., 60
Campbell, Ruby Virginia, 85
Campbell Town, 43, 60, 75, 84
Camp, Tharon, 266
Cannon, Rev., 59
Canterbury, Willie Oglesby, 304
Carlton, Frank E., 78
Carlton, Walter, 75
Carlton, W. W., 52
Carmichael, L. L., 33
Carpenter, Bishop Charles C. J., 57
Carr, Juanita, 241
Carr, Mack, 263
Carroll, Bill, 165, 206, 219, 234
Carroll, Boots, 208, 277
Carroll (Boots) Service Station, 270
Carroll, Charles, Jr., 255, 257
Carroll, Emmett, 208, 277
Carroll, J. Leonard, 122, 208, 240, 277
Carroll, R. G., 261
Carroll, Robbie, 296
Carroll, Will H., 50, 314
Carter, Jerry, 255–257
Carver, R. D., 53
Cass & Daugette Construction Company, 206
Cassidy Family, 60
Cassidy, Rev. John, 54
Cass Toy Factory, 251
Castagnoli, Aden, 45–46
Castella, E., 130
Castelli, Elizabeth, 79–80
Cather, James Pat, 77
Catlin, L. C., 59
Cedar Grove, 246, 282
Centreville, 83, 95, 97–99, 101–102, 105, 107, 117, 123, 132, 137, 225, 251
Centreville Press, 200
Chambers, Alberta, 153
Champion, Barney, 286
Champion, B. E., 237
Champion, Lorraine, 286
Charlie's Shop Shop, 270
Chase, W. A., 33
Chicago Bridge & Iron Works, 210
Children of the Confederacy, 161
Cinelli, Mrs. Frank, 265
Circus, John Robinson's, 38
City Restaurant, 148
City Shoe Shop, 181
Civil Works Administration, 204–205
Civitan Club, 251, 253, 266
Clarke, W. L., 41
Clark, Mr., 13
Clark, W. L., 41
Clements Family, 56
Clements, J. D., 162
Clements, Jerry, 175, 178, 179, 184
Clements, K., 42
Clements' Store, 181
Clerks' Union, 213
Cleveland, Allen, 113
Cleveland, C. L., 109
Cleveland, Matthew, 50
Clifton, Joe, 137
Clifton Undertakers, 205
Clinner, Herman, 255, 277
Clinner, Sylvia, 250
Coalburg Coal & Coke Company, 64
Cobern, Bill, 258
Cobern, Wayne, 258
Coca–Cola Bottling Works, 121, 140, 149, 163, 179, 200, 237, 262, 270
Demolished, 281
Coke Ovens Park, 14, 16, 310
Coketon, 124
Coleanor, 50, 95, 103, 145, 165, 192, 206, 211, 219
Cole, C. H., 98
Coleman, Anna, 50
Coleman, Scott, 147
Coleman, W. C., 87–88
Colier, Forest, 22
Colleen's Flowers, 298
Collier, C. C., 41
Collins, Bill, 127
Collins, Chester, 250, 254
Collins, Jane Greene, 113, 115, 180
Collins, Joseph T., 99, 104
Collins, William, 207
Comerford and Casey, 42
Comer, Gov. B. B., 98–99, 101, 103, 105, 126
Come, Tom, 115
Confederados, 113
Conley, J. L., 66
Continental Club, 278
Contri, Rock, 116–117, 187
Convict Labor, 124–125, 127
Cook, Anna (Barns), 51
Cook, Don, 296
Cook, Martin, 51
Cook, Nealey, 51
Cook, Shirley, 51
Cooper, Howard, 179
Cooper, W. H., 158
Corey, W. E., 101
Coston & Sons, 210
Cottingham & Company, 67
Cottingham and Suttle Store, 170
Cottingham, Arthur, 296
Cottingham, E. N., 41, 85
Cottingham, Luther, 296
Cox, Bill, 236
Cox, Donise, 63
Cox, W. A., 236
Cox, W. D., 40
Crabtree, Dr., 243, 247
Crawford, A. H., 75
Crawford, D. W., 126
Crawford, George, 101, 107, 191
Crawford, J. Campton, 58
Crawford, Martha, 58
Crawford, Morris, 287
Crawford, Rufa, 58
Crawford, Sheriff, 80
Crawford, W. T., 137
Creel, Lela Mae (Mills), 51
Crider, Herman, 269
Crimes, Rodney, 296
Crim, H. F., 137
Crocker, Tim, 306
Cross, Charley, 28
Crowder, Carol Zoe, 248
Crowder, Dinah, 270
Crowder, Dr. John, 247–248
Crowder, John, 193, 248, 256
Crowder, Wendell Wade, 248
Crowe, J. Turner, 187, 217, 242
Crowe Residence, 177
Crowe, Virginia, 177
Crownsmen, 255, 257
Culture Club, 145
Cummins, Charles, 36
Cundiff, Buleah, 204
Cundiff, Mit, 49, 84, 137
Cunningham, George, 219
Cunningham, Herschel, 219
Cunningham, Robert, 290
Curtis, Dr. Joseph, 86
Curtis, J. F., 113
Curtis, Mrs. Dr., 117
Cutts, Ashantis, 306
Cutts, Ruth Johnson, 46
Cutts, S. C., 59
Dagotown, 90, 128
Dailey, Bama (Ashworth), 51
Dailey, G. E., 247, 261
Dailey, George E., 227
Dailey, James, 237

Dailey, Jewel Jones, 166, 206
Dailey, Malcolm, 211
Dalrymple, William, 203
Daniel, M. J., 7
Daniel, Mrs. Mamie E., 153
Daniel, S. G., 7
Davenport, J. L., 49
Davenport, Mrs. A. W., 226
Davidson, A., 50
Davie, Bill, 121
Davie Building, 187
Davie, Dr. N. T., 133, 136, 143
Davie, J. B., 42, 58, 82, 85, 95, 98, 102–103, 106, 123–124, 130, 183, 253
Davie, J. W. "Bill", 103, 254
Davie, Marshall, 193
Davie, Marshall, Jr., 212, 253, 286
Davie, Mary H., 205, 253–254
Davie, M. C., 94, 117, 119, 122, 145, 151–153, 164, 176, 179, 182, 190, 193, 197, 214, 225, 253
Davie, Mrs. J. B., 103
Davie Residence, 205
Davie, Robert, 204, 211
Davie's Store, 120, 121, 145, 149, 174, 176, 179, 181, 186, 188, 200, 240, 248, 253, 270
Davis, Edward, 36
Davis, Fred H., 262, 264, 281
Davis, Steve, 258
Dean and King, 42
Deason, Billie, 308
DeBardeleben, Henry F., 7, 9, 11, 64
Debs, Eugene V., 70
DeCraft, Don, 281
Deerman, George, 194
Deerman, Jerry Claude, 301–302
Deerman, Jo, 301
Deerman, Lewis, 175, 179, 211
Deerman, L. H., 193
Deerman, Patricia Clinner, 54, 145, 255
Deerman, Paul, 306
Deerman, Vic, 305, 309
Deerman, W. D., 247, 254, 261, 269
Deerman, Willard (Owens), 51
DeGraffenreid, Ryan, 263, 305
Deloney, Supterintendent, 101, 171
Del Vecchio, Dr. P. A., 291
Dennis, Benjamin, 56
Dennis, Mark, 41
Dennis, Robert, 237
Dent, O. Z., 99
Desmond, B. B., 177
Desmond, Clara Logan, 269
Desmond, George, 137
Desmond, George O., 273, 282
Desmond, Mary Logan, 206
Desmond, Tulon, 15, 143–144, 147, 194, 197, 210–211, 218, 219–220, 225
Desmond, Tulon E., 15, 144
Dexter, Benjie, 122
Dexter's Livery Stable, 23
Dickey, William, 43
Dickinson, W. D., 101
Dinkins, Claude, 40
Dinkins, W. B., 137
Disciples of Christ, 60
Dison, T. T., 184, 192–194
Dodge, Lillian, 45
Donaldson, Edward, 52
Donaldson, Isabella, 52
Donaldson, Mary J., 52
Dooley, Emma, 50
Dooley, I. S., 50, 59, 147, 153
Dorsey, Captain, 136
Douthit Family, 55
Douthit House Hotel, 42, 86–87, 106, 140, 148, 157, 172, 174, 178, 181
Douthit, Maude, 172, 183, 198
Douthit, W. H., 26, 42, 71, 86, 87
Downing, F. M., 134
Downing, Frank, 118
Downs, Eddie G., 268, 270–271, 278–279, 284, 286
Downs, Rev. James S., 55
Driver, W. B., 146
Dudley, 9
Dudley, Eleanor, 249
Duncan, Anna Blake, 50
Duncan, Annie, 58
Duncan Family, 37
Duncan, J., 33
Duncan, J. M., 189
Duncan, Louise, 206
Dunlap Family, 55
Dunnabie Residence, 148, 193
Dunnabie's Store, 145, 147–148, 181
Dunn, George, 58
Dunn, Pearl, 133
Durette, Emmett, 60, 165
Durette Family, 59
Dyer, Berl, 129
Dyer, Carl, 51
Dyer, Harry Cecil, 229
Dyer, Myrtle, 51
Dyer's Blacksmith Shop, 149
Dyer, Vernon, 51
Eastern Star, Order of, 50, 272–273
Echols, Rev. Robert, 92
Edmonds, Ben L., 119, 133, 135, 174, 193
Edmonds, Ben L., Jr., 204
Edmonds, B. J., 19
Edmonds, DeWitt, 19
Edmonds, Dr. Eugene, 149
Edmonds Dry Cleaners, 148, 181, 183, 186–187
Edmonds Family, 60
Edwards Furnace, 2
Edwards, Giles, 13
Edwards, Lydia, 13
Elam, V. D., 76–78
Elite Cafe, 181
Elks, Benevolent & Protective Order of, 49, 85, 116, 118, 119, 136, 140, 146–147, 149, 157, 164, 183, 205, 207, 210, 249, 251, 262, 268, 274–275
Elks Lodge Ritual Team, 205, 210
Elliot, Lottie, 122
Ellis, Elmo Israel, 46, 61–62, 150, 178, 183, 205, 208, 211, 218–219, 250, 283, 299
Ellis, George, 137
Ellis, George E., 202
Ellison & Thompson, 77
Ellison, Charles, 37
Ellison, Dr. Rhoda Coleman, 1, 3, 5–7, 29, 38, 47, 61, 77–78, 93, 101, 124, 128, 137, 145, 303
Ellison, John T., 101
Ellison, T. B., 33
Elloff, David, 62
Elloff, Mildred, 62
Elrod, C. K., 239
Emblem Club, 251
Emos, Dan, 201
Ensley, 197
Eoline, 11, 25, 88
Erwin, Lt., 68
Eubanks, Charles, 162, 163
Eubanks, Herman, 162
Eubanks, J. H., 153, 163
Eureka Company, 15
Evans, Charles, 277
Evans, Dave, 191, 222, 231, 240, 242
Evans, Dave, Sr., 191
Evans Mine, 155
Evans, William, 155, 191
Everleigh, Charles, 66
Everse, Martin, 16
Everse, Sandra McCool, 131, 211, 269, 307, 314
Excelsior Coal Company, 30
Export Coal Company, 20
Faggard, D. J., 41
Faith Baptist Church, 55, 285, 310
Fancher, Aaron, 237
Fancher, Herman R., 221, 231–232
Fancher, John W., 228
Fancher, John W., Jr., 228
Fancher, J. W., 213
Fancher, Margaret Parker, 133
Fancher, Martha, 241
Fancher, Milton, 218, 228
Farlee, C. T., 58
Farley, Fannie, 58

Farmer, Lewis, 231
Farmer, Lewis, Jr., 230
Farnetti, Frankie, 226
Farrah, H. H., 19
Farrah, Louie, 19
Farrar, Annie Merle, 108
Farrar, Dr. Joe Morgan, 61
Farrar Family, 60
Farrar, Henry H., 197
Farrar, H. H., 218
Farrar, James, 72
Farrar, Jean Israel, 61, 160, 285, 299
Farrar, Joe, 280, 285, 292
Farrar, Morgan, 285
Farrar, Roselyn, 228
Faulk, Billie, 70
Faulk General Store, 181
Faulk, I. M., 187
Faulk (I. M.) Novelties, 149
Fayette, Henry, 194
Ferrire Family, 260
Ferrire, Frank, 55, 206, 212, 226, 247, 253, 256, 268–273, 276, 278, 280–282, 284, 286, 288, 290–291, 293, 296–297, 302, 304–305, 307
Ferrire Store, 246
Fiddlers' Convention, 152
Findley, Fayette, 37
Findley, Robbie, 308
Findley, Walker (W. W.), 268, 270
Finnen and Foster Grocery Store, 249
Finnen and Jones Grocery, 270
Finnen, Arthur, 60, 93, 129, 203, 311
Finnen, Billy, 283–284
Finnen, Billy Roy, 255, 257, 283–284
Finnen, Bobbie Jo, 63
Finnen, Brian, 283–284
Finnen, Darrell, 36, 267
Finnen, Ed, 247
Finnen Filling Station, 207
Finnen, Henry Thomas, 230
Finnen, James Harold "Bozo", 220, 225
Finnen, John, 54, 60, 122, 205
Finnen Meat Market, 149
Finnen, Nell, 220, 239–241
Finnen, Pat, 19, 36, 76
Finnen, Tom, 36, 220
Finnen, W. E., 261
Finnen, Will, 60, 194
Finnen, William, 203
First Alabama Cavalry, 136
First State Bank of Bibb County, 298, 307
Fisher, Louie, 201
Fitts Brothers, 194
Fitts, James R., 277
Fitts, Lester, 246
Fitts, Libby, 207
Fitts, Mary Elizabeth, 246–247
Fleming, George, 294, 314
Flinn, J. H., 42
Florine, Phillip, 120
Flowers, Walter, 287
Floyd, James P., 204
Folsom, Gov. Jim, 239
Foman, Hector, 51
Forbes Piano Company, 181
Forman, Sarah, 51
Forty–fourth Alabama Infantry, 87
Foster, Victor Y., 247, 249
Fountain, Eugene, 224
Fountain, H. E., 234
Fountain, J. C., 235
Fountain, Lola, 234
Fountain, Victor Donnis, 234, 235, 236
Fournier, Frank, 67
Four Points, 97
Francis, Rev. J. C., 142
Franklin, J. M., 90, 91
Franks, Clarence, 237, 266–267, 277, 279
Franks, Enos, 237, 250
Franks, Howard, 311
Franks, Joyce, 310
Franks, Loiel, 228
Franks, Robert, 266
Frazier, Nat, 153
Frederick, Beth Fitts, 247
Frederick, Brown, 122, 145
Frederick (Brown) Pool Hall, 148
Frederick, G. C., 60
Frederick, Harmon W., 35, 186, 204, 236, 275
Frederick, Huey, 189, 215–216, 255, 262
Frederick, Lawrence, 231–232
Frederick, L. B., 277
Frederick Motor Company, 184, 189, 190, 207, 253
Frederick Motor Company Building, 207
Frederick, Roy, 200
Frederick's Chevrolet, 161
Freedman's Bureau, 8–9
Freeman, G. W., 52
Freeman, M. D., 33
Free Will Baptist Church, 60, 204
French, Preacher, 60
Friendship Baptist Church, 59
Frog Level, 47, 94, 138, 273, 282
Frost, Jarome, 51
Frost, Mary Ellen, 51
Fuller, Charles, 36
Fuller, Charley, 19
Fuller, Dr. H. M., 225
Fuller, Nelson C., 126
Fuller Residence, 199
Fuller, Will, 161
Fulton, Clay, 22
Fulton, John, 15
Fuqua, Dr. E. M., 127
Gafford, Frank, 103
Gafnea, Susan Riley, 194, 201
Gallagher, Dan, 137, 143
Gallagher Family, 137
Gallagher, Neil, 143
Gallent, James, 237
Gardeugli, Liugi, 115
Gardner, James, 12
Gardner, Lilly, 76–77
Garner, Uncle Tom, 133
Garnsey, 95, 103, 145
Garvich, Miss, 152
Gatlin, B. H., 147
Gatlin Family, 60
Gentry, E. H., 237, 304
Gentry Hardware, 149
Gentry, Margaret, 260
George, Jack, 219
George, John, 22
George, Louis P., 29
George, Monroe, 22
George, Roy, 285
George, Sammy, 219
Georgia, Mobile & Ohio Railroad, 11, 25, 196
Gholson, John, 21
Gibbons, Rose, 54
Gifford, Frank, 106
Gilbert, Albert, 15
Gilbert, Ballard, 269, 277
Gilbert Family, 60
Gilbert, John, 40, 60, 93, 123
Gilbert, Martha, 305
Giles, Ronnie, 237
Gillespie/Collins Funeral Home, 15
Gillespie, Ed, 210, 214, 247
Gillespie, E. G., 192–193
Gillespie Funeral Home, 205
Gillespie Furniture Store, 163, 175, 181, 183, 186
Gillespie, James, 137
Gillespie, Mary Etta (Mrs. Ed), 210, 214, 241
Gillespie, Matthew Leo, 137
Gillespie Residence, 214
Gillespie Springs, 93–94, 273
Ginnett, Pat, 98
Gleaton, Clinton, 51
Gleaton, Ora Lee Yeager, 51
Gleaton, Robert, 237
Glover Branch, 162, 190
Glover, J. E., 42, 58
Glover Spring Branch Bridge, 149
Goggins, J. C., 231
Goggins, Randall, 282, 290
Golson, J. F., 137, 195
Golson, John, 135, 249
Golson, John F., 90
Golson, J. R., 98
Golston, John F., 210
Good Roads Association, 123
Goodwin, Louis Charles, 47
Goodwin, L. W., 50
Goodwin, Rucker, 200
Gordon, Tom, 36

Gorgas, Josiah, 3
Govie, Sam, 237, 287
Grady, Benje, 306
Grammer, A. J., 51
Graves, Gov. Bibb, 124
Gray, Melton H., 146
Gray, Ross F., 234–235, 290
Graziono, Geuseppi, 29
Green, A. A., 107
Green, Cooper, 260, 266
Greene Family, 56
Greene, Jim, 21–22
Green, Evelyn, 133
Green, John, 137
Green, Mariana, 135, 138
Greenpond Presbyterian Church, 59–60, 199
Green Shoe Shop, 181
Green, Virginia, 13
Gresham, 1
Gresham, Gen. Walter Quinton, 1, 2
Gressham, Mrs. E. H., 1
Griffin and Company, 122
Griffin, Carolyn, 286
Griffin, Dr. W. W., 148
Griffin, Henry, 52
Griffin, Mrs. W. T., 144
Griffin, Olera, 152
Griffin, Ovelle, 152
Griffin, Roy, 144
Griffin, W. T., 134, 137, 141, 145
Grissett, Virginia Dyer, 229
Grote, John, 7
Guarantee Cash Grocery Company, 155, 183
Gunlock Hill, 21, 54–55, 246
Gunnels, H. C., 103–105
Gurnee, 10, 30–31, 56
Gurnee Junction, 145
Gurnee, W. S., 7
Habib, Dr. Roshdy, 293
Hall, Floria Lawrence, 51
Hallman, Vickie, 302
Hall, Travis, 255, 258
Hally, Frank (Holley), 39
Hamilton, Robert, 134
Handley, A. E., 89
Hardemon, Frank, 63, 266
Hardemon, James, 236
Hardemon, Shirley, 63
Hargrove, 9, 31, 103, 137, 139
Hargrove, Bob, 133
Hargrove, Chester, 227
Harkey, Tula, 43
Harkin, John, 66
Harkins, Frances Jennett, 314
Harkins, Lucille, 153
Harmon, Daniel, 136
Harmon, Jack, 60, 251, 282
Harmon, John, 88
Harmon, Judy, 260
Harmony Baptist Association, 52
Harper, Estelle, 51
Harper, W. H., 261
Harris, Billy, 217, 219
Harris, Helen (Leonard), 31
Harris, H. H., 138
Harris, Lillian Cinelli, 265
Harris, Munroe, 126
Harris, Rev. Jerry, 60
Harris, Trion, 43
Harris, T. T., 22, 41, 54
Harris, W. D., 213
Harris, W. D. "Billy", 222
Hartley, Billy, 263
Hartley, Boyd, 263
Hartley, Columbus, 81
Hartley, Margaret Lane Edmonds, 115, 240, 263
Hartley, Mrs. Boyd, 251, 262, 270
Hartley, Sam, 43
Hartley, Samuel, 49
Harvey's Drug Store, 78, 120, 139
Harvey, W. A., 82, 117, 135
Harvey, William A., 27
Harvey–Wright Drug Store, 148
Hawken, H. C., 138
Hayden, Rev., 59
Hayes, Claude, 247
Hayes, Dovie, 42
Hayes Hamburgers (Uncle Mel), 148, 149
Hayes, James Newton Erasmus, 269
Hayes, L. J., 98, 137, 152
Hayes, Mel, 75, 161
Hayes, M. Y., 54, 72
Hayes, Robert, 100
Hayes, Ronnie, 250
Hayes, Wallace, 237
Hayes, W. Y., 202
Haynes, Charles B. (Charlie), 247, 251, 254, 296
Hays, L. J., 82, 117
Head, Gertrude, 204
Heard, Bernice, 207
Heard, Effie, 46
Hebron, 134, 282, 291, 305
Heflin, Howell, 307
Heikel, Adam, 135
Helena, 11–12, 266
Henderson, Elaine Stewart, 262
Henderson Family, 21
Henderson, Fred, 194
Henderson, Frederick, 255
Henderson, Freeman, 51
Henderson, Henry, 197
Henderson, Preacher, 129
Henderson, Reanie, 51
Henderson, Rev. Richard, 51
Henderson, Rev. Sam, 51
Henderson, S. A., 52
Henry Ellen, 9
Herring, George, 245
Herron's Garage, 174
Hewlett, Rev. Roy, 153
Hickman, Faye Pope, 297, 303, 307–308
Hickory Hill Cemetery, 113, 231
Hicks, Bobby, 258
Hicks, David, 255–256, 258
Hicks, Emma, 310
Hicks, Gene, 255
Hicks, Jack, 258, 286
Hicks, Mrs. Edith, 287
Higginbotham, Billie, 257
Higginbotham Field, 259
Higginbotham, Morris, 252, 255–257, 259–260, 285
Higginbotham, Oscar, 237
Hill Creek, 94, 134, 224
Hill Creek Ball Park, 166, 203, 207
Hill Creek Commissary, 149
Hill Creek Miners' Union, 207
Hill Creek Mine, 127–129, 139, 155, 164, 192, 197, 203, 213, 222, 231, 239, 242, 244
Hill Creek Mining Company, 128, 161, 175, 179, 187, 206
Hill Creek Park, 93
Hilltop Needle Club, 135
Hill, Will, 50
Hirsh, Capt. Edouard, 141
Hobdy, J. B., 108
Hoggle, Renee, 306
Hoit, James, 219–220
Holdridge, W., 43
Holdrige, Prof. W., 42
Holifield, Clara Sue, 213
Holifield, Mrs., 21
Holland, Bail, 22
Holland, Carrie, 231
Holland, Charity, 244
Holland, Claude, 222–223, 236
Holland, Donald, 258
Holland, Neil, 231
Holland, Robert, 41
Holland, Will, 231
Hollifield, Watson, 22
Hollis, Rev. Samuel, 58
Holsamback, Alice, 62
Holsombeck, Roy, 255–256, 258
Holt, Barbara, 306
Hooter, Joseph, 66
Hope Hill Baptist Church, 58
Hornsby, C. E., 279
Horse Creek, 78
Horton, Leo, 250
Hosmer, J. V., 75–76
Hosmer, Lester, 204
Hosmer, Thomas Wayne, 279
Houton, Lewis, 62
Houton, Marie, 62
Howard, Claude, 108
Howard, J. J., 54
Howell and Jones, 128
Howell Coal Company, 99
Howell, Willard, 99
Hubbard, Charles, 134
Hubbard, Dr. Ben, 259, 261
Hubbard, Mike, 296

Hubbard, Rufe, 82
Hubbard, W. G., 52
Hubbard, William, 255
Huckabee, C. C., 174
Hudson, Miss S., 83
Huey, Ben, 108
Huey, Mrs. Ben, 138
Huey's Cafe, 215, 270
Huey's Place, 211, 216
Hughes, Bettye Kitchen, 140, 208, 226
Hughes Family, 56
Hullett, Rev. Rudolphus Strider, 52
Hunnicut, N. W., 219
Hutchins, Claudie M., 234–235
Hutchins, Delana, 306
Hutchins, Preston, 237
Hyde, Sue Pickett, 241
I&N Quality Store, 110, 148, 160–161, 163, 169, 174, 178, 180–181, 183, 186, 199, 248, 270, 309
Influenza Epidemic, 141–142, 156
Ingram, James, 250
Irwin, Charles, 239
Israel, Annie, 61
Israel, Dave, 62, 175–176, 195, 210, 239, 297
Israel, Elmo (See Ellis, Elmo Israel)
Israel Family, 62
Israel, Frank, 178, 211
Israel, Frank S., 205, 210
Israel, Joe, 152, 163, 187, 200
Israel (Joe) Department Store, 148, 213
Israel (Joe) Dry Goods, 181
Israel (Joe) Residence, 149
Israel, Julius, 62
Israel, Libby Jean, 178, 198
Israel, Melvin, 62, 201
Israel, Oscar, 72, 123
Israel, Samuel B., 46, 61–62, 71, 152, 163, 176, 178–179, 189, 193–194, 199–201, 210, 240–241, 273
Israel, William, 123
Israel, William "Wolf", 46, 61–62, 71, 83
Italian Band, 142
Italian Catholic Cemetery, 55, 79, 310
Italian Hall, 116
Ivey, Bob, 200
Ivey, E. J., 58
Ivie, Jennie, 58
Jackson, A. S., 147
Jackson Clothing Company, 213
Jackson, Frank, 46
Jackson, Gonzlo, 118
Jackson, James, 285
Jackson, Monroe, 79–80
Jackson, P. S., 147
James, H. P., 264
Jeanette Fruits and Notions, 149
Jenkins, Dr. Ben H., 247, 250
Jenkins, Morgan, 49
Jennett, Ben, 143, 144
Jennett, Ben F., 137
Jennett, Ben, Jr., 223, 224
Jennett, Pat, 98, 137
Jernigan, Rev. Don, 302
Jewish War Relief Committee, 152
Johns, 139
Johns, Llewellyn, 12
Johnson, Alton, 291
Johnson, Crawford, Sr., 121
Johnson, D. A., 58
Johnson, Frances Ruth, 168
Johnson, Gail, 306
Johnson, J. A., 214, 236, 262
Johnson, Luther Elmer, 219
Johnson, S. L., 16–18, 20, 22, 25, 33, 38, 84, 123
Johntown, 123
Johntown Baptist Church, 218
Jones, Almon, 140
Jones, Annie Mae, 63, 292–293
Jones, B. F., 54
Jones, Bobby, 280
Jones Brothers Mine, 155
Jones, Carl, 292–293
Jones, Donnie, 285
Jones, Douglas, 237
Jones, Dr., 12
Jones, Dr. B., 49
Jones, Dr. C. W., 113, 149, 163–164, 166, 181, 186, 193, 197
Jones, Dr. Walter P., 32
Jones, Earl, 197
Jones, Edward H., 153
Jones, E. E., 118
Jones, E. P., 261, 266, 268–269, 277–278, 281, 286, 290, 296–298, 302–303
Jones, Ethel, 306
Jones Family, 56
Jones, Floyd, 210
Jones, Glen, 296
Jones, Gov. Thomas G., 67–69
Jones Hardware, 270
Jones, Herman, 122
Jones, H. H., 126
Jones IGA Store, 293
Jones, J. A., 84
Jones, James F., 72
Jones, Jesse, 145
Jones, Jewell, 206
Jones, J. F., 183
Jones, John C., 5
Jones (J. W.) Drug Store, 149
Jones, Kristi Deerman, 211
Jones, Lela, 84
Jones, Maurice, 240, 244, 247
Jones, Monette, 293
Jones, Mr. & Mrs. Roy, 290
Jones, Myrtle Cole, 274, 293, 303, 307, 309–310
Jones, Paul, 273
Jones Residence, 149
Jones, Roy, 258, 290, 295
Jones, Ruth Carr, 285
Jones, R. W., 42
Jones, Shianne, 290
Jones Timber Company, 273
Jones, Todd, 211
Jones, Tom, 293
Jones, W., 58
Jones, W. A., 15, 84, 192, 200, 237, 241, 262, 281
Jones, W. F., 58
Jordan, Henry "Cakeman", 276, 278
Justice, Fred, 259, 260
Kaars, William, 66
Kaler, Claire (Clara), 153
Kaler, Melvin (Morris), 153
Kaler, Zavel, 153
Katz, Nathan, 201
Keith Barber Shop, 181, 184, 186–187
Keith, Mims, 21
Keith, Miss Willard, 153
Keith, R. B., 117
Keith's Barber Shop, 148
Kelley, John, 263
Kelley, Martha Fancher, 158, 221, 228, 230, 232, 237, 241, 263, 268, 296–297
Kellog, Mrs., 50
Kemp, Lige, 17
Kennedy Assassination, 156–157, 267
Kennedy, Josiah, 260
Kennemer, J. R., 139, 151
Kerr, Diane, 104, 266
Keyes, Annie, 179
Keyes, Eunice, 256
Keyes, Lyda M., 282, 287–289, 293
Keyes, Mrs. Bill, 217
Kilby, Gov. Thomas, 151–152, 157
Kindig, Charles, 148
Kindig, Charles Clarence, 148
Kish, John, 237
Kitchen, A. N., 240, 269, 277–278, 282, 288–289, 308
Kitchen Cleaners, 270
Kizziah, Harvey, 263
Klondyke Mine, 99, 128, 155, 192, 197, 206, 213, 220, 222
Klotzman Building, 213
Klotzman, Dave, 210
Klotzman Dry Goods, 148
Klotzman Family, 62, 200
Klotzman (Joe) Dry Goods, 181
Knights and Ladies of Honor, 85
Knights of Labor, 27, 54, 66, 70

Knights of Pythias, 49–50, 85, 119, 132, 198
Knights of Pythias Hall, 146, 149, 155, 161, 195, 207, 261
Knights Templar, 119
Kornegay, Frances, 307
Kren, Frances, 228
Kren, John, 183, 192, 249, 274
Kren (John) Grocery, 149, 181
Kren, Margaret, 206
Krenshaw, Mr., 173
Kren's Store, 183, 186
Kren Store, 193
Krentzman Family, 61
Krentzman, I. B., 103, 106, 119, 122
Krentzman, John B., 61
Krentzman, Jonathan, 82
Kronenberg, Israel, 61
Kuhns, Mamie Foster, 156, 202
Ku Klux Klan, 81, 146, 161, 240–241, 259
Lacey, Mamie B., 50
Ladies Needle & Social Club, 119
Lagrone, L. C., 117
Lagrone Market, 149
Lagrone, R. C., 72, 137, 230, 236
Lagrone, Rufus Cristo, 219
Laird, Winnie Lou, 153
Lamar, Dan, 127
Lamont, Joyce H., 77
Lamoreaux, Dr. Phillip, 1
Langdon, W. A., 26
Langston, Joab, 52
Langston, John M., 247, 254, 261, 266, 268, 278, 294, 296
Langston, Ramona, 241
Langton's Big Star Store, 270
Lang, W. S., 90–92
Lanier, M., 147
Latardi, Mr., 26
Latham, Sheriff, 67–68
Latham, Tom, 137
Lavender, W. W., 96
Lawley, Frank, 22
Lawley, Fred, 19
Lawley, Phelon, 237
Lawley, Rufus, 134
Lawley's Beauty Shop, 270
Lawley, Virgil, 278
Lawly, Olvis, 237
Lawrence, Annie Jo, 62
Lawrence, A. R., 230
Lawrence, Clifton, 237
Lawrence Family, 24, 59
Lawrence, Irene, 58, 249, 262
Lawrence, Jack, 24, 60, 179, 218
Lawrence, Margaret, 62
Lawrence, Raymond, 230
Lazanovick, Mipalo, 101
Lazzare, S., 116
League of Brothers Synagogue, 61, 130, 201
Lebourg, Albert, 122
Lee, E. D., 138
Lee, Mildred, 109
Lee, Mrs. E. D., 140
Lee, Rev. Dallas P., 109
Lees, Jim, 286
Lees, Mary Ann, 299
Lemley, Carolyn, 128
Lennigan, M. E., 33
Leonard, Francis, 227
Leonard, Johnny, 207, 208
Levin, L., 72
Lewis, D. M., 106, 118
Lewis, Dora, 148, 183
Lewis (Dora) Millinery, 181
Lewis, D. W., 103
Lewis, E., 33
Lewis, J. E. "Ed", 120
Lewis, John L., 238, 242–243
Lewis, Theo, 51
Liberty Baptist Church, 58–59, 286
Liddle, Alden, 258
Liddle, Billy, 250
Liddle, Bobby, 250
Light, Ester, 153
Light, Oscar, 163
Light (Oscar) Grocery, 149
Lightsey, Debbie, 63
Lightsey, Glenn, 314
Lightsey's Pond, 132
Liles, Fred, 212
Lily Shoals, 249
Lint, Charles, 36
Lions Club, 218, 272, 275, 280
Lippeatt, Billy, 295, 296
Lippeatt, Emery, 247, 263, 278, 293
Lippeatt, Floyd, 258
Lippeatt, Mary Marchant, 229
Little, A. B., 240
Little, Hop, 263
Little Italy, 21, 33, 47–48, 79, 111, 115, 130–131, 149, 161, 246, 265
Little Italy Band, 115–116
Little, Mrs. R., 12
Lloyd, Dave, 33
Lloyd, Dr. Stewart, 20
Logan, C. D., 124
Logan, W. H., 94, 98, 117, 137, 149
Logan, W. W., 174
Lolly, E. D., 58
Longhorn Trestle, 11
Louisville and Nashville Railroad, 10, 12, 68, 145
Louisville and Nashville Railroad Depot, 170
Lovejoy, Frances, 314
Lovejoy, Jimmy, 83
Lowery and Thomas, Bakers, 41
Lowery, J. B., 87
Lowery, J. G., 58
Lowerytown, 33, 156, 222, 224, 246, 305, 313
Lowrey, W. B., 87
Lucas, C. W., 206
Lucas, J. J., 58, 137
Lucas, Joe, 263
Lucas, Milton, 258
Lucas, Mrs. J. J., 119
Lucas, Wayne, 298
Lucile, 11, 25, 27, 97, 106, 124, 192, 282
Prison Tragedy, 125–127
Ludlow's Place, 159
Lutenberger, Mike, 122
Lyell, Sir Charles, 1, 32
Lyles, Fred, 206
Lynch, Edward "Punch", 227
Lynch, Fred, 227
Lynch, Samuel, 66
Macon, Bertha, 138
Mahaffey, Eddie, 210
Majors, Keith, 306
Majors, Robert, 6, 53, 121
Malouf, Dr. George, 243
Malvolti, Giacomo, 115
Manring Family, 60
Manring Hotel, 42, 87, 149, 174, 181
Manring, Mrs., 184
Marchant, Clara Jean, 273
Marchant, Dottie, 229
Marchant, Fred, 229
Marchant, Lloyd Palmer, 219
Mars, 123
Marsh, Walter, 258
Martin, Amos, 51
Martin, Annabelle Dunnabie, 147
Martin, Anna Dean Hutchins, 235
Martin, Charles, 237
Martin, Clarence, 197
Martin, Curtis, 51, 194
Martin, Dr. Charles, 106
Martin, Huell Lane Frederick, 189, 215
Martin, K., 59
Martin, Lucian "Dump", 15
Martin, L. W., 12
Martin, Martha, 51
Martin, Marvin M., 129, 149, 195
Martin, Mike, 276
Martin, Miranda, 276
Martin, Moodie, 18
Martin, Reed, Jr., 276
Martin, Reed, Sr., 15
Martin, Rev. C. B., 192
Martin, Stewart, 296
Martin, Thomas W., 126
Martin, W. B., 42, 54
Marvel, 122, 127, 137, 145, 192, 206, 244, 247, 251, 260
Masonic Lodge, 148, 155, 161, 174, 176, 179–181, 183, 187, 189, 213, 240, 260, 293

Masonic Order, 49–50
Mason, Roy R., 282–283
Massey, Rev. E. G., 59
Mattich, Tom, 115
Mattioli, Clelio, 115
Mattioli, Ernest, 115
Maxwell, J. W., 76
Maxwell, Mrs. Fred, 56
Mayday, Mary, 117
Mazer, Sidney, 153
McBride, H. G., 40
McBryde, P., 66
McCafferty, Father, 54
McCannon, Frank, 147
McCartney, Sam, 84
McCarty, Dr. William, 210
McCary, C. M., 126
McClellan, Cecil, 255
McClendon, Wesley, 69
McCool, Ben, 161
McCool, Billy Vic, 250
McCool, Emmett, 263
McCool, George, 92
McCool, Huey, 181
McCool, James Donald, 225
McCool, Margrette, 308
McCool, Marlon, 36
McCool, M. C., 98
McCool, Shirley, 248
McCool, T. A., 82, 193, 197, 203
McCool (T. A.) and Sons Grocery, 149
McCool, Tip, 211
McCoy, John, 126
McCulley, Elsie, 62
McCulley Family, 60
McCulley Hill, 21, 128, 282, 298
McCulley, Lester, 129
McCulley, Raymond W., 254
McCulley, R. W., 247
McCumber, Thomas, 39, 42, 170
McDonough, Joe P., 133
McElroy, Frank, 200
McElroy, James, 200
McElroy, Thomas, 22
McElroy, Walace, 236
McElroy, Widow, 22
McFarland, Mr., 126
McGinnis, W. D., 118
McGowen, Edith (Thomas), 32
McGrawtown, 81, 93, 128–129, 137, 205, 260
McGuire, Lee, 92
McGuire, P. F., 278
McIntosh, Louvella, 108
McIntosh Sisters Clothing, 270
McKenzie, Johnny, 296
McKenzie, Perry, 237
McKenzie, W. R., 274, 279
McKinney, A., 161
McKinney Auto Parts, 159, 290
McKinney, Julia, 159
McKinney, Milton, 277
McKinney, Raymond "Chink", 219
McKinney's Parts & Service Station, 270
McLauren, H. B., 241
McMaster, J. M., 5
McMasters, J. M., 98
McMath, Jerusia, 49
McMillan, Eddie, 288
McNair, J. R., 108
McNatt, S., 66
McQueen, James W., 12–13
McSpadden, Dealia Reach, 51
McSpadden, Joe Wheeler, 200
McSpadden, R. E., 211
Meadow, Dr. A. E., 12, 37, 54, 85, 113, 118
Meadow, Dr. J. M., 43, 49, 85
Meagher, J. F., 134
Medders, Mrs. T. G., 274
Medders, Thomas Lee, 258
Medlin, Tandy, 98
Melita, Boresea, 101
Melodeers, 210
Melsoni, Charles, 250
Melsoni, Franklin, 258
Melvin, Margaret, 108
Mercedes–Benz, 313
Merrill, Pelham J., 166–167, 178, 195, 198, 249, 250
Mertz, John, 54
Milam, Charles, 117
Mildred's Flowers & Clothing, 270
Miller, B. M., 126
Miller Gang, 78
Miller, Otto, 60, 162, 200, 202
Miller, Pearl, 62
Miller, Sam, 62
Miller, Wanda Pope, 134, 148, 171, 192, 225
Milner, John, 128
Miners Federation, 27
Minor, Lewis, 15
Mitchell, Jordan, 201
Mitchell Lodge, 49
Mitchell, Miss Jack, 202, 204
Mitchell, R. A., 132
Mitchell, Rev., 59
Mitch, William, 203, 207
Mixon, A., 147
Mize, B. J. W., 261
MIze, Charles, 200
Moffat, 11, 124, 192, 197
Moffatt Brothers Mine, 155
Monk, Dorothy, 63
Montevallo, 7, 9–10, 55, 83
Montgomery, Dr. Ethel, 247–248, 268
Moody, Mary Ann, 291
Moody, Morris, 258, 278, 287
Moon, Charles Roland, 226, 236
Moon, Howard, 206, 226, 230–231
Moon, James D. "Dan", 212, 222, 226
Moon, James D., Sr., 226
Moon, John Thomas, 226
Moon, Mary Clyde, 271
Moore, D. C., 247, 252
Moore, James, 212
Moore, J. G., 36, 43, 64–65, 70, 97, 124
Moore, Louis, 282
Moore, L. S., 218, 244
Moore, R. G., 193, 213, 260, 296
Moore, R. G. "Pug", 200
Morgan, Irby, 71
Morgan, Mrs. J. S., 131
Morgan, Samuel, 23
Mornin, Ottouis, 115
Morris, Mike, 296
Morrison, Anna, 31
Morrison, Annie Mae, 108
Morrison Boys, 78
Morrison, D. L., 126
Morrison Family, 55
Morrison, Floyd, 240, 245, 247, 261–264, 269, 273, 276, 280–281
Morrison, Gladys, 167
Morrison, Mrs. Frank, 179, 210
Morrison, Oscar, 115–117
Morrison, Sam, 40
Morrison, Sylvester, 81
Morrison, W. G., 7
Morrison, W. T., 193, 249, 260
Morris, Sam, 201
Morse, Kate, 108
Morse, Lonnie, 138
Morse, Lydia, 108
Morse, Pam, 314
Mosley, Jacqueline, 62, 303
Mosley, J. D., 62, 294
Moss, D. W., 58
Moss, Stan, 294
Mount Carmel Baptist Church, 42, 51–52, 226–227, 229, 280, 287
Mount Carmel Cemetery, 15, 45, 92, 137, 199, 229, 231, 243–244, 269
Mount Nebo Primitive Baptist Church, 52
Mud Creek Baptist Association, 52
Mulga, 127
Muller, Rev. S. A., 130
Murphee Drug Store, 174, 179, 181, 183, 186, 202, 259, 291, 302, 313
Murphee, Pat, 163
Murphy, Henry Grady, 120
Murphy, William J., 127, 204, 222, 252, 260, 297, 303, 305, 307
Muscoda, 168
Muse, J. D., 60

NAACP, 287
Nabors, John, 84
Nall, Mitchell, 59
Namath, Joe, 286, 292
Nash, Mick, 36
Nathews, Ike, 110
Neal, George M., 108
Neicci, Frank, 115
Nelson, Elija, 219
Nelson, Marie, 302
Nelson's Field, 99
New Brockton, 87
New Castle Coal & Iron Company, 64
New Hope AME Church, 59
New Zion AME Church, 59
Nichols, A. L., 52
Nichols, Harman W., 122
Nicholson, E. L., 58
Nicholson, N. C., 58
Nicholson, W. E., 219
Noble, Samuel, 7, 14
Nolen, Dr. Percy, 239, 259–260, 266, 268
Norris, Zelda Jordan, 276
North Bibb County Voters League, 286
Novgrad, Nat, 201
Novoson, Marc, 296
Nucci, Frank, 117
Nucci, Lester, 117
Nunnally, H. C., 95
Nunn, J. I., 197
Oakley, Annie, 50, 121
Oakley, Charles L., 109
Oakley, Jim, 39, 200
Oakley, Sheriff, 100
Odd Fellows Hall, 22, 39, 42, 70
Odd Fellows, Order of, 49, 50, 85, 105, 119, 122
Odess Department Store, 148, 174, 180–181
Odess, Morris M., 61, 117, 132, 142, 152
Oglesby, Ann, 251
Oglesby, Carl, 179
Oglesby, Rev. Sabert Gavelston, 59, 62
Ogletree Grocery, 148
Ogletree Residence, 149
Old Timers Game, 308–309
Opera House, 95, 104, 106–107, 110, 117, 132–133, 135, 139, 149–150, 153, 159–161, 163, 174–175, 178, 181
Opportunity School, 195, 198
O'Riley, J. P., 50
Our Savior Episcopal Church, 55–57, 140, 152, 206
Owens, Ina (Grady), 51
Owens, Mary Clyde Moon, 231, 271
Owens, Walter, 279–280, 302, 305
Owens, Walter, Drive, 59
Owen, William M., 52
Owings, Dr. William, 286
Owls, Order of, 119
Oxmoor, 15
Ozbirn, Dr. Thomas W., 300
Pandora, The, 119, 133
Parker, Colleen Weaver, 298
Parker, Edward, 50
Parker, Felix, 50, 290, 302, 304–305
Parker, James Coleman, 218
Parker, Joy, 267
Parker, L. P., 216, 277
Parker, Margaret, 206
Parker, R. C., 121
Parker, Rebecca Payne, 46
Parker, Rufus A., 250
Parker, Willie Jo, 46, 50, 58–59, 146, 227, 304
Park of the Four Orders, 50, 104, 122, 136, 140, 146, 153, 155, 165
Parks' Barber Shop, 283
Parks, Carey, 226
Parks, Catherine Blake, 226
Parks, Comer, 226
Parks, C. P., 226, 251
Parks, Frank, 226
Parks, John, 205, 226
Parks, Mrs. Comer (Blake), 228
Parks, Robert, 226
Parsons, Grayce, 228
Pass, Rev. J. H., 203
Pate, Howard C., 268, 275, 284, 286, 296
Patridge, Ed, 245
Patridge, E. L., 241
Patterson, Ann, 63
Patterson, Pat, 63
Patterson, Willie, 29
Patton, C. J., 239
Patton, Gov. Robert M., 124
Peace Baptist Church, 59, 153, 193
Peacock, Andrew Johnson, 113
Peacock, Dr. L. E., 17, 76, 82, 93, 113–115, 119, 133, 137, 142, 149, 161, 163–164, 181, 186, 192, 199, 214, 229, 281
Peacock, Mrs. L. E., 56–57, 115, 135
Penhale, Ken, 4, 10, 11, 86
Perry, Bud, 22
Perry, Cora, 58
Perry, Edward, 127
Perry, Glenn O., 208
Perry, J. W., 279
Perry Sewing Machine Office, 181
Perry, W. H., 125
Perry, Walter W., 41, 49–50, 58, 82, 132, 135, 148, 155, 219
Persinger, Jesse Cole, 28, 53, 54
Peyton, Jim, 138
Philharmonic Society, 37, 60, 115
Phillips, Bruce, 12
Phillips, Miss Issa, 43
Philpot, Earl, 195, 299
Pickett, David, 57, 62, 286, 302
Pickett Family, 32
Pickett, Sue, 32, 145, 275–276, 278, 280, 283, 286–287, 293–294, 296, 301–303, 308
Pierson, Dover, 247
Pierson, J. D., 241
Pierson, L. B., 237
Pierson's Hardware, 213, 270
Pilgrim Shoe Repair, 149
Pioneer Mining & Mfg. Company, 15
Piper, 2–3, 50, 95, 98, 103, 137, 145, 165, 192, 200, 206, 211, 251, 283, 298
Piper, O. H. R., 249
Pipe Springs, 58
Pitts, W. N., 66
Plotmisk, M., 288
Poginno, Joe, 99
Polglaze & Basenberg, 273
Polglaze, Leah Shamblin, 45
Polglaze, Richard, 45
Polglaze, W. M., 45
Pollyanna Crochet Club, 135
Polovich, Jake, 138
Pont, Mrs. Ed, 153
Ponzi, Charles A., 111–113
Poole, Rev. & Mrs. Mose, 275
Poole, Rev. Mose, 276, 280
Poole, Thomas DeWitt, 275–276, 283
Poole, Victor, 279
Poole, Willie Mae, 275
Pope Brothers Cleaners, 149, 170–171, 175, 181
Pope, Frank, 171
Pope, Harvey, 171, 192, 305
Pope, Herbert, 226
Pope, Jerry, 309–311, 313
Pope, Macon, 170–171, 174, 305
Pope, M. F., 134
Pope's Drug Store, 145, 148, 155, 163, 181, 187
Pope, Willis, 225
Porter, Ed, 126
Porter, George, 126
Porter, J. V., 54
Posey's Jewelry, 181
Pounds, Kathleen Yeager, 51
Pounds, Lester, 246
Pounds, Willie C., 237
Pow, Allan, 66
Pow, Jerry, 311
Powell, Joseph, 136
Powers, Ada, 12
Powers, D. J., 12
Powers, Emma, 12
Prather, Charles H., 226
Prather, Herbert H., 226
Prather, Hollis Edwin, 226
Prather, Howard, 15, 167

Prather, Howard "Booty", 226
Prather, James, 15
Prather, Jeremy, 15
Prather, Jim, 81
Prather, Maude, 226
Prather, Zeb, 15
Prather, Zebulon V., 226
Pratt, Betty, 241
Pratt Coal and Coke Company, 9
Pratt, Francis, 273
Pratt, Leonard C., 27
Pratt, Lillie, 43, 117
Pratt, Minnie, 27
Pratt, R. H., 43, 86
Pratt's Ferry, 123
Pratt, William, 280
Pratt, William Luther (W. L.) 5, 190, 202
Praytor, Rev. Robert E., 7, 26, 30–31, 33, 57–58, 311–312
Preston, Kenny, 296
Price Building, 139, 183
Price, Della, 294
Price, Joe, 236
Price, S. W., 126
Primitive Ridge, 33, 52, 84, 122–123, 137, 204, 224, 282, 305
Pritchett, Etta Reach, 40
Pritchett Ice Company, 143, 149
Pritchett, Matilda, 53
Pritchett, Pitman (P. P.), 53–54, 163, 192, 202, 249
Propst, Ryan, 307
Prospectors Club, 119
Prosperity Society of Blocton, 47
Pruett, David, 166, 169
Public Works Administration, 206, 210
Puffie, Puff, 115
Pullman Plant, 259
Putnum Thermo–Gas, 270
Quinn, Alberta, 267
Quinn, Alice, 225
Quinn Barber Shop, 149, 181, 186
Quinn, John, 40
Quinn, Nell, 271
Quinn, W. L., 130
R & R Grocery, 270
Rachel, W. B., 58
Rakestraw, Edna (Hartley), 51
Rally, Milton, 49
Randall, Dr. George W., 82, 103–106, 117, 119, 133, 135, 137, 140, 142, 146, 149, 157, 167, 177, 181, 192, 207, 212
Randall, Henry C., 177, 219, 239, 266
Randall, J. W., 247
Randall, Martha White, 177
Randall Medal, 119
Randall Residence, 177, 182, 193
Randle, Wilcox, 210, 242, 247, 249
Randolph, 86
Rationing Board, 225, 237
Ray, Billy, 161, 211
Ray, Dr., 12
Ray, Dr. Daniel W., 293
Ray, Effie, 256, 293
Ray, Harry, 293
Ray, Jim, 237
Ray, J. U., Jr., 26, 69, 113
Ray, Rev. John L., 58
Ray, W. R., 134, 137, 161, 192–193, 203–205, 210, 231
Reach, Elizabeth, 251
Reach, Hulet, 129
Reach, J. C., 7
Reach, John, 237
Reach, Luther, 51
Reach, Mary Bell, 250
Reach, Mattie, 51
Reach, Murphy, 277
Reach's Gas Station, 270
Reach Spring, 93
Reach, Yvonne, 250
Red, Bud, 80
Red Dot Coal Company, 224, 242
Red Eagle, 11, 137, 139, 155, 161, 282
Red Feather Coal Company, 125–126, 155
Red Men, Order of, 49, 85, 119, 198, 211, 261
Reese, Jesse "Jabo," 309, 313
Reeves, S. W., 40
Rensselaer Polytechnic Institute, 9
Reynolds, E. D., 82–83, 97, 103, 117, 119, 128
Reynolds, H. C., 82–83
Reynolds, H. C. & W. B., 22, 84
Reynolds, H. E., 40, 83, 97, 107
Reynolds, W. B., 82
R. F. D. Funeral Home, 159, 169
Rhodes, Herman, 306
Rhodes, Scott, 306
Rice, Bill, 283
Rice (Charles) Market, 181
Rice, George D., 80, 95
Rice, Johnny, 283
Rice, Nannie, 283
Rice, Rev., 59
Rice's Market, 186
Richardson, James, 33
Richardson, Judy Farmer, 231
Richardson, Norman, 296
Ricketts, Olene, 153
Ricorolo, Larcuzini, 116
Ridgley Lodge, 49
Riley, Cal, 126
Riley, Henry D., 234
Riley, Love, 22
Riley, Mart, 17
Riley, Martin, 119
Riley, Milton, 49
Riley, Percy, 213
Riley's Cafe, 149, 201
Riley, Walter, 234, 237
Riter, Morton, 201
River Bend, 117, 123, 137, 291
Robertson, Hayward, 237
Robertson, J., 33
Robertson, James, 37
Robinson, Bessie, 57, 89, 112, 154–155, 243–244
Robinson, Carl, 252
Robinson, Dr., 12
Robinson Family, 56
Robinson, Jo Ann, 224
Robinson, Margaret, 250
Rocky Church Community, 84
Roden Coal Company, 247, 253
Roden, Mrs., 249
Rogers, Harold, 237
Rogers, Ray, 306
Rogers, W. L., 118
Roland, Billy Mack, 18, 69, 308–309
Roland Funeral Home, 227
Roland, H. M., 241
Roland, Horace, 18
Ronilo, Bill, 275
Rosboro, Fred R., 60
Roseberg, Rabbi, 61
Rosser, Dr., 85
Rossi, Leno, 211
Ross, Jonette McCool, 258
Rotenbury, Eugene, 243
Roten, E. M., Jr., 260
Roten, George, 260
Royal Supply Company, 181, 183
Roycroft, A. B., 84
Roycroft Building, 176, 184
Roycroft, Hayes, 122
Roycroft (Hayes) General Store, 149
Roycroft, James, 236
Roycroft, James M., Jr., 176
Roycroft, Jimmy, 220
Roycroft, Ken, 237
Russell, Laura, 308
Russell, Paul E., 246
Sachs, A. S., 82
Sacred Heart School, 55
Salvadori, T., 261
Samsel, John, 213
Sansing, Billy, 237
Sansing, Joe, 19
Sansing, Nancy Ray, 203, 209
Sansing, Walter, 255–258
Santana Sportswear, 266, 268–269, 271, 296
Saramayou, Pete, 132
Schlosser, F. P, 42
Schober, Joe, 29
Schooler, Miss C. B., 43
Schultz Creek, 76, 132
Schultz Creek Baptist Church, 117

Scott, Evelyn, 168
Scottsville, 86, 97, 123, 134, 139, 157, 161
Scottsville Cemetery, 230
Scudder, Mildred Lee, 110
Scurlock, Carey, 171, 196, 303–304, 307
Scutchfield, Dr. F. Douglas, 286
Seagle, Mike, 296
Seales, Bill, 21
Seals, Jno., 72
Seals, Mrs. Glen, 217
Sedberry, G. F., 96
Self, H. H., 54
Selman, Rev., 75
Sewell, Ellen, 62
Sewell, Louie, 50
Shadwick, Denise, 288
Shadwick, Leslie, 293
Shadwick, Mr. & Mrs. Leslie W., 288
Shadwick, Regina, 292, 293
Shamblin, Melba, 221
Shannon Brothers Hardware, 148–149
Shannon Hardware Company, 122, 145
Shannon, Jim, 163
Shannon, Pat, 119
Shannon, Patrick, 84, 98
Shannon, Peter, 98
Shattuck, Charles, 255
Shaw, John Forest, 145
Shaw, Mike, 198
Shaw, Mrs. J. M., 156
Shaw, Willodean, 248
Shelby, Richard, 304
Sheppard, Frank, 29
Sheridan Family, 56
Sherrod, Benjamin, 179
Shiloh National Park, 8
Shiny Rock Cemetery, 58
Shows, Woodrow, 63
Shuttlesworth, Sam B., 261
Silver, Ben, 153
Silver, Calmon, 153
Silver Family, 181
Silver, Sam, 282
Silver's Economy Store, 152
Simpson, Glenda, 52
Simpson, R., 288
Sims, Bertha, 133
Sims, Billy, 288
Six Mile, 86
Slaton, G. Odette, 213
Slaton, Patsy, 248
Sloan, Barry, 296
Sloan, John, 306
Sloss, James W., 9, 11
Sloss–Sheffield Steel & Iron Company, 12, 64
Smalley, Jack, 260
Smelly, Doc, 138
Smelly, Edward Russell, 219
Smith, Allie (Harper), 51
Smith, Billy and Nina, 245, 252, 265, 278
Smith, Cherry, 6
Smith, C. S., 69
Smith, David, 296
Smith, Dr. Eugene Allen, 1, 6, 26, 32
Smith, Dr. Kennedy, 278
Smith, Edwin, 238
Smitherman, M., 58
Smitherman, Sheriff Glenn, 288
Smith Hill, 16, 21–22, 33, 41, 83, 92–93, 98–99, 103, 111–112, 116, 127–128, 135, 139, 142, 149, 152, 194, 206, 224, 245, 282, 285, 302, 305
Fire, 171
Smith Hill Dispensary, 83, 95–97, 171
Smith Hill School, 43, 133
Smith, Jack, 237
Smith, Joe, 19, 36
Smith, John, 238
Smith, John R., 122
Smith, Ken, 250
Smith, Kennedy, 6, 267
Smith, Kenneth, 260
Smith–Kuffner Hardware Company, 179, 181, 183
Smith, Lucille, 238
Smith, Morgan, 195
Smith, M. T., 192
Smith, Nina Franks, 6, 228
Smith, P. A., 14
Smith, R. D., 7
Smith, Rufus, 21, 111
Smith, Rufus D., 145
Smith, Sara Frances, 250
Smith Super Saver Grocery, 270
Smith Town, 5, 41, 85
Smith Undertakers, 181
Smith, Uriah, 4–5, 7, 61, 72, 83–85, 145
Smith, William "Billy", 6, 175
Smith, Willie, 83
Smith, Y. S., 106
Smith (Y. S.) General Store, 149
Smyth, Davene Pickett, 310
Snelson, John William, 275–276, 283
Snelson, Mr. & Mrs. Elisha, 276
Snelson, Sarah, 276
Snider, Ellison, 42
Socialist Party, 70
Southern Railroad, 100
Sparks Family, 60
Sparks, W. D., 117
Spencer, Rev. J. W., 53
Spenser, R., 58
Squire, Joseph, 1, 3, 10–11, 16, 218
Stabler, Claude, 45
Stacey, G. H., 204
Stack, N. R., 40
Stacks Hill, 21, 33, 58, 291
Stacy, G. H., 262, 271, 314
Stamps, E. M., 117
Stamps, Leige, 60, 93
Stamps, Rev. E. M., 60
Stamps, Shirley McCool, 248
Stamps, Will, 22
Standard Bottling Works, 121, 163, 281
Stanley, Ben, 244
Stanley, Ben O., 249–251, 254, 256, 260, 262
Stanley Residence, 149
Stanley's Grocery Store, 149, 170, 175, 181
Star, Alf, 206
Stedham, Elizabeth, 138
Steele, Eunice, 248
Stemetz, Kathleen, 228
Stephens, Paul, 277
Steward, J. W., 122
Stewart, Charles, 280
Stewart, Fred E., 262
Stewart, James, 136, 236
Stewart, James F., 151
Stewart, Joe, 247
Stewart, Johnny, 117
Stewart, J. W., 56, 117, 132, 149
Stewart, Mrs. James, 134
Stewart, Mrs. J. W., 106
Stewart, Nathaniel Edward, 97
Stewart, Paul, 258
Stewart Spring, 272
Stewart, W. A., 251
St. Francis Catholic Church, 21, 54–55, 130, 135, 137, 140, 160, 203, 246, 288
St. Francis Catholic School, 134
Stokes, Mrs. J. B., 249
Stone, Cornelius, 22
Stone, Jim, 21
Stone, Leonard, 21
Stone, Sandra, 306
Stone, Suzette, 306
Strand Theatre, 195–196, 205, 231, 252, 299, 302, 313
Street, Mary Curd, 58, 240
Street, Mrs. O. D., 250
Street, O. D., 26
Strickland, E. Paralee, 52
Strickland, J. H., 40
St. Thomas Aquinas Catholic Church, 55, 246, 285, 310
Sturdivant, Mrs. J. O., 117
Suell, Ed, 237
Sullivan, Nora, 51
Sumner, James Arthur, 269
Suttle, Elmore, 207
Suttle, John W., 85
Suttle, Julien LaFayette (J. L.), 5, 41, 85–86
Suttle, Julien LaFayette, Jr., 86
Suttles, George "Mule," 314
Suttletown, 5, 16–17, 23, 85
Suttletown String Band, 37
Sutton, Norman, 279
Sweeney, Charles, 252
Swift, Dr. Marjorie, 255

Swindal, Bernice S., 224
Swindal, Col. James B., 156–157, 224
Swindal, Mrs. S. Y., 156
Talliali, Charlie, 29
Tankersley, J. R., 39
Tannehill Furnaces, 2
Tannehill Ironworks State Park, 15–16
Tate, G. G., 137
Tatum, Vance, 213
T.C.I., 5, 8, 14–16, 64–65, 67, 69–70, 84–85, 87, 90–91, 93, 101, 122, 124, 244
T.C.I. Blocton School, 44
T.C.I. Hospital, 119, 120, 132, 136
T.C.I. Park, 39, 50
T.C.I. School, 101–102, 106–108, 136, 146, 166
Instructors, 168
Spring Festival, 168
Tennyson Literary Society, 154–155
Terry, V. P., 191
Thames, Lt. E. D., 136
Theatorium, 110, 132–133, 140, 145, 149, 152, 160–163, 169, 173, 181
Thomas, Bill, 36
Thomas, D., 33
Thomas, Darius Allen, 32
Thomas, Dr. M. C., 119, 120, 133
Thomas, Eunice, 108
Thomas, F. E., 50
Thomas, John R., 41
Thomas, J. R., 22
Thomas, Peter B., 32–33, 64, 66, 70
Thomas, Russell, 218
Thomas Town, 21–22
Thomas, W. A., 33
Thomas, W. D., 42
Thomas, Willard, 108
Thompson, A. V., 160
Thompson, A. W., 139
Thompson, Carrie, 49
Thompson, J. F., 95, 126
Thompson, Len, 224
Thompson, Paul, 296
Thompson, Seth, 49
Thompson, Will, 92
Thompson, W. O., 126
Thrasher, Ada Reach, 51
Thrasher, Bart, 78
Thrasher, Beatrice (Dyer), 51
Thrasher Boys, 78
Thrasher, Charles, 203
Thrasher, Dr. Helen Raye, 252, 266
Thrasher, Gladys, 206
Thrasher, Goldie, 179
Thrasher, J. W., 9
Thrasher, Kenny, 296
Thrasher, Mr. & Mrs. George, 266
Thrasher, Robert, 137
Thrasher, Sallie, 262
Thrasher's Mill, 134
Three Arts Club, 119, 145
Thursday Afternoon Study Club, 145
Tibbs, La Rue, 250
Tibbs, Mary Jo Jennett, 214
Tidwell, Ira, 181
Tidwell, R. E., 108
Tillery, George W., 7
Toumey, Michael, 1, 6
Tovich Fruit Stand, 148, 181
Triangle Dry Goods, 181
Trigg, Dr., 12, 17, 133
Trigg, Dr. A. W., 119
Trigg, E. S., 113
Trott, T. W., 59
Trussville, 15
Tucker, Dr., 161
Tucker Family, 56
Tuggle, Mrs. T. C., 22
Tuggle, T. C., 83, 95, 110–113, 116, 119, 132, 145, 148, 152, 153, 161, 173, 192, 195, 236
Tuggle Theatre, 185–186, 189, 194–195
Tulon's Place, 169, 179, 187, 190
Turner, Charles, 7
Turner, Lula A., 52
Turner, Mrs. J. M., 251
Turner Residence, 177, 193
Tuscaloosa Baptist Association, 52, 57
Twentieth Alabama Infantry, 86
Tyco Mine, 155
Tyler, A. L., 7, 14
Tyler, W. F., 39
Underwood, Mr., 13
Union Baptist Church, 59, 276
United Confederate Veterans, 86, 119, 134–135, 139, 146, 152, 157, 161
United Daughters of the Confederacy, 117, 119, 131, 135, 146, 152
United Mine Workers, 30, 66, 90–91, 93, 99, 139, 151, 158, 203, 207, 238
United States Steel Corporation, 5, 101, 107, 151, 191
Universal Motor Company, 160
Upchurch, J. P., 195, 213
Vance, Sheriff, 76
Vanderford, Dorothy, 240
Vanderford, June, 240
Vanderford, L. S., 205, 224–225, 227–228, 238–240
Vandergraff, Frank, 139, 147
Van Hoose, James, 56
Vanity Beauty Shop, 263, 270
Vaughn, S. E., 37
Verner, Prof. W. H., 43
Verner, S. F., 43
Vernon, James, 250
Viarigge, A., 265
Vickery and Deason, 89
Vickery, T. J., 37, 84, 117
Victory, Owanna Burt, 292
Waddell, Burt, 200
Waddley Band, 146
Wade, Jack, 255
Wadsworth, J. B., 49, 71, 82, 92, 219
Wagstaff, Rev. W. M., 59, 193
Wales, 25
Walker, Ed, 79
Walker, Kenny, 296
Walker's Bake Shop, 23
Walker, William, 40, 49
Wallace, Ben, 84
Wallace, Frank C., 134
Wallace, Gov. George, 264, 267, 271, 276, 279, 291
Wallace, Gov. Lurleen Burns, 85, 276
Wallace, J., 288
Wallace, J. H., 75–76
Wallace, Mabel, 108
Wallace, Margaret Lightsey, 86
Wallace, Mary, 108
Wallace, Maud, 146
Wallace, Miss S. V., 43
Wallace, T. C., 74, 136–137
Wallace, Thaddeus C., 86
Wallick, Dave, 201
Wallwork, J. W., 33, 104, 115–116
Walter, Mrs. Jesse, 152
Walters and Gamble Ladies' Wear, 148
Walters and Gamble Millinery, 181, 184, 187
Walters, J. E. J., 41, 77
Walters, Mrs. R. E., 41
Ward, Ed, 147
Ward, John S., 137
Ward, R. S., 241
Warren, Harris, 167
Warren, P. H., 239
Washington, Calvin, 296
Waugh, James, 33
Waugh, Sam, 33
Weaver, John Richard, 60
Weinstein & Israel, 71, 73
Weinstein, H., 61
Wells, Blanche, 204
Wells, F. E., 50
Wells, J. H., 147
Wells, Mem, 46, 146
Wells, W. H., 54
Welsh Congregational Church, 37, 60
Werer, Willy, 115
Wes–Bloc Manufacturing Company, 260, 262, 266
Wesleyan Methodist Church, 60, 204
Wesley, Daisy, 227
Wesley, Marvin, 291
Wesley, Thomas, 291

West, Annette, 241
West Blocton
Incorporation, 72, 74, 82
West Blocton Apostolic United Pentecostal Church, 63
West Blocton Bakery, 153, 181
West Blocton Band, 160
West Blocton Barons, 287
West Blocton Baseball Association, 122
West Blocton–Bessemer Bus Line, 159
West Blocton Bible Methodist Church, 60
West Blocton Chamber of Commerce, 189–190, 192, 198
West Blocton Church of Christ, 62–63, 260
West Blocton Church of God of Prophecy, 62–63
West Blocton Discount Drug Store, 282, 291
West Blocton Dispensary, 82–83, 88, 95–97, 102
West Blocton Drug Store, 259
West Blocton Elementary School, 209, 279
Lunchroom, 244
New Building, 206
West Blocton Fire Department, 172, 174, 194, 214, 216, 244, 266, 277
West Blocton First Baptist Church, 58–59, 83, 161–163, 227, 239, 243, 253, 274, 277–278, 282
West Blocton Garage, 122
West Blocton Health Center, 270–271, 286, 291, 297
West Blocton Heights, 59, 93, 139, 152
West Blocton Heights School, 46, 259, 278
West Blocton High School, 88–89, 102, 122, 134, 153–154, 162, 202, 204, 206, 225, 241, 261–262
Band, 206–207, 219–221, 243, 249, 275
Baseball Team, 249, 295–296
Basketball Team, 164–165, 193, 247, 286, 291
Dedication of New Building, 271
Football Team, 165–167, 193, 195, 199–200, 211, 222, 237, 249, 252, 255–256, 258, 260, 275, 278, 286, 294, 295
Girls' Basketball Team, 166
Indoor Restrooms Added, 247
Integration, 273–274, 278–280
Majorettes, 248
New Building, 204–205, 264, 268, 270
School Colors, 193
Tennis Team, 205
West Blocton Hunting Club, 167
West Blocton Improvement Committee, 16, 55, 61, 85, 244, 280, 309, 312
West Blocton Junior High School, 207
West Blocton Library, 272–273
West Blocton Loan Company, 237
West Blocton Methodist Church, 209, 218, 228–230, 238
West Blocton Municipal Building, 298, 300
West Blocton Pharmacy, 205
West Blocton Post Office, 110, 260–261
West Blocton Savings Bank, 138–139, 148, 149, 152, 159, 174, 181, 185–188, 197, 198, 237, 241
West Blocton School, 151
West Blocton School Improvement Assoc., 132, 152
West Blocton Second Baptist Church, 195, 204
West Blocton String Band, 194
West Blocton Supply Company, 240
West Blocton Theatre, 195, 197, 207
West Blocton Town Hall, 313
West Blocton Water System, 93–94, 197, 210, 254, 262, 265, 272–273, 282, 285, 291–292, 302
West, Buddy, 220
West, Charles Glen, 250, 255, 257
West, Dorothy, 306
West, Douglas, 306
West, Ed, 122
West End, 87, 122–123, 135, 179, 205, 210, 212, 231
West End Grocery, 270
Western Auto Store, 240
West, George H., 227
West, Jack, 237
West, Johnny, 280
West, J. T., 29
West, Kent, 227
Westonian, 205, 236
West, Robie Ann, 306
Whaley Furniture Company, 270, 278
Whatley, James, 255
Wheeler, H. E., 103, 104
Whidden, Len F., 17
Whitecap, 81
White, C. C., 139, 174
White, Dr. W. R., Jr., 298
White, Harold, 237
White, Mrs. C. C., 152
White, Tom, 12
White, Tommy, 255–259
White, Verta Lee, 62
White, Winston, 243, 260
White, W. R., 213
Whiteside, Birdie, 168
Whitsett, Rheinda, 294
Whitten, Annie Moore, 43
Wildo, George, 50
Wilkins Family, 55
Williams, Alex, 290
Williams, Benjamin Polk, 57, 62
Williams, Charles Albert, 246
Williams, C. M., 147, 152
Williams, Dewey, 283–284
Williams, Mamie, 59
Williams, Mattie (Price), 51
Williamson, C. P. Company, 9
Williams Spring, 272
Williams, Virginia, 306
Williams, W. B., 92
Williams, W. H., 52
Wilson, Dr. Cunningham, 91
Wilson, Lois, 162
Wilson, N. L., 67, 69, 152
Wilson, T. H., 241
Wilson, Thomas, 69
Winkenhoffer, Mr., 206
Winkle, Jack, 201
Winovick, Wasilije, 100
Winston, Jessie, 108
Witherspoon, Rev. J. W., 59, 80, 84
Woco Pep Service Station, 148
Wood, Alice, 49
Wood, Elizabeth "Baby", 226
Wood, Fred, 166
Wood, George W., Jr., 305
Wood, Huey, 109, 140, 153, 157, 164, 212
Wood, Jake, 200
Wood (J. F.) Motors, 207–208
Wood (J. F.) Plymouth & Dodge, 253
Wood, J. Fred, Jr., 297
Wood, Jim, 200
Wood, John, 118
Wood, John A., 109
Woodmen of the World, 119
Wood Motor Company, 260
Wood, Mrs. R. H., 262
Wood, Rev. F. M., 49
Woods (Albert) Billiard Parlor, 181
Woods, Carrie, 229
Woods, Cross, 229
Woods, James, 247
Woods, John, 22
Woods, Martha,, 229
Wood's Pool Hall, 186
Woods, Rev. F. M., 69
Woodstock, 9–10, 13, 26, 53, 78, 123, 137, 159, 266, 303, 314
Woods, Willie, 59
Wooley, Rev. D. Z., 142
Wooten Family, 60
Worthington's Grocery, 181

WPA, 211, 218
Wright, Charles, 159
Wright, Charley, 143
Wright, Eloise H., 153
Wright, R. B., 69
Wright Residence, 149
Wright, Ruben H., 45
Wright's Amusement Hall, 94–95, 98, 123, 136, 138, 172
Demolished, 159
Wright, W. H., 133, 137, 149, 152, 159, 164, 193
Wright, W. H. "Pinkie", 179–180, 251, 277, 297
Wylam, 152
Wynnes, A. W., 183
Wynn Family, 56
Wyper, Andy, 117, 160
Wyper, John, 117
Wyper, Will, 117, 140
Yeager, Harvey, 82
Yeager, Lucille, 256
Yeager (O. K.) Service Station, 270
Yeager, Oliver K., 261, 268–269, 277, 284, 286, 296
Yearby, Rev. J. T., 58
Yellow Front Store, 162, 219, 229
Yellowhammers, 119, 135, 146
Youngblood Mines, 127
Young, Brooks, 220
Young, Charles, 36
Young, George, 179, 222, 224, 242
Young, George E., 288
Young, George H., 113, 177, 182, 184–185, 198
Young, George W., 139
Young, J. R., 189
Young, Kenneth D., 214
Young, Lennie, 199
Young, Mrs. Kenneth D. (Lillian), 145
Young, Mrs. W. R., Jr., 204
Young, W. R., 22, 128, 137–138, 191–192, 198–199
Young, W. R., Sr., 281
Zavelo Jewelry, 148
Zavelo's Shoe Fixery, 148
Zeanah, Oscar M., 237, 239, 243, 247, 255–256, 259, 283, 287

The Cahaba Trace Commission and the author gratefully acknowledge the following people whose monetary contributions assisted in making the publication of this book possible.

Ollie M. Adams
John & Tina Nolen Adkins
Alabama Historic Ironworks Commission
David & Gina N. Aldrich
Jon & Marie Anderson
Charles L. Allen
Kae C. Allen
Steve & Nell Battle
Bruce & Janet Beerman
Bibb County Heritage Association
Nell Finnen Bleidt
Blocton Lodge No. 710, B.P.O.E.
Elizabeth J. Brannon
Celeste E. Burnam
Ann Pickett Buttrey
Libby Jean Caplan
Debbie McCool Cobern
Renette Israel Corenswet
Turner & Betty Crowe
Malcolm & Jewel Dailey
Robert & Pearl Davie
Mary Ruth Pope Dennard
Tulon E. Desmond
Elmo I. & Ruth Ellis
Joe Engel
Martin & Sandra McCool Everse
Margie J. Gentry
William E. Gladstone
Dora H. Going
Faye Pope Hickman
Sue Pickett Hyde
William & Sally Israel
Michael Kavanaugh
Jean Young Ivey
Dudley & Diane Kerr
Mary Sue Kofner
Patsy Pickett Linton
Milton & Julia McKinney
Roy Mason
Wanda Pope Miller
James D. & Barbara Moon

Louis & Gloria Nolen
Walter & Mary Clyde Moon Owens
Joe & Juanita Arnold Owings
James & Mary Parker
Howard & Geraldine Pate
Ed & Lane Segrest Patridge
Glenn O. Perry
Macon Pope, Jr.
Daniel W. Ray
Nelda Ray
Mary Ann Arnold Roe
Jonette McCool Ross
Kenneth & Helen Doss Roycroft
J. B. & Betty Scurlock
Kenny Scurlock
Sam I. Silver
Shirley McCool Stamps
Roger Stein
Marvin & Frances Stewart
Julien and Jeanne Suttle
Dr. Helen Raye Thrasher
Town of West Blocton
Mike & Julie Waldrop
Roy Webb, Jr.
Willie R. White, Jr.
George Howard & Mary Kay Young
James W. Young

ABOUT THE AUTHOR

Charles E. Adams was born in West Blocton and attended public schools there before entering the University of Alabama where he earned B.S. and M.A. degrees and a certificate in graduate history. After teaching in Bridgeport and Robertsdale High Schools and serving in the Air Force during the Korean War, he became Conference Director in Continuing Education at the University of Alabama, a position he held for twenty-eight years. He has written a history of the West Blocton United Methodist Church and has contributed articles to newspapers, magazines, and historical anthologies on the history of his hometown. He resides in Tuscaloosa.